# THE GEOGRAPHY OF
## WISCONSIN

# THE GEOGRAPHY OF
# WISCONSIN

John A. Cross and Kazimierz J. Zaniewski

THE UNIVERSITY OF WISCONSIN PRESS

The University of Wisconsin Press
728 State Street, Suite 443
Madison, Wisconsin 53706
uwpress.wisc.edu

Gray's Inn House, 127 Clerkenwell Road
London EC1R 5DB, United Kingdom
eurospanbookstore.com

Printed in the United States of America
This book may be available in a digital edition.

Library of Congress Cataloging-in-Publication Data
Names: Cross, John Alden, 1951– author. | Zaniewski, Kazimierz J., author.
Title: The geography of Wisconsin / John A. Cross and Kazimierz J. Zaniewski.
Description: Madison, Wisconsin : The University of Wisconsin Press, [2022] |
Includes bibliographical references and index.
Identifiers: LCCN 2021021555 | ISBN 9780299335502 (hardcover)
Subjects: LCSH: Wisconsin—Geography.
Classification: LCC F581.8 .C76 2022 | DDC 917.75—dc23
LC record available at https://lccn.loc.gov/2021021555

# Contents

# Illustrations

# Preface

At the time we began working on this project, our goal was to provide an up-to-date comprehensive overview of Wisconsin's geography that could be used by college students and be of interest to the general public, literally anyone who wants to know more about this state. We intended to include the results of the 2020 census. The drafting of the chapters was largely complete by spring of 2020, but the onset of the Covid-19 pandemic not only delayed the expected completion date for the census, but its disruption plus other issues affecting the Census Bureau may result in numbers that are not consistently comparable with those collected in previous years.

The 2020 census already faced political controversy given President Trump's desire to exclude undocumented persons (Liptak 2019; Mervis 2020). Furthermore, concerns had been raised about the Census Bureau's efforts to safeguard confidentiality by using a new differential privacy mathematical algorithm that "will degrade the quality of information" (Mervis 2019, 114), particularly for minor civil divisions and smaller units. The pandemic shutdown hit shortly before the official census day of April 1, when many students had vacated their dormitories and campus homes (Wines and Bazelon 2020), and office personnel were working at home or from their second homes, often many miles distant. It remains unclear how many workers will return to their previous homes if employers permit continued telecommuting and how the hospitality and travel industry will recover. The door-to-door follow-up by census takers among nonrespondents to the census was delayed, and the nonresponse rate appears more problematic than in previous censuses. All of these circumstances cast doubts regarding the accuracy of the delayed 2020 census results. Thus, particularly when dealing with county, city, village, and town populations, this book reports the 2010 census. Yet the census is among the least of the consequences of the pandemic.

Wisconsin's economy, along with that in almost every part of the world, was devastated by the pandemic, which necessitated shelter-in-place orders, the closing of many businesses, and limitations placed upon the gathering and movement of people. While many effects were immediately apparent, such as with the closing of schools, the collapse of the tourism business, the canceling of most airline flights, and the closure of Wisconsin's taverns, the resulting unemployment, business failures, and financial losses will be felt in the state for years. Thus, in perusing this *Geography of Wisconsin*, the reader is provided an account of America's Dairyland on the eve of the coronavirus pandemic.

While the medical crisis should not influence the state's geology, vegetation, or climate, it has had, and will likely continue to have, a major impact upon Wisconsin's human geography, particularly economic geography. Shopping patterns will undoubtedly change, as growth of online commerce may persist. The pandemic highlighted spatial differences in healthcare services and economic well-being that are described in the book, while leaving many residents of Wisconsin hoping for a return of the state's recreation and tourism industries described in the final chapter. For these persons, the book provides a detailed inventory of Wisconsin's economy and attractions on the eve of the pandemic. Whether these activities resume unaltered or in some new configuration or spatial setting remains to be seen. Hopefully, the many annual festivals and celebrations that were canceled during 2020 will resume in the future, although changes will undoubtedly occur. Yet the past is always a key factor to understanding the present and the future. Thus, this book should help the reader appreciate the post-pandemic Wisconsin.

We want to thank several individuals for their assistance with this project. The editorial staff of the University of Wisconsin Press, beginning with Gwen Walker and ending with Nathan MacBrien, have been most helpful. Managing editor Adam Mehring has skillfully coordinated everything. Jane Curran provided superb copyediting. In addition to Dennis Lloyd, director of the press, other staff deserving credit include Jennifer Conn, who coordinated the cover design; Terry Emmrich, who as production manager worked with the designer and oversaw printing; and Casey LaVela, who as sales and marketing manager has her staff promoting the book. Furthermore, several individuals have been particularly helpful in providing or explaining statistics that were used to prepare the maps or that were reported in the text. These persons include Michael S. Halsted, Harbors & Waterways program specialist at Wisconsin Department of Transportation (data for Wisconsin ports); Ernie Perry, Mid-America Freight Coalition, College of Engineering at UW–Madison (data for Wisconsin ports); Brian D. Anderson, forest inventory analyst at Wisconsin Department of Natural Resources (data on forests); David Haugen, forester with the U.S. Forest Service's Northern Research Station, St. Paul, Minnesota (detailed forest harvesting data by species); plus a variety of individuals with the Wisconsin Department of Agriculture, Trade and Consumer Protection and the Wisconsin Department of Natural Resources who have answered the authors' questions over the years.

In addition, the book greatly benefits from quotations from a variety of scholars who spent their careers studying Wisconsin. Appreciation is expressed to the American Geographical Society, the American Association of Geographers, the Milwaukee Historical Society, the Wisconsin Geological and Natural History Survey, the Wisconsin Historical Society, the Wisconsin Library Association, which holds the copyright to the Work Projects Administration's *Wisconsin: A Guide to the Badger State*, and Professor Marc Levine of the University of Wisconsin–Milwaukee's Center for Economic Development Publications for providing permission to publish the longer quotations, to utilize digital data, and to reproduce historical photographs within this book. The lead author taught a geography of Wisconsin course at the University of Wisconsin–Oshkosh for over a decade, and he wishes to acknowledge the contributions of his many students, expressed by their comments and questions during classes and the information they discovered and included in their research papers. Furthermore, the authors talked with innumerable academics and public employees regarding their research and activities, which provided insights that both were expressed in the classroom and informed this work.

The text was written by John Cross, who also took all of the unattributed color photographs that illustrate *The Geography of Wisconsin*. He greatly appreciates the efforts of his wife, Joann, in reading and critiquing the entire manuscript and accompanying him on many trips to take the photographs. The maps were prepared with MapViewer, MaPublisher, and Illustrator software by Kazimierz Zaniewski, who was the lead author of *The Atlas of Ethnic Diversity in Wisconsin*, published by the University of Wisconsin Press in 1998.

# THE GEOGRAPHY OF
## WISCONSIN

*Chapter One*

# Introduction to Wisconsin

H AVE YOU EVER WONDERED how you would explain where Wisconsin is located to a person who lives in a distant foreign land? What would you tell the person whom you might encounter on a trip to Tahiti or New Delhi, India? Telling that person that Wisconsin is home to the Green Bay Packers is unlikely to help, inasmuch as in most of the world soccer has a much greater following than American football. Let's grab a basketball or a beach ball to help locate our state.

## LOCATION OF WISCONSIN

Even though the basketball does not have the continents and oceans printed upon it, one can easily define the equator as being equally distant from the two poles, as we spin the ball on its axis. Wisconsin is astride the 45°N parallel, halfway between the equator and the North Pole (Figure 1.1). This fact alone tells us a fair bit about the state. It is sufficiently far from the equator to lie outside the tropics, yet not so close to the poles that our climate is dominated by cold air year-round. Indeed, because of Wisconsin's midway location, our weather is considered temperate, with a cold winter, yet a warm summer.

If we sketch out in chalk the location of North America on our ball, we can show the person Wisconsin's relative position in the continent. Wisconsin is located well within the interior, far distant from any body of salt water (Figure 1.2). Although we are closer to the Atlantic Ocean than the Pacific Ocean, the closest body of salt water is located to our north. The southernmost part of Wisconsin is located about 850 miles as the crow flies from the Gulf of Mexico. The northeasternmost part of Wisconsin is nearly 520 miles from southernmost part of James Bay, itself an appendage of Hudson Bay. This continental position not only strongly influences the greater annual range in temperatures that Wisconsin experiences relative to coastal regions, but it also delayed the European settlement of the state. The oldest buildings in Milwaukee are two centuries younger than the oldest structures in Boston, New York City, Williamsburg, or St. Augustine.

If you have an outline map of the United States, you can show your friend that Wisconsin is in the North Central part of the nation. While it does not technically share a border with Canada given the eastward jog of northern Minnesota's off-shore boundary, if you were to travel by boat straight

FIGURE 1.1. Plaque indicating place in Wisconsin that is equally distant between the North Pole and the equator. Because the Earth is not a perfect sphere, this location precisely equally distant in miles from the equator and the North Pole is located 8 minutes and 45.7 seconds of latitude north of 45°N. This monument is along U.S. Highway 141 between Crivitz and Beaver in Marinette County. Another monument is precisely displayed at 45°N in Door County.

northeast from Wisconsin's Lake Superior shoreline, you would make landfall in Ontario, Canada. Wisconsin's northern position can be further demonstrated by showing that Wisconsin's southernmost border is farther north than Canada's southernmost shore along Lake Erie. Even Toronto, Canada's largest city, is located closer to the equator than is the southern tip of Wisconsin's Lake Winnebago. Given the state's location closer to the Atlantic than the Pacific, the U.S. Census Bureau classifies Wisconsin as being in the East North Central region.

In locating Wisconsin, one may be tempted to describe its position in the hemisphere. Wisconsin is literally in the center of the northern half of the Western Hemisphere. Not only does the 45°N parallel run through Wisconsin, but the 90°W meridian crosses Wisconsin, indicating that Wisconsin is located halfway between the prime meridian, which runs through Greenwich, England, and the international date line, which separates Alaska from Siberia. The 45°N and 90°W lines intersect in Marathon County, highlighting Wisconsin's centrality. The geographic center of Wisconsin is just 1,536 feet east of the 90°W line, one mile east of Auburndale, in Wood County (Rogerson 2015).

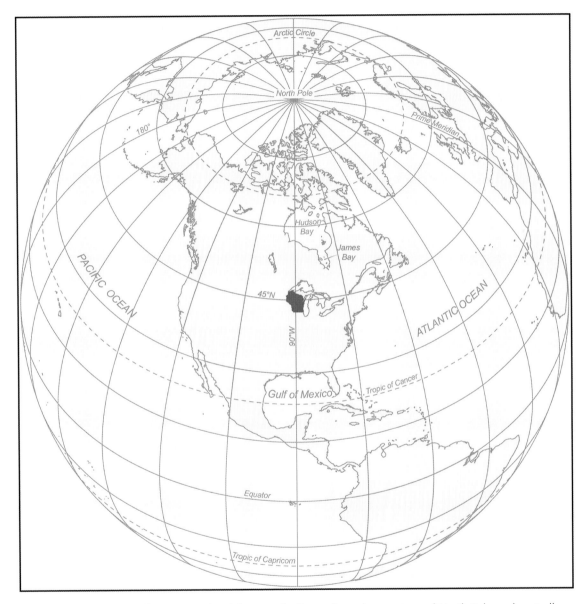

FIGURE 1.2. Wisconsin's location on the globe, equally distant from the equator and North Pole, and centrally located within North America.

Relative to other states, Wisconsin is north of Illinois and east of both Minnesota and Iowa—or at least the northeasternmost corner of Iowa. It is located south and west of Michigan's Upper Peninsula. We can also describe the various region names—besides the Census Bureau's East North Central designation—of the United States that include Wisconsin. Wisconsin is one of the Great Lakes States. It is part of the Midwest. Physiographers consider it part of the Central Lowlands. Historically, it was part of the Northwest Territory.

## Boundaries of Wisconsin

Some of the most prominent features that define Wisconsin's location are the bodies of water that surround the state. To the east, Wisconsin is bordered by Lake Michigan, and legally the border with Michigan is a line in the center of Lake Michigan equally distant between the shorelines of both states. To the north, Wisconsin is bordered by Lake Superior, with the legal boundary in the lake waters offshore drawn equally distant between Wisconsin and Minnesota. The Mississippi River forms the western boundary of Wisconsin from Prescott downriver, with the precise border following the center of the main channel of the river. Upriver from Prescott the St. Croix River marks the border for 125 miles. From that point the boundary line runs straight north to the St. Louis River, which serves as the boundary between Minnesota and Wisconsin from that point to where it flows into Lake Superior.

Portions of the boundary between Wisconsin and Michigan's Upper Peninsula are also delineated by rivers. For nearly forty miles, beginning at its mouth along Lake Superior, the Montreal River marks the border. The Menominee River and its Brule River tributary form the boundary between the two states upriver from the Bay of Green Bay to Brule Lake, the headwaters of the Brule River. When Congress first established the boundary between the Territory of Wisconsin and the State of Michigan, it set the boundary erroneously assuming that the two rivers had a single source at Lac Vieux Desert (Kellogg 1918; Martin 1930). Actually, Lac Vieux Desert lies within the Wisconsin River watershed. Thus, the border between the states extending between Brule Lake and the Montreal River is a slightly angled diagonal line, slightly changing its angle at Lac Vieux Desert.

The maritime boundary running through Green Bay between Wisconsin and Michigan was twice litigated in the U.S. Supreme Court during the twentieth century (Martin 1930, 1938). First, in 1923 Michigan claimed that it had sovereignty over a greater share of the waters of Green Bay, including Washington Island, Chambers Island, and several lesser islands. Michigan also claimed 463 square miles around Lac Vieux Desert and the Montreal River. The maritime squabble centered on the location of "the 'most usual ship channel' from Green Bay to Lake Michigan" (Martin 1930, 117), which Wisconsin argued was north of Washington Island, while Michigan argued it was the Porte des Morts Passage, separating the island from the Door Peninsula. Following three years of review, which considered over 3,000 pages of documents, 299 maps, and testimony provided at hearings in both states (see Martin 1930 for a highly detailed and informative discussion), Wisconsin prevailed, given its long possession of the territory. Yet the wording of the 1926 U.S. Supreme Court decree led to additional litigation, as it indicated that those islands between Rock Island and the tip of Michigan's Garden Peninsula were part of Wisconsin and stipulated confusing fishing grounds, leading to a 1936 Supreme Court ruling that finally settled Wisconsin's border (Martin 1938).

The southern border of Wisconsin was surveyed in 1831 following the designation by Congress that the northern border of the state of Illinois should be located at 42°30′ N latitude. As one can see on detailed maps today, the actual border that was surveyed deviates from that parallel, an artifact of surveying tools that were far less sophisticated than what is available today. Even a triangulation survey of Wisconsin's border in the 1880s showed that the surveyed state line is over a half mile farther north of where it should have been along the Mississippi River, but farther south in the area of Beloit and eastward (Thwaites 1888).

## Naming of Wisconsin

The lands and waters that comprise Wisconsin had previously been parts of other named lands. The name "Wisconsin" itself was first written by the French as "Ouisconsin," a rendering of a Native American name that was first written as "Meskousing" by Father Jacques Marquette, most likely for the river that flows through the center of the state (WHS 2017). The word itself has been variously attributed to describing a river with a thousand islands, with many tributaries, with high hills, with a rushing channel, with muskrat houses, or with holes in its banks occupied by birds' nests (Vogel 1965). In his comprehensive explanatory dictionary of Wisconsin place-names, Robert Gard (2015, 359) notes, "Most accounts say the word came from the Objibwe as Wees-kon-san, meaning the 'gathering of the waters.'" While the exact derivation of the name is uncertain, the Wisconsin Historical Society (WHS 2017) argues that it is a Miami Indian name that means "river running through a red place."

The lands that became Wisconsin were considered part of New France. With the defeat of the French during the French and Indian War, or the Seven Years War, as it was termed in Europe, Quebec began to be administered by the British. As part of the Quebec Act of 1774, Wisconsin was formally included within the territory of the Province of Quebec. Although the Treaty of Paris in 1783 officially transferred that part of the Province of Quebec south and west of the Great Lakes—including Wisconsin—to the United States, in reality Wisconsin remained under French Canadian or British control until the conclusion of the War of 1812. The British North West Company had fortified trading posts at Fort La Pointe (Madeline Island) until 1815, as well as Fort St. Louis (Superior), Fort Folle Avoine (Danbury), and Prairie du Chien (St. Feriole Island).

Officially Wisconsin was designated as part of the United States' Northwest Territory from 1787 to 1800 and Indiana Territory between 1800 and 1809, yet its small white population—many of whom were of mixed white–Native American ancestry—was dominated by French Canadians employed by the British-owned North West Fur Company. Thus, when the governor of Indiana Territory appointed a justice of peace at Green Bay in 1803, "the law of his court was French, like his accent, though he wore a British red coat and owed allegiance to the American Government" (WPA 1954, 39).

It was not until the War of 1812 that the first American military forts were established in Wisconsin, when the area was part of Illinois Territory. When Illinois gained its statehood in 1818, Wisconsin was appended to the Territory of Michigan. It remained a part of that territory until Michigan gained its statehood in 1836. While it was part of Michigan Territory, lead mining developed in the southwestern area of Wisconsin, and the first five counties were established in what was to become Wisconsin.

## The Territory of Wisconsin

The Territory of Wisconsin was created by Congress in 1836. Initially, it included all of the lands that comprise today's state of Wisconsin plus all of the region between the Mississippi and Missouri Rivers north of the state of Missouri and south of Canada—with the alignment of the U.S.-Canadian border not settled until the ratification of the Webster-Ashburton Treaty in 1842. Because of the concentration of Wisconsin's population in the lead-mining district, Belmont was selected as the first capital of Wisconsin Territory in 1836. In actuality, the buildings (Figure 1.3) that are preserved at First Capitol State Historic Site are located about two and a half miles northwest of the current site of Belmont, inasmuch as the community relocated to gain railroad access in the 1860s. In 1837 Burlington—that is, the Burlington in Iowa, not the one in Wisconsin—became the second territorial capital, and in 1838 it became the capital of Iowa Territory, after the region west of the Mississippi River was separated from Wisconsin Territory (Cravens 1983). Madison became the capital of Wisconsin Territory in 1838, having been selected because the site between Lakes Mendota and Monona was centrally located between the lead-mining communities of southwestern Wisconsin and the expanding agricultural

FIGURE 1.3. First Capitol State Historic Site, three miles northwest of Belmont in Lafayette County, preserves the 1836 Territorial Legislature's Council House (right) and the lodging house for legislators (left), also the home of the territorial Supreme Court's chief justice. The buildings are presently a museum.

settlement extending west from Lake Michigan. The actual site of the then-yet-to-be-established village of Madison, which existed in name only, was owned by former federal district judge James Doty, one of the territorial legislators, and Stevens T. Mason, governor of Michigan. The first meeting of the territorial legislature met in the unfinished—and unheated—capitol building in late fall 1838. Madison has remained as Wisconsin's capital city since that time, initially as territorial capital and state capital since 1848.

## SURVEYING OF WISCONSIN'S LANDS

Lands in Wisconsin were surveyed for sale to both settlers and speculators following guidelines set forth in the Northwest Ordinance, beginning in 1831 when Wisconsin was still part of Michigan Territory. The Public Land Survey system, first utilized in Ohio, was eventually used not only throughout the five states that grew out of the Northwest Territory but in much of the remainder of the United States.

### Townships and Sections

Lands were surveyed into townships that are squares that measure 6 miles by 6 miles (Figure 1.4). Each township comprises 36 sections, of which each measured—or should have contained, assuming there were no surveying errors—one square mile or 640 acres. The townships were laid out and numbered north of a base line, which in Wisconsin is the boundary between Wisconsin and Illinois. Townships are numbered east or west of the fourth principal meridian, which is a line extending due north from near Hazel Green. Today, the southern part of that line marks the eastern boundary of Grant County. Townships, running 6 miles north-south and 6 miles east-west, are given a unique identification, such as T7N, R9E. Thus, this township, which defined the political Town of Madison, is in the seventh tier of townships from the Illinois state line, extending from 36 to 42 miles north of the border. It is located in the ninth range of townships east of the principal meridian. Thus, it is located 48 to 54 miles east of the border between Grant County and both Iowa and Lafayette Counties.

The survey township was laid out in six tiers of sections, with six sections per tier. The sections are numbered, starting with the northeasternmost section, designated section 1. The numbering runs consecutively to the northwesternmost section, numbered section 6, with section 7 being located immediately to its south. In this second tier of sections from the north, the sections are numbered 7 to 12, with section 12 being located immediately south of section 1. They continue as if following a rope running back and forth along the tiers of sections, such that the southwesternmost section in a township is section 31, while the southeasternmost section is section 36. Section 16 was reserved from sale, such that the later sale or lease of its lands could support the establishment of public schools for the children of those who settled in the township.

The surveying of the sections and the townships had a profound impact upon the landscape of Wisconsin and its political and educational systems. Survey markers indicated the corners of the sections, and one in Madison is undoubtedly one of the most impressive in the nation (Figure 1.5). The marking of the sections of the townships imposed a grid upon the landscape (Figures 1.6 and 1.7).

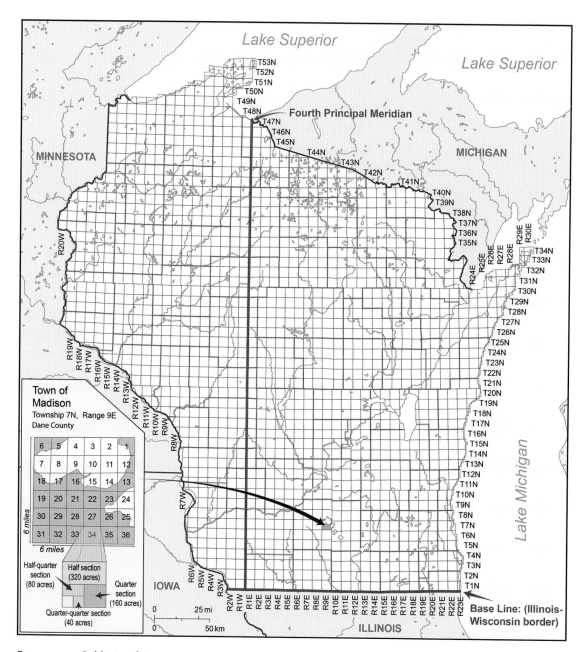

FIGURE 1.4. Public Land Survey System townships and counties in Wisconsin, with inset showing the township in which Madison is located. Map drafted using shapefiles for townships, sections, and quarters from Wisconsin Department of Natural Resources Open Data website (https://data-wi-dnr.opendata.arcgis.com/) for townships, sections, and quarters.

FIGURE 1.5. Wisconsin State Capitol building, located in Madison. The dome of the State Capitol is above the corner of sections 13, 14, 23, and 24.

Property lines ran north-south and east-west, with rural properties typically displaying square or rectangular shapes. Similarly, fences show this arrangement, and woodlots or areas of uncut forest on many farms are often square or rectangular, particularly in areas of flat land or gently rolling terrain. The most conspicuous display of the survey grid is seen in the arrangement of rural roads, which typically are along the property lines separating the survey sections. Given that farms often occupied a full quarter section—and 160-acre quarter sections were the most common allocation of homesteads in Wisconsin, although some individuals acquired a half of a quarter section—sometimes the rural roads run through the middle of a section, one-half mile east or west from the section lines or one-half mile north or south of the section lines.

Following the Public Land Surveys, which began after treaties had been negotiated with various Native American groups (see also chapter 7) and before most settlers had arrived in Wisconsin, lands were offered for sale directly to settlers who had already occupied property. For others, lands were available for sale at various land offices, of which eleven were established in Wisconsin. With Congress's passage of the Homestead Act in 1862, homesteaders received their lands without any payment other than a filing fee, but they were required to live on the property for five years and to cultivate a portion

FIGURE 1.6. Aerial view of dairy farmsteads along section line roads in the Town of Withee in Clark County, demonstrating the grid pattern upon the landscape created by the Public Land Survey.

FIGURE 1.7. Smoky Hollow Road, a section line road, leading north toward a church in the Town of Arlington, Columbia County. Note the offset to the west in the road, created by unevenly measured sections.

of the land. This requirement that settlers occupy their lands, rather than living in a nearby village such as was common in many farming regions of Europe, resulted in a dispersed settlement. Assuming that farms each covered 160 acres, while four farmhouses might be located at an intersection where four sections cornered, these farmsteads would by necessity be at least a half mile from their nearest other neighbors.

Given that Congress stipulated that the lands of section 16 of each township were set aside to help finance public schools, and because most children walked to school, one-room schools offering grades one through eight were erected at several locations scattered across each township. While the one-room school is largely a feature of the past, many of the buildings remain (Figure 1.8), having been converted into homes, businesses, or storage facilities. These schools necessitated one of the most local government units in Wisconsin, the local school district. Within many townships separate districts were established for each school. Seventy-five years ago Wisconsin had over 5,000 such school districts. With consolidation and the eventual requirement that each district operate a high school, by 2019 they numbered 421 (WLRB 2019). In addition, Wisconsin has 16 technical college districts, whose board members are chosen by either presidents of the school boards within the district or the county board chairs whose counties are within the district (WLRB 2019).

FIGURE 1.8. Former Lost Creek School, now used as a residence, within the Town of Pepin in Pepin County.

## Towns and Counties

Towns often constitute the smallest minor civil division in Wisconsin, officially termed civil towns (not the "townships" nomenclature used in most states that have such divisions of a county). Most towns in Wisconsin comprise all of the non-incorporated portions (which if incorporated constitute villages or cities) of a Public Land Survey township, or several survey townships, or at least many of the sections of a survey township, although river and county boundaries are sometimes responsible for exceptions. The town is the lowest level of administrative unit within Wisconsin (Figure 1.9), comprising part of a county, but it does not include the lands of any city or village that it might surround. While some towns exercise considerable authority, including having their own police forces, most town governments limit their services to maintenance of local roads, seeing that they are plowed to remove snow in the winter, exercise some land use control, and provide a volunteer fire department, which is sometimes done in conjunction with an adjacent town (Paddock 1997). As of 2019 Wisconsin had 1,249 towns, having lost four since 2017 and an additional two between 2015 and 2017 (WLRB 2015, 2017, 2019). Their number declines as town lands are merged into cities or villages.

FIGURE 1.9. Former town hall of the Town of Hartland, Shawano County. Although a larger modern town hall has been erected nearby, this building is typical of the structures that housed town records and accommodated periodic town meetings.

In general, Wisconsin cities are larger than villages, but there are many exceptions. Wisconsin statutes for nearly a century have stipulated minimum population thresholds that communities must have to incorporate as a city or village. Yet previously established cities have been "grandfathered in," just as communities that met the minimums, but that subsequently lost population, maintain their incorporated status. At present, the minimum size set for incorporation is "150 residents for an isolated village and 2,500 for a metropolitan village located in a more densely populated area" (WLRB 2015, 233). Incorporation of new cities requires a minimum size of 1,000 in rural areas and 5,000 in more densely settled areas. The Town of Caledonia in Racine County was the state's largest town in 2000, having a population of 23,614, yet in 2006 it incorporated as a village. However, not all urbanizing towns seek incorporation as a village or city. For example, the Town of Grand Chute, which had a population of 20,919 in 2010, has not gained such incorporation, inasmuch as the city of Appleton's incorporated area splits the town into noncontiguous segments. In contrast with the heavily populated Town of Grand Chute, 12 towns in Wisconsin had fewer than 100 residents at the time of the 2010 census. Three had fewer than 50. The Town of Wilkinson in Rusk County had 40 residents, while the Town of Cedar Rapids, also in Rusk County, had 41. The Town of Popple River in Forest County counted 44 residents.

The presence of towns throughout most of the midwestern states (with the exceptions of southern Illinois, where counties were not divided into towns, and in several sparsely populated areas of northern Minnesota), the New England states, and those areas settled by New Englanders, including New York and Pennsylvania, distinguish these areas from the southern and western states. In those other regions the county is the smallest unit of government, other than those municipalities that seek incorporation as cities, villages, or towns (and there the term implies a small community, not a subunit of a county). The existence of towns in Wisconsin has undoubtedly provided an element of local government that has likely discouraged many small communities from seeking incorporation, which would result in many formally designated small cities, such as one encounters in states such as Missouri. In total, Wisconsin had 602 incorporated places, including 190 cities and 412 villages in 2019 (WLRB 2019).

At the time of statehood in 1848, when Wisconsin was admitted as the thirtieth state of the Union, Wisconsin had 29 counties. The majority of these counties were located south of the Wisconsin and Fox Rivers. Those in northern Wisconsin included far greater land areas than those to the south (Figure 1.10). With settlement and clearing of the forests, the huge northern counties were partitioned into smaller counties, such that by 1901 Wisconsin consisted of 71 counties. Menominee County, created in 1961 from parts of Shawano and Oconto Counties, is the most recently formed county in Wisconsin (WCG 1998). Its borders coincide with the boundaries of the Menominee Indian Reservation, and unlike all of the other counties, it is not divided into separate towns, as the county and town have coincident areas.

Lands within Indian reservations are subject to tribal sovereignty as set forth in the specific treaties with the various Indian nations that were enacted during the nineteenth century. Thus, they are often not subject to local and state laws or zoning, but they are subject to federal laws. In a similar manner the federal government, rather than town, county, or state officials, oversees law enforcement on federal lands, most noticeably the national forests, national wildlife refuges, and various other U.S. government lands.

FIGURE 1.10. Map of Wisconsin published by J. H. Colton and Company in 1855. Although this map, published just seven years after statehood, fairly accurately portrays Wisconsin's border with Michigan, it was to be litigated three-quarters of a century later. Note that many of these counties were subdivided over the next century, and the names of several counties changed. Map from lead author's collection.

Wisconsin has over 600 other special purpose districts, including metropolitan sewerage districts, agricultural drainage districts, sanitary districts, sewer utility districts, water utility districts, mosquito control districts, and professional sports team stadium districts, among others. Some cover small areas involving few persons, while the metropolitan sewerage and professional sports team stadium districts cover large highly populated areas and are able to levy property taxes or impose sales taxes to fund their operations (WLRB 2019).

## WISCONSIN'S PLACE IN THE NATION AND WORLD

What would you tell a person from another part of the world—or even the United States—about the characteristics of the state where you live? While Wisconsin leads the nation in the production of certain products, such as cheese, and is the home to one of America's favorite National Football League franchises, the Green Bay Packers, in many respects Wisconsin is a middle state. It was the thirtieth state to be admitted to the Union, with 20 states admitted later. Wisconsin's geographic land area—54,310 square miles—places it exactly in the middle, relative to the other states. Twenty-four states have a larger land area than Wisconsin, while 25 have a smaller area. Wisconsin ranked twentieth in population at the time of the 2010 census, with 5,686,986 persons enumerated. Its 2020 resident population was 5,893,718, still the nation's twentieth largest.

Comparison of a state's size with foreign countries can also provide a useful perspective. Wisconsin's land area is over twice that of the Republic of Ireland. Its lands are about 10 percent greater than England and 28 percent larger than Cuba. Wisconsin occupies more land than Austria and Switzerland combined, and it is a little less than half of the size of Italy. Wisconsin's land area is only 2,600 square miles less than that of Bangladesh, which is the world's ninety-second-largest country—a middle sized country. Yet, Bangladesh has a population that is nearly 160 million greater than that of Wisconsin. When comparing Wisconsin's population to that of European nations, Wisconsin is similar to both Denmark and Scotland.

Lacking both sea-level locations and any mountain range, the mean elevation of Wisconsin, approximately 1,050 feet, is average among the states. Indeed, 26 states have a lower average elevation, while 23 have a higher value (Wikipedia 2019). Yet, only 9 states have a smaller elevation range between their highest and lowest locations. Furthermore, the highest elevation in Wisconsin, at 1,951.5 feet, is lower than the highest elevation within 38 of the states. Six counties have elevations exceeding 1,900 feet, while those along Lake Michigan share the lowest elevation, which averages 579 feet (WSCaO 2017).

We can also look at Wisconsin's official state seal, which is displayed on the floor of the capitol rotunda and on the state flag, for clues about the state's character and economy. Wisconsin's state seal was designed in 1851 and was slightly modified 30 years later. It is largely focused on Wisconsin's coat of arms, which has been featured on the state's flag since 1863, when it was created to accompany Wisconsin troops on Civil War battlefields (WLRB 2015). Even though it had received statehood only a decade and a half before the Civil War began, only seven states provided more soldiers to the Union Army. A close examination of this seal illuminates the many changes that have occurred in the state over the nearly one and three-quarter centuries since its creation.

At the top of the seal is the badger, whose inclusion as the state animal has more to do with other elements displayed on the seal than with the importance of the animal itself. Two men flank the shield, upon which four symbols are featured, representing farming, mining, navigation, and manufacturing (clockwise from the uppermost left symbol). The man on the right holding the pickaxe is a miner, while near his feet is "a pyramid of 13 lead ingots [that] represents mineral wealth" (WLRB 2015, 950). The early miners were often likened to badgers, either because they burrowed mining pits into the ground like the animals, or because many miners created dugout shelters in the ground in which to live like badgers. To the left of the lead is a cornucopia, from which a collection of fruits and vegetables grown by the state's farmers are spilling. The man to the left represents a sailor.

As we will see in chapter 5, mining was a leading industry during Wisconsin's territorial days and during the first decades of statehood. Its inclusion, along with the farmer's plough and the cornucopia, summarized the two most important economic activities of Wisconsin at the time the seal was designed. Both shaped Wisconsin's historical geography. Navigation was far more important during that time than today, and sailing ships upon the Great Lakes played a crucial role in supplying the early cities along Lake Michigan before the arrival of railroads, as is described in chapter 10.

Today, Wisconsin is quite different from what it was a century and a half ago. Economically, *Forbes* magazine in 2016 ranked Wisconsin as the twenty-seventh "Best State for Business" (Badenhausen 2016), while it was ranked twenty-first in 2019 (Forbes 2019). Its median household income in 2014 was the twenty-first highest among the states. The Census Bureau's five-year American Community Survey for 2011 to 2015 indicates that Wisconsin ranked twenty-seventh in the proportion of its adult population over 25 years of age that had earned a four-year college degree (USCB 2019a). While these statistics indicate that by some criteria Wisconsin is an average state, when one considers the high school graduation rate, Wisconsin is tied with Hawaii for eleventh place. Wisconsin ranks fourteenth in the life expectancy of its residents.

## WISCONSIN'S LANDSCAPES AND THIS BOOK'S ORGANIZATION

The landscape of any state or nation is shaped by its unique combination of physical features, including landforms, vegetation, climate, and soils, and by its human inhabitants. The size of that population and its influence are reflected in the cultural and economic activities that have molded the use of the physical environment. Thus, what we see on the landscape today not only shows how contemporary residents are utilizing the land, but also provides numerous clues about how earlier populations utilized the land, influenced by their needs, technologies, cultures, and desires. One cannot understand the contemporary landscape without considering an area's historical geography, given that so much of what we see is a legacy of the past. Indeed, although the red dairy barn displayed on the state's license plates is not as commonly seen today as it was a half century ago, some barns that we see today date from the nineteenth century. This book is designed to give the reader a better understanding of what can be seen when looking at a Wisconsin landscape.

The physical features of Wisconsin that we see, and the foundation upon which humans built their farms and communities, are described in chapters 2 through 4. Chapter 2 focuses on describing the physical landforms of Wisconsin and explaining the role of geology and geologic processes in their formation. Chapter 3 describes the climate of Wisconsin and reviews the pattern of vegetation associated with a locale's climate and soils. Chapter 4 illustrates that hazards associated with weather and climate not only strongly influence the state's farmers but also disrupt transportation.

Resource exploitation underpinned the early settlement of Wisconsin. Furs were the first commodity that was sought, yet the opening of lead mines in the early nineteenth century initiated the development of Wisconsin in earnest, while a century later mining for iron ore focused attention on the northernmost counties. The influence of mining on the early historical geography of Wisconsin, in contrast with its contemporary role, is the focus of chapter 5. Chapter 6 focuses on the state's forests, noting that the clear-cutting of the forests of central and northern Wisconsin, beginning a century and a half ago, sparked the development of several of the state's cities, set the stage for farmers to tackle the challenge of growing crops on the cutover lands, and shaped the state's contemporary forest industries.

Wisconsin's farmlands were largely settled during the nineteenth century, at the same time that the United States was experiencing a massive wave of European immigration, a topic that is explored in chapter 7. Because over two-fifths of Wisconsin is farmland, two chapters explore the contemporary spatial patterns of agriculture in Wisconsin, looking at crop production in chapter 8 and at livestock in chapter 9, particularly the dairying that distinguishes America's Dairyland. Just as the physical environment, the soils, the climate, and environmental hazards, discussed in chapters 2 through 4, shaped Wisconsin's primary production as described in chapters 5 through 9, the availability of transportation, whose varieties and networks are reviewed in chapter 10, serves the farms, mines, sawmills, and manufacturing facilities scattered across the state. Transport facilitates the movement of the state's population to work, to shopping, and to recreation. Wisconsin's manufacturing industries are explored in chapter 11. Industrial jobs, concentrated in urban centers that are well linked by transportation lines to sources of raw material, labor, and markets, are described in chapters 11 and 12.

Rural Wisconsin has experienced many changes since immigrants poured into Wisconsin during the nineteenth century. Many of today's immigrants have focused on the state's urban centers. Population change has been uneven across the state, resulting in many demographic characteristics that are differentially expressed across Wisconsin, as explored in chapters 12 and 13. The state's urban centers, which provide the focus of the state's tertiary economic activities or services, are also described in these two chapters. Chapter 13 also explores rural and urban disparities in income, the availability of health care and social services, and political orientation. Finally, the outdoor sports and tourism industry, so important to many rural areas of central and northern Wisconsin, the importance of professional sports teams in several of the state's metropolitan centers, and the state's many festivals provide the focus for chapter 14.

Numerous maps are provided throughout the book to guide the reader. While Figure 1.11 shows the location of the counties, county seats, and principal places, many sites mentioned in the book are too small to include on a general location map. Such places are often located by county in the text, yet the reader or traveler who wishes to see their locations displayed on a detailed large-scale map may wish to refer to a recent edition of DeLorme's (2020) *Wisconsin Atlas and Gazetteer*.

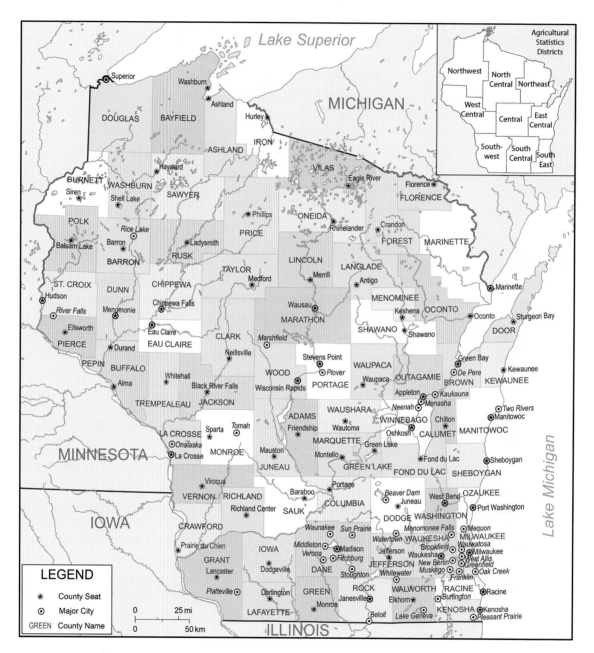

FIGURE 1.11. Reference map for Wisconsin, showing locations of counties, county seats, other major cities, and major rivers. Agricultural districts often depict Wisconsin's regions. Map drafted using U.S. Census Bureau Cartographic Boundary Files-Shapefiles (https://www.census.gov/geographies/mapping-files/time-series/geo/carto-boundary-file.html).

*Chapter Two*

# Wisconsin's Physical Landscape

## *Its Geology and Landforms*

THE FIRST VIEWS of Wisconsin's physical landscape that a traveler sees when approaching the state vary tremendously, depending upon one's location. A person crossing the surveyed base line that separates Wisconsin from Illinois will notice the vast flat plain that constitutes so much of central and northern Illinois continues into the southern half of Wisconsin, only sporadically broken by river valleys or undulations of glacial moraines. A traveler along the Mississippi River will note steep bluffs, rising 500 feet or more, exposing sedimentary strata towering over the river terrace (Figure 2.1). A person approaching Wisconsin's Lake Michigan shore is likely to encounter sandy beaches, backed by low escarpments of eroding glacial debris or by sand dunes (Figure 2.2). In contrast, the view of Wisconsin from Lake Superior is of a far more rugged shoreline, marked by steep sea cliffs in places, such as in the Apostle Islands or along the Bayfield Peninsula. Ridges, 1,000 feet higher than Lake Superior, are visible in the distance. A person venturing from Michigan's Upper Peninsula into Wisconsin south of the Gogebic Range will note that in many areas the soils are thin, with exposures of granites and gneiss—igneous and metamorphic rocks well over a billion years older than the sedimentary rocks seen in southwestern Wisconsin, which are nearly a half-billion years old (Figure 2.3).

Physical landforms that constitute much of the landscape of Wisconsin—ranging from what an early geologist called the Mississippi Cañon (McGee 1891) on the west to the Niagara Escarpment in the northeast—clearly demonstrate the role of geology, stratigraphy, and geologic processes in shaping the landscape. Yet, continental glaciations during relatively recent geologic time smoothed much of the topography in three-quarters of the state and created the Great Lakes.

This chapter first reviews the geologic history of Wisconsin, including the age and character of the underlying bedrock, the types of rocks comprising the bedrock, and the stratigraphy that characterizes the sedimentary bedrock. Differences between the rocks and landforms within the outlier of the Canadian Shield exposed in northern Wisconsin and the sedimentary formations elsewhere influence the fertility of the soils and the distribution of mineral resources. Particular emphasis is placed on discussing how Pleistocene glaciations shaped Wisconsin's landforms and excavated the Great Lakes.

FIGURE 2.1. Bluffs along the highway and railroad facing the Mississippi River, southwest of Lynxville in Crawford County.

FIGURE 2.2. Grasses atop sand dunes along eroding Lake Michigan shore in Kohler-Andrae State Park, Sheboygan County.

FIGURE 2.3. Sheer cliffs of Precambrian metamorphosed mafic intrusive rock along Menominee River from Niagara in Marinette County. This river marks part of the boundary between Wisconsin and Michigan's Upper Peninsula and a fault separating two rock units.

The advance and retreat of the glacial ice resulted in the creation of a physical landscape that is distinctly different from what one sees in the Driftless Area of southwestern and west-central Wisconsin, which was never overridden by the ice sheets.

## BEDROCK GEOLOGY OF WISCONSIN

The rocks that comprise the state's bedrock vary greatly between northern and southern Wisconsin—both in their type and age (Mudrey, Brown, and Greenberg 1982; WGNHS 2005). Although in places the bedrock is exposed at or near the surface, in most locations it is buried beneath the soil. In places soils formed with the weathering of the bedrock below, yet bedrock is often below unconsolidated glacial and lacustrine deposits that date from the Quaternary. This geological time period includes the Pleistocene and the Holocene, the time of the glacial ice sheets and the epoch that followed their melting. The Pleistocene dates from 2.6 million years ago until the Holocene began 11,700 years ago, but Wisconsin's bedrock is far older. The oldest bedrock in Wisconsin is Upper Archean gneiss in Iron and Vilas Counties near Michigan's Upper Peninsula. Gneiss itself is a type of metamorphic

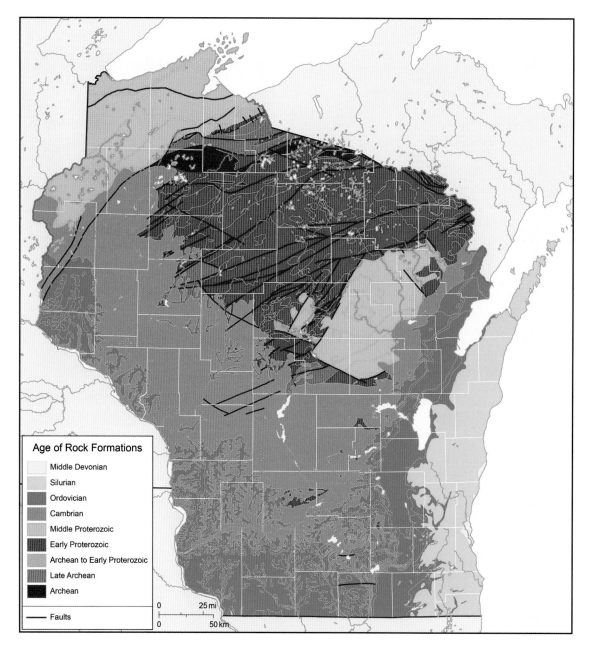

FIGURE 2.4. Bedrock geology map of Wisconsin. Map redrafted based upon Mudrey, Brown, and Greenberg (1982) and U.S. Geological Survey, Wisconsin Geologic Map Data (https://mrdata.usgs.gov/geology/state/state .php?state=WI). Used with permission of Wisconsin Geological and Natural History Survey (University of Wisconsin–Madison, Division of Extension).

rock, created by the alteration of the crystalline structure of even older granites by heat and pressure within the earth's crust, such that the minerals are arranged in alternating layers of white and dark crystals.

A large variety of Precambrian era igneous and metamorphic rocks comprise the bedrock throughout much of northern Wisconsin (Figure 2.4). Archaen gneisses are the oldest, while the youngest are Keweenawan rocks, consisting of igneous lavas, gabbros, and some sandstone, which formed around 1.1 billion years ago. These Keweenawan rocks, which constitute the bedrock in most of Douglas and Bayfield Counties plus the northern portion of Ashland and Iron Counties, are separated by ancient faults from even older rocks to their southeast. Within the broad area from Sawyer County and the central parts of Ashland and Iron Counties to central Portage and Waupaca Counties to the south, the predominately igneous and metamorphic rocks typically range in age from 1.7 to 2.8 billion years, Lower Proterozoic in age. Some granitic rocks that are approximately 1.5 billion years old illustrate the intrusion of slightly younger rocks.

Much of this igneous and metamorphic rock, contorted and displaced by both intense folding and faulting, is the legacy of mountain building that occurred 1 to 2 billion years ago. Extensive areas of basalts and rhyolite, both types of extrusive volcanic rocks associated with the ancient mountain building, constitute the bedrock from Rusk and Taylor Counties on the west to Marinette and Forest Counties, along Michigan's Upper Peninsula on the east. To the southeast of these Lower Proterozoic volcanic rocks, the bedrock is dominated by granites, which are Middle Proterozoic in age. These granites represent intrusive igneous rocks, formed by upward movement of a large mass of magma that then solidified below the surface of the land, forming what geologists call a batholith. Subsequent erosion of the earth's surface exposed the granites of the Wolf River Batholith, which constitute the bedrock from eastern Marathon and Waupaca Counties through Oconto County.

The ancient igneous and metamorphic crystalline rocks of northern Wisconsin constitute the southern outlier of the Canadian or Laurentian Shield that dominates Canada's bedrock from Lake Superior to Baffin Island. As is characteristic of continental shield rocks in many parts of the world, their geology is complex, and the rock is highly mineralized, including many significant deposits of metallic ores. In addition, numerous faults indicative of crustal movements that occurred hundreds of millions of years ago crisscross the landscape, separating and displacing the various rock units visible on the bedrock geology map.

South of the Precambrian rock outcrops, these older igneous and metamorphic rocks are covered by layers of sedimentary rocks in the eastern, southern, and most of the western parts of the state. These sedimentary beds are over 1,000 feet thick in far southeastern Wisconsin. Formed by the deposition and cementation of materials eroded from earlier preexisting rocks, these sedimentary rocks include sandstone, shale, and dolomite. Sandstone and shale are consolidated layers of sands and clay with silt, respectively. Dolomite is somewhat similar to limestone but is calcium-magnesium carbonate rather than calcium carbonate. It is formed from chemical precipitation of carbonate minerals in ocean waters. It is more resistant to erosion than limestone, yet it is subject to chemical solution and cave formation.

Sedimentary rocks that constitute Wisconsin's bedrock were deposited in layers during the Cambrian, Ordovician, Silurian, and Devonian geologic time periods, roughly 540 to 360 million years ago, when the area was covered by sea water. With the uplifting of the rocks, younger rock units were not deposited, and the existing sedimentary rocks were exposed to erosion. The youngest of these rocks, Devonian-age shale that outcrops along Lake Michigan from Milwaukee through Ozaukee County, lies atop the older units. It has undoubtedly been eroded from other locations. Cambrian period sandstones and shale constitute a broad swath of central Wisconsin's bedrock south of the Precambrian shield rocks. In southwestern Wisconsin steep valleys cut through relatively horizontal layers of sedimentary rocks, exposing alternating layers of shale, dolomite, and sandstone, along with lesser amounts of limestone and conglomerate. The youngest rocks, around 450 million years old, are located at the highest elevations, while older rocks are encountered along the valley bottoms.

At several locations in central and south-central Wisconsin isolated outcrops or inliers of Precambrian granite and quartzite rock exist, representing old residual erosional remnants that are surrounded by younger sedimentary rocks. Examples of granite inliers are found at Berlin and Redgranite in Green Lake and Waushara Counties, respectively, where quarries once operated. Farther south is the Baraboo Range, running from central Sauk County into west-central Columbia County, which represents the exhumation of an ancient ridge of quartzite, a metamorphosed sandstone that is highly resistant to weathering and erosion. This quartzite is around 1.6 billion years old. The erosion of the surrounding sedimentary rocks over the past several hundred million years exposed the ancient ridge dissected by an incised valley, which was further modified by glaciation during the Pleistocene. Before we consider the consequences of glaciation, the role of bedrock, tectonics, and stratigraphy upon Wisconsin's physical landscape needs to be explored.

## TECTONICS, STRATIGRAPHY, AND WISCONSIN'S LANDFORMS

Erosion removed much of the rock created by volcanism and tectonic activity that created high mountains across Wisconsin well over a billion years ago. Sedimentary rocks were laid down upon those eroded igneous and metamorphic rocks in at least the southern two-thirds of Wisconsin nearly a half billion years later when the area was covered by a sea. Subsequent differential uplift of the southwest corner of Wisconsin and downwarping of central Michigan into a broad structural basin slightly tilted the rock layers. In contrast, tectonic activity within northern Wisconsin dating back over a billion years and continuing through Paleozoic times (Schultz 2004) resulted in steeply folded rocks associated with the downfolding of the area now occupied by Lake Superior and extensive faults, some of which today mark the boundary of Precambrian and more recent bedrock. Both of these areas display landforms today that owe their existence to their bedrock's differential resistance to erosion.

### Lake Superior Uplands

First, let's look at the landscape between the southern shore of Lake Superior and the crest of the Gogebic Range and the Douglas Range continuing west toward the St. Croix Valley. Landforms in his region illustrate the consequences of both faulting and folding associated with the creation of the Lake Superior syncline. This long narrow trough, into which layers of sediments accumulated, marks a rifting of the earth's crust. Subsequent volcanism resulted in vast and multiple outpourings of lava, such that the Keweenawan bedrock consists of a complex layering of sedimentary and igneous rocks, many of which have been altered and infused with mineral depositions. The lowest units are at least partially metamorphosed. These layered rocks are tilted the most along their southeastern exposure, such that in the Gogebic Range (often called the Penokee Range in Wisconsin) they are nearly vertical.

The greater resistance to erosion of the Penokee Range's iron-bearing rocks relative to nearby strata created the ridge, which rises to an elevation of over 1,800 feet. Its highest point is Mount Whittlesey, which reaches 1,872 feet (Schultz 2004). The Penokee Range does not constitute a divide between watersheds, as several rivers flow through gaps in the ridge. For example, the Bad River flows though the Penokee Gap, while the Montreal River runs through a wide gap that cuts across the Gogebic Range.

From Superior, which occupies a lowland that is partly attributable to glaciation that deepened Lake Superior that was followed by glacio-eustatic rebounding following the melting of the Pleistocene ice sheet, one looks northwest toward the conspicuous Duluth Escarpment in Minnesota. Separating Duluth from Superior is an ancient fault, with the bedrock upon which Superior is located being down-dropped relative to the upland behind downtown Duluth. Geologically, Superior is situated on a graben, separated by faults not only from Duluth but from higher lands toward the south. The Douglas Fault is marked by the Douglas Copper Range, which cuts across northern Douglas County

FIGURE 2.5. Big Manitou Falls, Wisconsin's highest waterfall, in Pattison State Park, Douglas County.

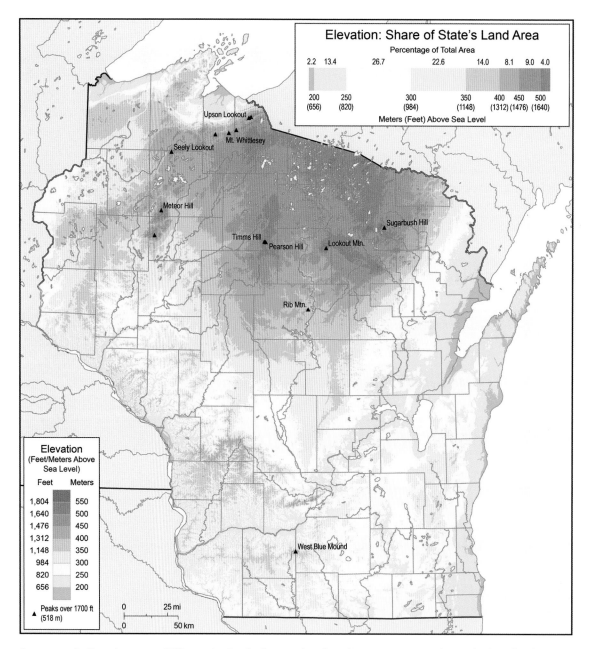

## Elevation: Share of State's Land Area

Percentage of Total Area

| 2.2 | 13.4 | 26.7 | 22.6 | 14.0 | 8.1 | 9.0 | 4.0 |

| 200 | 250 | 300 | 350 | 400 | 450 | 500 |
| (656) | (820) | (984) | (1148) | (1312) | (1476) | (1640) |

Meters (Feet) Above Sea Level

Upson Lookout
Seely Lookout
Mt. Whittlesey
Meteor Hill
Timms Hill
Pearson Hill
Lookout Mtn.
Sugarbush Hill
Rib Mtn.
West Blue Mound

### Elevation
(Feet/Meters Above Sea Level)

| Feet | Meters |
| --- | --- |
| 1,804 | 550 |
| 1,640 | 500 |
| 1,476 | 450 |
| 1,312 | 400 |
| 1,148 | 350 |
| 984 | 300 |
| 820 | 250 |
| 656 | 200 |

▲ Peaks over 1700 ft
(518 m)

| 0 | 25 mi |
| 0 | 50 km |

FIGURE 2.6. Elevation map of Wisconsin. Drafted using data from https://www.opendem.info/download_contours.html; Open Digital Elevation Model (OpenDEM Project).

into Bayfield County. Both 165-foot Big Manitou Falls (Figure 2.5) in Pattison State Park and Amnicon Falls in Amnicon Falls State Park are situated where the Black and Amnicon Rivers, respectively, tumble from the resistant volcanic rock to weaker sandstones exposed in the graben. Farther east, the rise south of Ashland runs parallel with the Douglas Fault, and the waters of Chequamegon Bay mark the site of eroded weaker sedimentary rocks. Even farther south is the Lake Owen Fault, with the bedrock between it and the Douglas Fault constituting the uplifted St. Croix Horst.

South of the Lake Superior Uplands extends a broad area of smoothed, yet undulating, igneous and metamorphic rocks, covered in places by glacial deposits. In this region we find the highest elevations in Wisconsin (Figure 2.6). Timms Hill, with an elevation of 1,951.5 feet in Price County, is the highest spot in Wisconsin. Nearby Pearson Hill, just a half mile distant and less than a foot lower, is the state's second-highest elevation. Farther south, just southwest of Wausau, is Rib Mountain, which is a far more prominent topographic feature. An outcrop of Precambrian quartzite, a highly erosion-resistant metamorphic rock, Rib Mountain, with an elevation of 1,924 feet, rises nearly 700 feet above the surrounding landscape, causing early geologists to erroneously claim it was Wisconsin's highest location (Martin 1965, 18). Rib Mountain and the Baraboo Range in southern Wisconsin are examples of a monadnock, an outcrop that was subsequently surrounded by younger sedimentary rocks that were later eroded, exposing the ancient ridge.

### The Niagara Escarpment

Sedimentary rocks that comprise the bedrock of the eastern half of Wisconsin are tilted slightly downward toward the east. Southwestern Wisconsin and adjacent southeastern Minnesota and northeastern Iowa constitute an upland composed of these sedimentary rocks, through which the Mississippi River carved a canyon. Numerous exposures of the relatively flat-lying sedimentary strata are visible in the bluffs facing both sides of the river. Farther toward the east, the sedimentary layers over the Wisconsin Arch begin to dip more conspicuously toward the Michigan Basin, where the Silurian dolomite that is the dominant bedrock from Door County south through Kenosha County is found over a mile beneath the surface in south-central Michigan, buried beneath later rock layers that are absent in Wisconsin.

Sedimentary rocks vary in their resistance to erosion. Shale, such as found in the Maquoketa Formation, is the least resistant to erosion, being a relatively poorly consolidated clay deposit. Thus, it is easily eroded by both flowing water and glacial ice. Where that rock unit outcropped we have the Bay of Green Bay, the Lower Fox River Valley, Lake Winnebago, and Horicon Marsh. Overlying this shale to the east are several layers of Silurian dolomites, which are among the most erosion resistant sedimentary rocks in Wisconsin, although they are often fractured and vulnerable to ground water contamination. The dolomite acts as a caprock, protecting the underlying weak rock from erosion. Where water flows across the boundary between the caprock and the shale, such as at Wequiock Falls to the northeast of the city of Green Bay, one can see the boundary between the unlike sedimentary rocks. As the weak shale is eroded before the falls, dolomite blocks break off. It marks but one spot along the Niagara Escarpment that "extends at least 650 mi (1,046 km) from eastern Wisconsin through the Upper Peninsula of Michigan, through Manitoulin Island and the Bruce Peninsula into southern Ontario and into the Niagara area of New York" (Luczaj 2013, 1).

FIGURE 2.7. Sven's Bluff Overlook atop Niagara Escarpment, overlooking the Bay of Green Bay in Peninsula State Park, Door County. Eight miles due east, marshes line the Lake Michigan shore.

The Niagara Escarpment is the most conspicuous topographic feature in the eastern third of Wisconsin, constituting the surface water drainage divide between the Fox Valley and Lake Michigan (Mickelson and Socha 2017). Because the layer of dolomite is dipping toward the east, the western exposure of the resistant rock unit is marked by an escarpment that corresponds to the steep edge of the cuesta. The asymmetrical character of the cuesta can clearly be seen in parts of Door County (Figure 2.7). Just north of Ellison Bay, the cliffs of Ellison Bluff rise over 200 feet above Green Bay. In contrast, on the eastern shore of the Door Peninsula, such as near Newport State Park, the dolomite gently dips below the waters of Lake Michigan. Where wave-cut cliffs exist, many are no more than 4 to 6 feet in height.

The Niagara Escarpment parallels the eastern shore of Lake Winnebago, with steep bluffs seen in places, such as at High Cliff State Park (Figure 2.8). Farther south the heights atop the cuesta, located nearly 300 feet above the level of Lake Winnebago, are the site of an extensive wind farm. The cuesta continues south of Fond du Lac and faces the eastern edge of Horicon Marsh. While the rock units that comprise the escarpment have been traced into Illinois, erosion and glacial deposition have muted the escarpment's topographic exposure. Nevertheless, the escarpment can be seen in places as far south as Waukesha County (Luczaj 2013).

FIGURE 2.8. Cracked blocks of dolomite exposed along the crest of Niagara Escarpment in High Cliff State Park in Calumet County. Ice-covered waters of Lake Winnebago are visible in the distance.

## GLACIATION SHAPES WISCONSIN'S PHYSICAL LANDSCAPE

Wisconsin's landscape has been profoundly influenced by multiple continental ice sheets that moved across most of the state over the past million years, scouring extant rocks and soils, reshaping the effects of earlier glacial advances and retreats, burying preexisting river valleys, and depositing glacial debris in multiple forms and shapes. The geologic consequences of the last episode of glaciation during the Pleistocene are so exquisitely and abundantly displayed in Wisconsin that North American geologists named it the Wisconsin or Wisconsinan glaciation. The Wisconsin glaciation began roughly 100,000 years ago. It was marked by a series of advances and retreats, before leaving the state around 11,700 years ago (Mickelson and Attig 2017). The Wisconsin glaciation created most of the state's glacial features, but the glaciated areas south of Madison in Dane, Monroe, and Green Counties, as well as a broad swath of Wisconsin extending west from Marathon and western Waupaca Counties into Pierce and Pepin Counties, show evidence of previous glaciations, including both the Illinoian glaciation that preceded the Wisconsin glaciation and earlier glacial advances. But that area was not covered by ice during the Wisconsin glaciation. In total, about three-quarters of Wisconsin's land area is covered by glacial deposits (Syverson and Colgan 2004). An area of west-central and southwest Wisconsin never experienced flowing ice from any of the Pleistocene glaciations, and this land is designated the Driftless Area. Given that the Wisconsin glaciation so distinctly shaped much of the state's landscape, its numerous consequences and features will be explored—and indeed a hiker can see them firsthand by walking the 1,200-mile Ice Age National Scenic Trail (Mickelson, Maher, and Simpson 2011).

### Glacial Lobes

As glaciers moved across the surface of Wisconsin, soil and unconsolidated earth materials were pushed by the ice. Rock often became incorporated in the ice, and as it moved across exposed bedrock

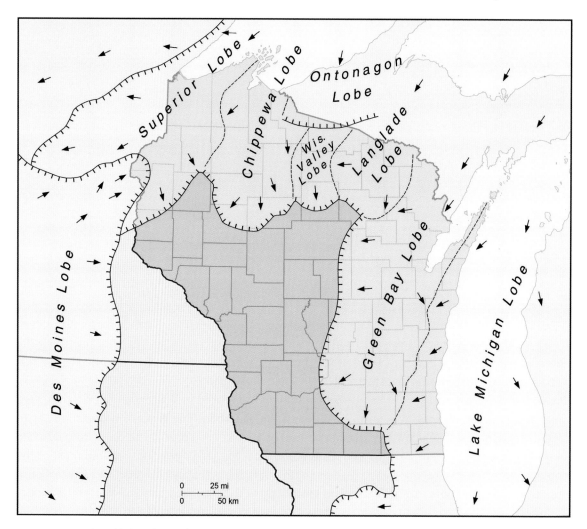

FIGURE 2.9. Glacial lobes during the Wisconsin glaciation within Wisconsin. Map drawn after a similar image in Clayton, Attig, Mickelson et al. (2006). Used with permission of Wisconsin Geological and Natural History Survey (University of Wisconsin–Madison, Division of Extension).

it scraped groves or scratches in that rock. These striations, which can be seen on exposed bedrock today, indicate the ice's direction of movement. The ice generally moved from the north, often following the alignment of preexisting valleys. The separate advancing tongues of ice are referred to as glacial lobes (Figure 2.9). The huge mass of ice that scoured the valley that is now occupied by Lake Michigan was the Lake Michigan Lobe. Just to the west, another lobe centered on the area west of the Niagara Escarpment, advancing from what is now Green Bay. It is termed the Green Bay Lobe. The primary lobes of the Laurentide Ice Sheet associated with the Wisconsin glaciation in northern Wisconsin, westward of the Green Bay Lobe, are the Langlade, Wisconsin Valley, Chippewa, and Superior Lobes. A small portion of Wisconsin was glaciated by a branch of the Des Moines Lobe, the Grantsburg sublobe, which moved from the west. Scouring was particularly effective in the area that became Lake Superior, whose greatest depth—the deepest of any of the Great Lakes—reaches 733 feet below sea level.

The movement of the glacial lobes smoothed Wisconsin's topography by scouring preexisting hills and rises—particularly if they were not outcrops of highly resistant rock—and by filling preexisting valleys. For example, when one considers the depth to bedrock (Figure 2.10), nearly a third of northern Wisconsin appears to have bedrock that is buried by at least 100 feet. In much of Bayfield County, the depth to bedrock is between 400 and 500 feet (Trotta and Cotter 1973), with the thickest glacial deposits corresponding to the location of an interlobate moraine between the Chippewa Lobe and the Superior Lobe, indicating the role of both glacial movement and subsequent glacial melting in depositing glacial materials. Within east-central and northeast Wisconsin one can trace an area of deep valley fill extending northeastward from Portage, where it joins today's Wisconsin River. This deep infill marks the preglacial route of the Wisconsin River, which made a long loop through today's Marquette and Green Lake Counties and two of its major tributaries, the Fox and Wolf Rivers. Thus, it has been argued that today's Green Lake and Lake Puckaway occupy parts of the former channel of the Wisconsin River, and that the preglacial Fox River drained in the opposite direction of its present course (Fleming 1985).

## Glacial Geomorphology

The most conspicuous geomorphic feature left by the glacial lobes is the ridge of glacial debris that marked the outermost reach of the ice. These terminal moraines or end moraines (Figure 2.11) are particularly impressive where the ice margin rested at a location for a lengthy time, with the rate of forward movement of the ice—including the rock material at its base, incorporated in the ice, and atop the ice—being balanced by the rate of its melting. The character of the land surface and bedrock over which the glacier moved, plus the thickness of the ice itself, influenced the amount of debris that could be scoured and deposited. End moraines in northern Wisconsin are particularly large, such as the five-mile wide Chippewa Moraine (Figure 2.12) that is marked by a variety of deep kettle lakes and high ice-walled hummocks near Plummer Lake (Mickelson, Maher, and Simpson 2011). Another conspicuous view of a terminal moraine is from State Highway 21 to the west of Coloma. There, from the Adams-Waushara County line, a person looking east from an extensive outwash plain sees the Johnstown moraine, rising over 100 feet in a distance of a half mile. This moraine marks the limit of the

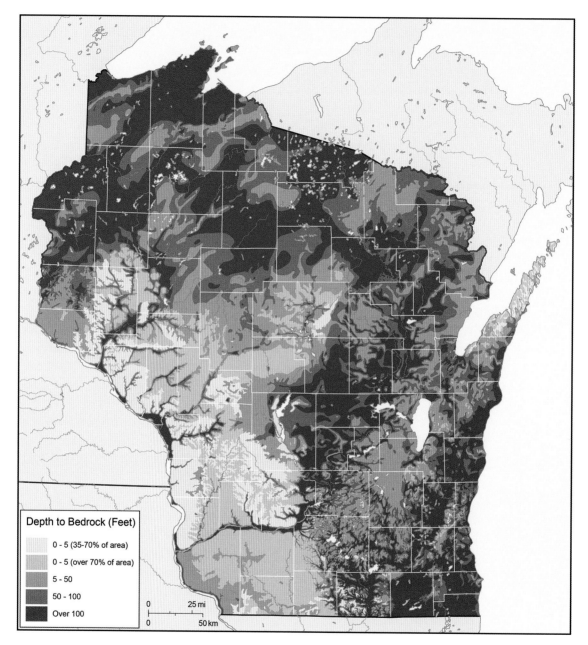

FIGURE 2.10. Depth to bedrock map. Map redrafted based upon Trotta and Cotter (1973), with data from Wisconsin Department of Natural Resources Open Data website (https://data-wi-dnr.opendata.arcgis.com/datasets/bd2c357cc13a47569ee625e49bab3a0e_0). Used with permission of Wisconsin Geological and Natural History Survey (University of Wisconsin–Madison, Division of Extension).

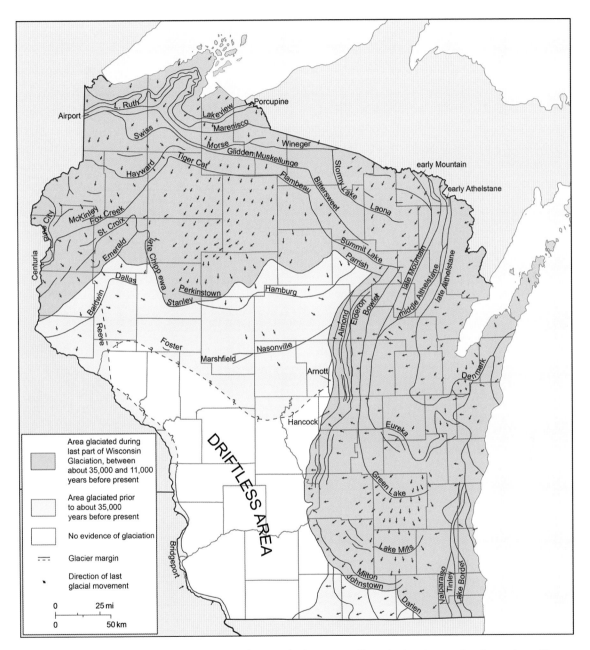

FIGURE 2.11. Glaciation within Wisconsin, indicating the locations of terminal and recessional moraines. Map drawn after a similar image in Clayton, Attig, Mickelson et al. (2006). Used with permission of Wisconsin Geological and Natural History Survey (University of Wisconsin–Madison, Division of Extension).

Green Bay Lobe, and its undulating moraine ridges are nearly three miles wide. In contrast to these conspicuous terminal moraines, throughout "much of southern Wisconsin, moraines are only about 50 feet high and a quarter to a half mile wide" (Mickelson, Maher, and Simpson 2011, 35). In a few cases, such as in northern Dane County, no end moraine is visible.

Terminal or end moraines are typically the most prominent type of moraine, while recessional moraines are low to moderate ridges of glacial debris that mark the location where a retreating ice sheet (its rate of melting exceeding the rate of ice movement) might pause for a number of decades. Although it was located behind the end moraine, and possibly other recessional moraines, the longer the advancing ice bringing glacial debris was balanced by ice melting at its location, the greater the accumulation of sand, gravels, and boulders in the moraine. Along Interstate Highway 41 near the Winnebago–Fond du Lac County line one crosses the Eureka Moraine, which over a distance of a mile rises and falls 60 feet. It likely marks a moraine of red till that buried an earlier moraine (Hooyer and Mode 2008). Farther south near Hartford the same highway crosses the Kettle Moraine, which is an interlobate moraine that consists of a large and complex assemblage of glacial deposits and outwash features that developed between the Green Bay and Lake Michigan Lobes. The conspicuous Kettle

FIGURE 2.12. Chippewa Moraine with numerous kettle lakes in aerial view east of New Auburn, Chippewa County. In contrast with the rough terrain within the moraine, farm fields occupy the smoother outwash plain southwest (left) of the moraine.

FIGURE 2.13. Field strewn with glacial boulders, west of Whitcomb (Town of Wittenberg) in Shawano County.

Moraine, which in places displays nearly 300 feet of local relief, extends from Kewaunee County into Walworth County.

Glacial deposits, called drift or till, can be found throughout the area between the terminal and recessional moraines. Where stagnant ice simply melted, a wide variety of debris may be deposited. A distinction between glacial deposits and water- or wind-deposited sediments is that till is highly heterogeneous, varying not only by rock type but by size. Some of the till material is clay or silt sized, yet it may also contain large boulders. Many glacial boulders are erratics, rocks that have been transported considerable distances and are not from the local bedrock. Thus, it is not unusual for farmers in southern Wisconsin to encounter granite boulders that have come from the Precambrian shield a hundred or more miles to the north. In addition, certain areas of Wisconsin, such as parts of central Shawano County, are known for their abundant large glacial erratics (Figure 2.13), making it difficult for farmers to plow fields.

Moving glacial ice sometimes shaped the underlying glacial debris into streamlined hills, called drumlins. Wisconsin is one of the best places in the world to see such features, and geologists mapped over 2,000 drumlins in the area of the former Green Bay Lobe (Zakrzewska-Borowiecki 1975). Drumlins, which vary considerably in both size and elongation, are typically elliptical, aligned in the direction

FIGURE 2.14. Drumlins dominate the landscape west of Campbellsport in southern Fond du Lac County.

FIGURE 2.15. Field with McMullen Hill, a kame formed from glacial debris, within Kettle Moraine State Forest northwest of Parnell in southwestern Sheboygan County.

of the ice movement. They resemble hills, which in parts of eastern Fond du Lac and Dodge Counties resemble inverted spoons, with their steeper slope facing the direction from which the ice was moving (Figure 2.14). Within the Campbellsport Drumlins Unit of the Ice Age National Scientific Reserve, "the drumlins are generally 60–120 ft high, rounded, irregular, to elliptical. Elongation ratios commonly are less than 2:1" (Black 1974, 63). In contrast, within western Dodge County the drumlins are far more elongated, resembling high embankments, rising 20 to 60 feet above the intervening swales that extend a mile or more, typically in a north to south direction. Regardless of where drumlins are found, they occur in large groups, comprising drumlin fields, and it is not unusual to find contiguous or overlapping drumlins.

Along the margins of glacial lobes, both in large terminal moraines and interlobate moraines, such as the Kettle Moraine, a variety of glacial features were formed by uneven melting of the ice. Kettles are depressions formed where blocks of ice were surrounded by glacial debris or the outwash gravels from a melting glacier. After the ice melted a kettle hole or depression remained, now often occupied by a small lake, surrounded by glacial or outwash material. Kames—or, as some geologists more specifically note, moulin kames—developed where holes in the melting ice were subsequently filled by glacial debris transported by meltwater. Within the Kettle Moraine, which contains particularly large kames in abundance, some kames rise over 100 feet above their surroundings near Dundee (Figure 2.15). Larger ice-walled-lake plains resemble hills, yet lake sediment occupies the hilltop rather than the surrounding lowlands—particularly where the bed of the lake formed on glacial till. At times the lake formed on glacial ice, which itself was covered with considerable glacial debris and other sediment. When the ice melted, a far more hummocky terrain resulted (Clayton, Attig, Ham et al. 2008), which is particularly well developed and preserved in the Chippewa Moraine Unit of the Ice Age National Scientific Reserve east of New Auburn. There the terminal moraines "are hummocky complexes up to 20 km [13 miles] wide that contain supraglacial sediment and numerous ice-walled-lake plains" (Syverson and Colgan 2004, 305). Eskers are sinuous accumulations of water-transported sands and gravels that mark the location of streams that had flowed atop or in tunnels under stagnant glacial ice, with prominent examples (Figure 2.16) found in both the Kettle Moraine and the Chippewa Moraine.

## Glacial Outwash and Glacial Lakes

The torrent of water from the melting Pleistocene glaciers shaped the physical landscape far beyond the greatest areal extent of the glacial ice. This is particularly visible within central Wisconsin, where the Green Bay Lobe blocked preexisting rivers, causing large lakes to form, whose outflowing waters carved new channels. Thus, the drainage network seen today differs from what existed before the Pleistocene or even the Wisconsin stage of the Pleistocene. Much of the physical landscape of central Wisconsin is best understood from the perspective of a giant lake bed filled with lacustrine deposits, rather than just the glaciers that fed waters to it.

The meltwater transported huge quantities of glacial debris, but unlike glacial till that is an unsorted mixture of many sized materials, the outwash is sorted and stratified. Heavy boulders were transported short distances. As the velocity and depth of the flowing water decreased, heavier materials were

FIGURE 2.16. Trail along top of Parnell Esker in Kettle Moraine State Forest west of Parnell in southwestern Sheboygan County.

deposited first. Gravels and sands settled out next, while fine-grain silt and clay were carried the greatest distance. Outwash accumulations—particularly of the larger sands, gravels, and boulders—are the greatest near the margins of the melting glacier, and an advancing glacier incorporated this debris into its terminal moraine. Where the outwash was deposited upon stagnant or retreating ice, kames, kettles, and hummocky terrain, as we previously discussed, developed. In front of the moraine, outwash fans—similar to alluvial fans seen where a mountain stream flows into a broad valley—and outwash plains developed. The character of these deposits was influenced by whether outwash water could readily flow away from the glacier. In some locations topography or other glacial ice blocked the water, forming an ice-marginal lake or a proglacial lake, its larger version.

Within central Wisconsin the preexisting Wisconsin River channel was blocked by the Green Bay Lobe both near Portage and the Baraboo Range. Glacial Lake Wisconsin covered an area of over 1,800 square miles to a maximum depth of 150 feet (Schultz 2004), at which point the water accessed an outlet to the Black River, which drained into the Mississippi River. During its roughly 5,000 years of existence (Clayton and Knox 2008), extensive outwash plains extended west from the Green Bay Lobe. Shorelines with beaches developed along the lake's margins, from which wind-blown sands were deposited. The preexisting valley channels were largely filled with lake sediments, resulting in today's broad, poorly drained landscape. Even though the area was not glaciated, glacial erratics can be found, particularly near the former shorelines, where icebergs floating on the lake transported glacial material. The presence of scattered steep-sided sedimentary mesas and buttes, many too small to have withstood the advancement of any glacier ice, provides additional evidence that the land was not glaciated, yet the sandy outwash around their bases is from the glaciations. Sandstone buttes and mesas, such as those at and near Mill Bluff State Park, Roche a Cri State Park, and many similar features in Adams and Juneau Counties, were islands during the Wisconsin glaciation (Figure 2.17). Some buttes are so heavily marked by wave-action that they resemble sea stacks, such as Devil's Needle near Camp Douglas (Black 1974).

Glacial Lake Oshkosh not only covered the area occupied by today's Lake Winnebago, but at times, depending upon the exact location of the Green Bay Lobe, which fluctuated over time, it covered most of Outagamie and Winnebago Counties, eastern Waupaca and Waushara Counties, plus half of Brown County and the easternmost five or so miles of Oconto and Marinette Counties. The ice sheet's location also controlled the lake's drainage and depth, which at its maximum was about 65 feet deeper than today's shallow Lake Winnebago. At higher stages, Glacial Lake Oshkosh drained

FIGURE 2.17. Shiprock, a butte of sedimentary rock surrounded by glacial outwash, along State Highway 21 in the Town of Richfield, Adams County.

into the Wisconsin River near Portage. At other times it drained toward Lake Michigan, such as through a valley partly occupied by the Manitowoc River today when the northernmost segment of the Lower Fox River Valley was blocked by ice. Other outlet valleys from Glacial Lake Oshkosh have been identified along the Neshota, Kewaunee, and Ahnapee Rivers, which "carried water along northwest-southeast trending bedrock controlled valleys that cut across the Niagara cuesta" (Luczaj 2013, 25). Thus, large areas of the region have thick deposits of lake silts and clay that overlie the glacial till, plus evidence of glacial sluiceways.

Extensive glacial lake deposits are also found in northern Wisconsin, particularly within Douglas, Bayfield, and Ashland Counties, where the waters of Glacial Lake Duluth rose until outlets developed leading to the St. Croix River. Smaller glacial lakes also developed south of Glacial Lake Wisconsin proper, including Glacial Lake Baraboo, actually a bay of Glacial Lake Wisconsin, and Glacial Lake Middleton.

### St. Croix Dalles, Wisconsin Dells, and Baraboo Range

The narrow 100-foot-deep gorge that surrounds the Wisconsin River at Wisconsin Dells was rapidly carved along fractures in the sandstone when Glacial Lake Wisconsin abruptly drained when the ice

FIGURE 2.18. Giant glacial potholes on ledge overlooking the St. Croix River at The Dalles of the St. Croix River within Interstate State Park, just south of St. Croix Falls, Polk County.

dam blocking its drainage was breached. The Dells, averaging 150 to 200 feet in width but narrowing to 50 feet in one place, are cut across a former drainage divide as are several of its contemporary tributary streams. It is a classic example of the extraordinarily rapid geologic scouring that can occur during catastrophic flooding (Clayton and Knox 2008). This is not the only location in Wisconsin where glacial meltwaters carved steep gorges.

The St. Croix River gorge, at Interstate State Park, experienced a megaflood of waters from Glacial Lake Duluth and Glacial Lake Agassiz to its north. These waters entered the Moose Lake outlet from Lake Superior and then the Brule–St. Croix outlet around 16,000 years ago when glacial ice blocked the eastward drainage of Lake Superior (Mickelson, Maher, and Simpson 2011). The St. Croix Dalles, with its giant potholes carved in the basalt (Figure 2.18) 100 feet above today's river level, indicates the magnitude of the flood. On the Wisconsin side of the St. Croix Dalles, the "potholes are as much as 6 by 10 ft across and 15 ft deep" (Black 1974, 181). They are larger on the Minnesota side of the gorge. When glacial ice blocking Lake Superior melted, the direction of drainage of the Brule River reversed itself. It now drains into Lake Superior, with a portage connecting it to the current St. Croix watershed.

The Baraboo Range, an ancient partly exhumed Precambrian quartzite monadnock, provides another fascinating example of how the glaciation reshaped the physical landscape and altered the drainage pattern of the Wisconsin River. While the gorge through the South Range of the Baraboo Range dates to Precambrian times, as Cambrian-age sandstones were deposited atop the eroded quartzite within the gorge, during the Pleistocene rivers again flowed through the gorge (R. A. Davis 2016). Some geologists have argued that the Wisconsin River entered the Lower Narrows (where the Baraboo River flows today) and then crossed the South Range through the gorge now occupied by Devil's Lake before the Wisconsin glaciations (Schultz 2004). Others claim that Wisconsin River flowage by way of Devil's Lake Gorge may have occurred when the advancing Green Bay Lobe blocked the valley east of the Baraboo Range (Lytwyn 2010). Given that the Baraboo Range is located near the greatest advance of the Green Bay Lobe, the summit of East Bluff is unglaciated, but one arm of ice entered the gorge from the north, while another entered it from the southeast. Terminal moraines were formed blocking the northern and southern approaches to the gorge, with Devil's Lake forming within the gorge between the two moraines (Figure 2.19). Thus, Devil's Lake State Park sits astride the boundary between the glaciated region and the unglaciated Driftless Area.

## THE DRIFTLESS AREA

Southwestern and west-central Wisconsin, plus a small part of northwest Illinois, were never glaciated during the Pleistocene and are called the Driftless Area, implying it lacks glacial drift or till deposits. Various hypotheses have been made to explain this absence of glaciation, which is likely related to a combination of factors (J. C. Knox 2019a). These include the area's somewhat greater elevation (500 feet higher than along the Wisconsin River in central Wisconsin), the fractured sedimentary bedrock to the north and northwest of the region, and the distance from much thicker glacial ice comprising the Lake Superior and Lake Michigan Lobes. The great depth of the Lake Superior

FIGURE 2.19. East Bluff rises nearly 500 feet above the frozen surface of Devil's Lake in Devil's Lake State Park, south of Baraboo.

Basin helped retard "ice advance into central Wisconsin and perhaps funnel[ed] ice to the west and east, leaving the Driftless area unglaciated" (Syverson and Colgan 2004, 306).

It is unclear how much of the Driftless Area was actually unglaciated. Arguments have been made that glaciation occurred in part of the northern half of the traditional Driftless Area early in the Pleistocene, well before the Illinoian glaciations (thus before 780,000 BP), with glacial evidence being largely removed by subsequent erosion. It appears that early glacial ice from the west pushed a mile or two across the Mississippi River into the mouth of the Wisconsin River (Mickelson, Knox, and Clayton 1982; Syverson and Colgan 2004) and a similar distance across the bend in the Mississippi River near Glen Haven (Carson, Attig, and Rawling 2019).

The Driftless Area is conspicuous because of its lack of glacial till, moraines, and glacial erratics and the presence of steep, well-incised valleys, buttes, and mesas that could not have survived glaciation. However, glaciation had many effects on the region. While the Central Sand Plain that occupies the flat lake bed of former Glacial Lake Wisconsin "is generally considered to form the northernmost part of the Driftless Area" (Mason 2015, 365), the area to its west and southwest comprises the archetypical Driftless Area.

## Distinctive Topography of the Driftless Area

In parts of the Driftless Area the local relief, the vertical distance between valley bottoms and ridge crests, often exceeds 500 feet. Unlike most of Wisconsin, where fluvial erosion of the landscape has only been able to proceed during the 10,000 to 12,000 years since the melting of the ice sheets, within the Driftless Area such stream erosion has proceeded for millions of years. Thus, the drainage network is far better developed, with more intricate dendritic arrangements of rivers and tributary streams. Lakes are virtually absent, and marshland is far less conspicuous. Because the bedrock consists of relatively flat-lying sedimentary rocks of varying resistance to erosion, steep slopes and some cliffs have developed where resistant sandstones and dolomite bedrock prevails. Overall, the slopes are longer and steeper than where glaciation occurred, yet steep hillslopes have also been attributed to periglacial conditions, including solifluction, that prevailed when the region was surrounded on three sides by glacial lobes.

The portion of the Driftless Area between the Chippewa and Wisconsin Rivers is particularly noteworthy for its highly incised topography (Figure 2.20). Little level upland remains, as "the area is a maze of stream-eroded ridges and valleys" (Schultz 2004, 114). Where the ridges are capped with

FIGURE 2.20. Deeply eroded valley along County Highway G in the Town of Milton, Buffalo County, within the Upper Coulee Country north of Fountain City.

dolomite, cliffs and rough slopes may occur, whereas where slightly older sandstones are exposed, the ridges are "lower and more rounded" (Schultz 2004, 115). The narrow, steep-sided valleys that dominate this region are locally called coulees, with Coulee Country extending from the St. Croix River to the Wisconsin River. One of the distinctions between coulees and valleys that are called hollows south of the Wisconsin River is that coulees "are steep-walled, tributary valleys, with sandy beds that are occupied by water flow only intermittently [as] most of the time their stream beds are dry" (Mather 1977, 25). Most coulees extend only several miles before abruptly ending at a ridge.

South of the Wisconsin River the topography changes. While the north-facing slope is quite steep and marked by many tributary valleys, its crest coincides with the Galena-Platteville Escarpment. This escarpment marks the location of a cuesta that follows the exposure of a resistant layer of dolomite that gently dips southward. It acts as a caprock for the flat to gently rolling land to the south. Locally, this escarpment is the Military Ridge, so named for the road constructed in 1835 along this high point between Madison and just south of Prairie du Chien. In his seminal review of Wisconsin's physical geography, Lawrence Martin noted, "The Military Ridge is not a symmetrical ridge with even slopes on either sides. It is an exceedingly unsymmetrical feature, with a short, steep, northern slope toward the Wisconsin [River] and a long, gentle descent southward to the Illinois line." When a railroad was constructed along the Military Ridge in 1881, it was "the longest stretch of railway in the state without a bridge over a stream" (Martin 1965, 62), as it marked a divide. Today the Military Ridge State Trail runs on the former railroad grade, from which one can readily see the change in topography depending upon whether one looks north or south (Figures 2.21 and 2.22).

The highest elevation in southern Wisconsin is West Blue Mound atop Military Ridge, where the cuesta is capped by a layer of Niagara Dolomite. At 1,719 feet in elevation, it is nearly 500 feet higher than the surrounding plain several miles to the south, and its summit is 1,000 feet above the Wisconsin River to the north. South from the Military Ridge a much greater proportion of the land comprises relatively flat interfluvial upland areas between scattered river valleys.

## Glaciation's Effects on the Driftless Area

Although glacial ice never covered the landscape of the Driftless Area from central La Crosse and Monroe Counties southward—and there is uncertainty as to whether Pre-Illinoian glaciation overrode the area northward to a line extending roughly from the northern border of Pepin County into northern Wood County—the Late Wisconsin Glaciation between 10,000 and 30,000 years ago strongly influenced the physical landscape. During the peak of the glaciation, the climate of the region was far colder than today. Cold periglacial conditions resulted in accelerated rates of mass wasting of the region's hillsides, as development of permafrost in the ground enabled solifluction and other mass movement to erode the slopes (Mason and Knox 1997). With permafrost a variety of periglacial features appeared on the landscape, including boulder streams, rubble mantles, and talus cones. Ice wedge–cast polygons on the ground are visible when viewed from above, given that the materials that filled them often differed in their water-holding characteristics (Clayton, Attig, and Michelson 2001). Talus slopes along Devil's Lake are classic examples of a periglacial feature, just as are the nearby boulder trains that formed over permafrost.

FIGURE 2.21. Sloping pastures and forested stream valleys looking north from Military Ridge toward the Wisconsin River Valley. Photograph taken from Military Ridge State Trail in Iowa County just a short distance from Figure 2.22.

FIGURE 2.22. Cornfield on nearly flat land extending south from Military Ridge. Photograph taken from Military Ridge State Trail east of Dodgeville in Iowa County.

The greatest glacial influence upon the Driftless Area came from the vast rivers that transported flood waters and outwash from the melting glaciers. Not only was the Mississippi River a massive outwash river fed by the Lake Superior Lobe and at times by outflow from Glacial Lake Aggasiz, which covered a large area to the north and west of present Lake Superior, but at times the St. Croix River brought meltwater from the Lake Superior Lobe. Farther south, the Chippewa River conveyed meltwater from the Lake Superior and Chippewa Lobes. The Black River brought water from the previously mentioned high-water outlet along Glacial Lake Wisconsin, but it also channeled water from tributaries draining the Chippewa, Wisconsin Valley, and Langlade Lobes. The Wisconsin River drained floodwaters from Glacial Lake Wisconsin once the ice dam along the edge of the Baraboo Range was breached. Its waters came from the Wisconsin Valley and Langlade Lobes to the north and the Green Bay Lobe to the east. As previously noted, waters from Glacial Lake Oshkosh drained into the Wisconsin River when its northeastern outlets were blocked by ice. While not within the Driftless Area, the Rock River that drained south-central Wisconsin also served as a glacial sluiceway.

These massive flows created glacial sluiceways or spillways, carving valleys that are far larger than what their rivers could erode today. Some of the rivers transported several orders of magnitude more

FIGURE 2.23. Floodplain prairie occupying Upper Dunnville Bottoms from Red Cedar State Trail southeast of Dunnville, Dunn County. The Chippewa River is barely visible in the distance, and hills to the left mark the edge of the glacial sluiceway.

water during their peak megafloods. Today we can describe the Chippewa, Black, and Wisconsin Rivers as misfit rivers; their valleys are simply too large relative to the small-sized rivers flowing through them. For example, the Dunnville Bottoms (Figure 2.23), along the Chippewa River just west of where the Red Cedar River joins it, are two to three miles wide. Within these river valleys, as well as the Mississippi River into which they flowed, huge quantities of glacial outwash were transported and deposited, leaving "ribbons of valley-train drift across an otherwise drift-free region" (Schultz 2004, 178). Throughout Dunn and Pepin Counties, the accumulation of fill above the bedrock within the Chippewa River Valley exceeds 200 feet, and it likewise surpasses 200 feet along the Wisconsin River from Portage to Prairie du Chien. Its depth along much of the Black River ranges between 100 and 200 feet (Trotta and Cotter 1973).

The dumping of so much outwash material into the Mississippi River not only raised the river bed but also reconfigured the channel. Two examples are noteworthy. Where the Chippewa River flowed into the Mississippi Valley, its waters slowed, and a large delta formed that dammed the Mississippi River. Lake Pepin occupied the valley upriver to near St. Paul, Minnesota, and it extended into what later became Lake St. Croix (Blumentritt, Wright, and Stefanova 2009). Another delta began forming from Mississippi River waters flowing into Lake Pepin, which not only cut off the mouth of the St. Croix River near Prescott but has extended well south of Hager City. Today, Lake Pepin occupies the river channel, extending over 2 miles across the valley and 22 miles upriver. Farther downriver, the Black River likewise formed a large delta where it flowed into the Mississippi River. In this case, the debris spread into the valley that existed between Trempealeau Bluffs, which include both Trempealeau Mountain and Brady's Bluff that are now largely within Perrot State Park, and bluffs 3 to 5 miles distant to the north and east—the valley that had previously been occupied by the Mississippi River. With the Mississippi River waters rising about 150 feet because of the delta blockage, the river began flowing and excavating in a narrower valley that separated the Trempealeau Bluffs from the Minnesota Upland (Martin 1965). That valley is now occupied by the main channel of the Mississippi River. From the broad plain of the former river channel can be seen several conspicuous river terraces, particularly to the east of Trempealeau (Figure 2.24). The uppermost terrace indicates the highest deposition of glacial outwash materials when the glacial sluiceways were most active, while the lower terraces marked levels at which the river deposited its load later. Once the outwash was deposited, later river flowage began to erode the outwash, marking the various river stages.

River terraces mark both the Mississippi River Valley and the glacial sluiceways. Both Pepin and Trempealeau are located largely on terraces about 45 and 60 feet, respectively, above today's Mississippi River. East of Lake Pepin are several distinct terraces, with one near the Chippewa River displaying a 95-foot scarp. At La Crosse only one terrace is readily visible, "bordered by a steep west-facing scarp 30 feet high" (Martin 1965, 156). In nearby Onalaska four terraces "separated by steep terrace scarps" are identifiable, rising 25 to 35 feet, 50 feet, 80 feet, and possibly 100 feet, as that level may mark postglacial sand dunes (Martin 1965, 158). The terrace at Prairie du Chien, which runs from 25 to 50 feet, "seems to be a giant alluvial fan" (Martin 1965, 158) that formed from the deposition of outwash from the Wisconsin River. The variability in the terrace heights is a function of the differing volume

FIGURE 2.24. River terrace facing Mississippi River viewed from Great River State Trail east of Trempealeau.

of water and outwash loads that the Mississippi River received in its various reaches, strongly influenced by the characteristics of its various tributaries—several of which were major glacial sluiceways.

Terraces also exist along the misfit tributary rivers. Subsequent fluvial erosion from the much smaller postglacial Wisconsin River has left "a series of three westward-grading stairstep terraces parallel to the main channel" (Clayton and Knox 2008, 388) between Lone Rock and Boscobel, of which the most prominent are located 56 to 75 feet above today's river. Similarly, along the Chippewa River one finds Durand located on a terrace 40 to 60 feet above the river. If it were not for these terraces, relatively little flat land would be found in the valleys cutting across the Driftless Area.

Meltwater from the glaciers contained vast quantities of glacial silt. Where this was deposited in river valleys, large quantities were blown from the west in the winter during low water levels. Additional amounts blew from glacial outwash deposits in eastern Iowa. Within much of the Driftless Area adjacent to the Mississippi River this aeolian silt, named loess, was deposited in a 4- to 16-foot-deep layer atop the land. In a few locations, it is nearly 60 feet deep. Thus, "loess is by far the most extensive surface material of Driftless Area uplands" (Mason, Jacobs, and Leigh 2019, 62). Even as far east as western Kenosha and Racine Counties, loess accumulation exceeded 2 feet in places. Much of the loess within central and eastern Wisconsin is derived from glacial lake beds, outwash plains, and even moraines, and as such, it is sandier than deposits closer to the Mississippi Valley (Scull and Schaetzl 2011). Thus loess, which originated with glacial erosion and transport, was deposited by winds on both glaciated and unglaciated lands. Within the Driftless Area the thickness of the loess is typically the greatest on broad ridgetops and decreases "substantially in thickness or pinch[es] out entirely from ridge summits to the convex shoulder slopes below" (Mason 2015, 369). Loess deposits are highly vulnerable to gully erosion. Agricultural practices beginning in the nineteenth century "increased erosion and sedimentation rates to levels that were unprecedented in preceding post-glacial time" (J. C. Knox 2019a, 29).

The Central Sand Plain that formed east of the Wisconsin River, occupying the bed of Glacial Lake Wisconsin that was closest to the terminal moraines and sources of the glacial outwash, displays relict sand dunes that date to the end of the Pleistocene. The surface nearest the terminus of the Green Bay Lobe is dominated by sands, and in places winds blew the sands into both parabolic and transverse dunes. The largest dunes range from 6 to 60 feet in height (Figure 2.25), with the dune field covering about 56 square miles. They formed following the drainage of the glacial lake, possibly during a drier climate or when melting permafrost exposed the lake sediments to erosion (Rawling et al. 2008). Unlike areas along the Great Lakes where winds continue to reshape sand dunes, those in central Wisconsin have remained inactive for the past 10,000 years and are largely covered with vegetation.

FIGURE 2.25. Relict sand dunes in Central Sand Hills area northwest of Nelsonville, Portage County.

## The Great Lakes

The Great Lakes are another lasting legacy of the Pleistocene. The area occupied by Lake Michigan had been a river valley whose course was determined by exposures of weak bedrock layers around the Michigan Basin. The Lake Michigan Lobe of glacial ice scoured, widened, and deepened the pre-existing valley, so that the deepest lake floor is nearly 350 feet below the water surface. The shoreline features along Lake Michigan display the various stands of water since the melting of the glacial ice, the role of isostatic rebound, and normal erosional and depositional processes that occur along the shore. Because retreating ice once blocked the flowage of meltwaters to the north, the shore of what is now southern Lake Michigan was initially located farther inland than today, with lake waters draining into the Mississippi River system by crossing a low divide in Chicago. At other times water levels were considerably lower than at present, and river valleys extended into the lake. For example, the Bay of Green Bay now occupies a drowned valley that was formerly occupied by a series of shallow lakes.

During the Pleistocene, the mass of glacial ice depressed the land surface, which simultaneously raised the elevation of a broad swath of land beyond the ice. With the melting of the ice, the surface that it covered began rising, an isostatic process that has yet to be completed. Because rebound or subsidence at a given location determines how its water level relates to lake outlets, which are also subject to vertical displacement, different locations along the Great Lakes are subjected to either rising or falling lake levels. In Wisconsin, Milwaukee has subsided by 14.4 cm (5.7 inches) per century, relative to the outlet of Lake Huron, while Sturgeon Bay's rate is 3.8 cm (1.5 inches) per century. In contrast, the port of Duluth-Superior subsidence rate is 25.3 cm (10 inches) per century, relative to the outlet of Lake Superior (Mainville and Craymer 2005). Where subsidence is high, the shoreline is subject to flooding and erosion, resulting in cliffs and shoreline retreat, which occur at many locations along Lakes Superior and Michigan.

Several harbors along Lake Superior represent partly drowned river valleys, which are fresh-water estuaries, such as the St. Louis River near Superior. Although marshes exist behind the sand spits that protect the harbors at Superior, Port Wing, and Ashland, much of Wisconsin's Lake Superior shore is marked by steep cliffs and eroding headlands. In many places glacial deposits are being eroded, while elsewhere it is solid rock. In these locations one can find red sandstone cliffs of 10 to 30 feet, wave-cut arches, and sea stacks. Among the most scenic of these rocky shorelines is the Pictured Rocks section of the Apostle Islands National Lakeshore.

Nearly 500 feet above today's Lake Superior shoreline are abandoned beaches and wave-cut cliffs created along Glacial Lake Duluth when its waters drained into the St. Croix River. Because of uneven glacioisostatic rebound, these beaches are found "45 feet higher above Lake Superior near the Montreal River than at Superior" (Martin 1965, 463). Evidence of several other shorelines, marking intermediate levels between Glacial Lake Duluth and Lake Superior, also exist.

Wisconsin's shoreline along Lake Michigan lacks the cliffs that formed along much of Lake Superior, except along the Bay of Green Bay's eastern shore. Where the Niagara Escarpment follows the western edge of the Door Peninsula, cliffs may be seen rising 100 to 200 feet above the waters of Green Bay. Examples are found in Peninsula State Park and near Ellison Bay. In contrast, the western shore of Green Bay is low and marshy.

The shore along Lake Michigan proper has a few low cliffs, four to six feet in height, cut in sedimentary rocks near Newport State Park and north of Whitefish Dunes State Park—both in Door County, but most of the shore consists of cliffs of eroding glacial deposits or sandy beaches, backed by sand dunes in places. The largest complex of dunes on the Wisconsin side of Lake Michigan is at Whitefish Dunes State Park. Noteworthy dunes are also found at Kohler-Andrae State Park south of Sheboygan. Low dunes associated with former beaches, separated by marshy swales, are found in Point Beach State Forest, north of Two Rivers.

Most of Wisconsin's Lake Michigan shore is marked by glacial deposits that are being eroded. In places, recession rates of 10 feet or more per year have been observed, and as Martin (1965, 305) observes, "This keeps the bluffs steep [and] the beaches at the cliff base are narrow, for the lake currents carry the eroded material away." Erosion near Carthage College in Kenosha has breached the valley of Pike River, which once flowed into Lake Michigan near the site of today's harbor immediately east of downtown Kenosha, nearly two miles south of where it now flows into the lake.

South of Sheboygan abandoned shorelines, wave-cut cliffs, and beach deposits are seen 23, 38, and 55 feet above today's Lake Michigan, created when the lake waters were higher (Martin 1965). Farther north, evidence of former shorelines associated with Glacial Lake Algonquin can be found 29 to 40 feet above today's lakeshore, while another shoreline running from 14 feet to 22 feet above Lake Michigan is associated with Glacial Lake Nipissing, with the variations in elevation illustrating differential isostatic rebound.

## PHYSIOGRAPHIC REGIONS

In his magisterial works about the physiography of the United States, Nevin Fenneman (1928, 1938) placed Wisconsin into two of the nation's ten physiographic divisions: the Interior Plains and the Laurentian Upland. The Laurentian Upland, which largely corresponds with the Canadian Shield, was composed of only one physiographic province, the Superior Upland. The Interior Plains was divided into three provinces, with one, the Central Lowlands, including all of Wisconsin except for that portion within the Superior Upland. The Central Lowlands, so named because of its generally low elevation and relatively horizontal stratigraphy, was considerably smoothed by continental glaciations. The Central Lowland province is divided into six sections. Those in Wisconsin include the Eastern Lake Section, the Wisconsin Driftless Section, and the Till Plains. Characteristics of the Eastern Lakes include "glaciated cuestas and lowlands, moraines, lakes and lacustrine plains," while the Wisconsin Driftless Section was a "maturely dissected plateau and lowland invaded by outwash plain" (Fenneman 1928, 277). The Till Plains extended from Illinois into only the southernmost part of south-central Wisconsin, being distinguished as "young till plains; morainic topography rare; no lakes" (Fenneman 1928, 277).

Four decades later in his regional geomorphology of the United States, William Thornbury (1965) divided Wisconsin between two geomorphic provinces, the Superior Upland and Central Lowlands. The Wisconsin sections of the Central Lowlands are similar to Fenneman's classification, including the Great Lakes Section and the Wisconsin Driftless Section, although he argued that the Till Plain Section's northern boundary should be farther south.

The physical landscape of Wisconsin is divided into five provinces within Lawrence Martin's (1965) authoritative description of the state's physical geography. These include the Lake Superior Lowland, the Northern Highland, the Central Plain, the Western Upland, and the Eastern Ridges and Lowlands. Their boundaries do not entirely coincide with those suggested by Fenneman or Thornbury, but earlier editions of Martin's work clearly influenced the work of those two authors. Regardless of how we draw geographic boundaries on the physical landscape—and every geographer is likely to have his or her unique take on their best locations—there are several distinctive dichotomies that distinguish the physical landscape of Wisconsin. Areas formed on Precambrian bedrock—largely confined to the Northern Highland—are distinct from that formed on sedimentary rocks. Lands that have been glaciated display many features unlike terrain that did not experience Pleistocene glaciation. Of those lands that were not glaciated, those covered by glacial lakes—such as the Wisconsin River Valley in central Wisconsin—are quite different from the Driftless Area. Likewise, those lowlands that adjoin either Lake Superior or Lake Michigan contain many shoreline features, including wave-cut cliffs and former beaches, that mark not only the glacial lakes that formed when the glacial lobes were melting, but also contemporary coastal processes.

## RIVER DRAINAGE BASINS

River drainage basins, together with the divides between them, provide another way of physically identifying the regions of Wisconsin (Figure 2.26). Wisconsin has been described as "divided by a major watershed, which determines (a) that some of the streams flow east into the Atlantic Ocean by way of Lake Superior or Lake Michigan, and (b) that the remaining streams—the larger number—shall flow south into the Gulf of Mexico by way of the Mississippi River" (Martin 1965, 18). While the Great Lakes and the Wisconsin River play key roles in defining the state's largest drainage basins, the divides between them are not always clearly defined, nor do the divides clearly follow ridge lines.

The Wisconsin River is the largest river to wholly flow within Wisconsin, with the exception of a small area around Lac Vieux Desert that is in Michigan's Upper Peninsula. The volume of discharge at its mouth is three times that of the Fox River, yet the divide between the two watersheds is neither well delineated nor consistent. Separating the two rivers at Portage was "a flat, swampy plain, where the Indians and the early explorers, either portaged their canoes or floated them across at high water" (Martin 1965, 355). As recently as the early twentieth century, floods on the Wisconsin River sometimes drained into the Fox River, and concerns were raised about the Wisconsin River becoming captured by the Fox River, given its lower elevation and steeper gradient. While a levee at Portage now confines the Wisconsin River, and the portage canal connecting with the Fox River has been blocked, that levee was threatened with overtopping during flooding in 1993, which was only prevented by emergency sandbagging atop the levee's crest (WDNR 1993). Today, the divide between the Upper Fox River Basin and the Lower Wisconsin River Basin is located less than two city blocks from the Wisconsin River in Portage.

The second-largest annual water discharge in Wisconsin is that of the Chippewa River basin, which is nearly twice that of the Fox River. The divide between the Upper Chippewa and waters flowing into Lake Superior is not located along the crest of the Penokee or Gogebic Range, but farther south, as several rivers, including the Montreal River and the Bad River, have carved gaps through those ranges. Father west, the divide crosses the narrow portage between the Bois Brule River and the

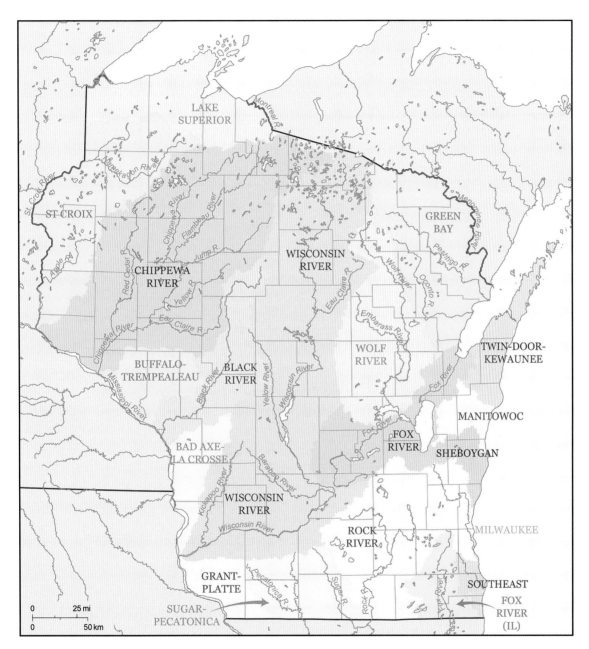

FIGURE 2.26. Rivers and major watersheds in Wisconsin. Rivers and lakes data from the U.S. Census Bureau's Cartographic Boundary Files (https://www.census.gov/geographies/mapping-files/time-series/geo/carto-boundary-file.html); watersheds data from Wisconsin Department of Natural Resources Open Data website (https://data-wi-dnr.opendata.arcgis.com/datasets/hydrologic-units-10-digit-watersheds).

St. Croix River, marked by a series of lakes in the former outlet of Glacial Lake Duluth. Given that the St. Croix, the Chippewa, the Buffalo, the Trempealeau, the Black, the La Crosse, the Wisconsin, and the Platte Rivers all flow into the Mississippi River, that river is by far the largest, but only its eastern bank is within Wisconsin.

Rivers that flow into Lake Michigan, other than the Fox River that is fed by the Wolf River with a watershed that is 76 percent greater than that of the Upper Fox River, have far smaller watersheds and discharges than the larger rivers flowing into the Mississippi River. The Niagara Escarpment separates these watersheds from those of the Fox and Rock Rivers from the Kewaunee River to the Milwaukee River watersheds, including the Manitowoc and Sheboygan Rivers. The Rock and Fox Rivers to the south drain into the Illinois River watershed.

Extensive areas of Wisconsin were covered by marshes when European settlers first arrived. Considerable marshes remain, although far-reaching drainage systems were constructed to drain many wetlands by both ditches and underground pipes. Several large swamps were dammed to create larger lakes with controlled depths, such as the Fox River at Lake Winnebago, reducing the extent of the marshes with a wider and slightly deeper lake, and the Rock River at Horicon Marsh. While the dam built at Horicon in 1842 is long gone, the water level in Horicon Marsh is now controlled by a combination of drainage channels and dikes. Horicon Marsh, which covers 51 square miles, is the nation's largest freshwater cattail marsh. East of the Niagara Escarpment several extensive marshes exist in Manitowoc and Sheboygan Counties, occupying lake beds formed following the final retreat of the Wisconsin glaciers. These include both Sheboygan and Manitowoc Marshes, each of which covers over 15 square miles. Far larger swamps exist within central Wisconsin in the area of the former Glacial Lake Wisconsin. The Great Swamp of Central Wisconsin, covering an area of nearly 500 square miles between Black River Falls on the west, Wisconsin Rapids on the north, Camp Douglas on the south, and the outwash plain from the terminal moraine of the Green Bay Lobe on its east, has been substantially altered by cranberry growers.

In total, Wisconsin has 24 water basins that are identified by the Wisconsin Department of Natural Resources (WDNR 2015a). Several major rivers, such as the Wisconsin and Chippewa Rivers, are divided into two or three separate watersheds, representing flowage patterns. The rivers themselves can often be divided into various reaches, such as above or below areas of rapids, which are often indicative of the importance of bedrock outcrops in shaping the river courses, courses that are often different today from what they had been before the Pleistocene glaciations.

## GEOLOGY IS BUT ONE ELEMENT IN WISCONSIN'S PHYSICAL LANDSCAPE

The geology of Wisconsin is but one of the factors influencing the state's physical landscape. The character of the bedrock, along with glacial or alluvial deposits, strongly influenced the soil that formed. Topography and drainage also affected soil formation. Yet, climate is also important in soil formation, both directly and indirectly through the vegetation growing on the soil. Vegetation itself is also highly dependent on the slope and drainage of the land and the texture and chemical characteristics of the soil, plus the local microclimate. Thus, the next chapter will focus upon describing Wisconsin's climate and vegetation and then on relating all three elements that created the state's soils.

*Chapter Three*

# Climate, Vegetation, and Soils of Wisconsin

WISCONSIN IS CLASSIFIED BY CLIMATOLOGISTS as having a humid continental climate, with the southern part of the state having the "warm summer" version (Köppen-Geiger classification Dfa), while the northern portion has the "cool summer" form (Dfb). Regardless of which Köppen-Geiger classification applies, both indicate that Wisconsin has a long cold winter (Figure 3.1), with the biggest regional differences being the length and warmth of the summer. While Wisconsin does not have the variety of climates that are found in many states, particularly those that have both high mountain ranges and lowlands, or both ocean shorelines and lands far inland, its climate does differ depending upon one's location. First let's consider those conditions that are common to all of Wisconsin.

## CLIMATE CONDITIONS CHARACTERISTIC OF WISCONSIN

Climate represents the long-term average weather conditions. Typically, weather data collected daily for thirty years is averaged, with climate described by average or mean values for temperature and precipitation. Data is averaged both annually and by month, so that the average high and low (maximum and minimum) temperatures and the average (mean) temperature for each month and for the year can be displayed. Likewise, average precipitation data, both monthly and annually, are calculated. In using climate normals, which are averages, one must remember that half of the values are greater than the mean and half are smaller. Weather conditions often depart from normal climate means, yet when they depart significantly, weather and climate hazards result. These are explored in chapter 4.

As mentioned in chapter 1, the 45°N parallel cuts across Wisconsin. Being halfway distant between the North Pole and the equator, Wisconsin's climate is influenced by both polar and tropical conditions. However, these are highly constrained because Wisconsin is about 3,000 miles from both the equator and the North Pole. Given its latitude, Wisconsin receives approximately twice the number of hours of sunlight in the summer than what it gets in the winter, with the differences most pronounced in Wisconsin's northernmost area. Furthermore, in the winter the sun is low in the sky, while in the summer the sun is high in the sky and its heating is much more intense, given that the sun's radiation

FIGURE 3.1. Smoke rising from chimney on a bitterly cold subzero day in late January 2019 on a snow-covered Amish farmstead within Brush Creek Valley, astride the Monroe-Vernon County line west of Ontario.

travels more directly through the atmosphere. Thus, summer temperatures are far warmer than in the winter.

Wisconsin has a continental climate, given that it is far from any ocean. While proximity to the Great Lakes does have local influences on climate in Wisconsin (Eichenlaub 1979) as we will explore later, during most of the time these lakes—particularly Lake Michigan—are downwind from Wisconsin, minimizing their influence. Climatologists have long recognized that land and water heat and cool differently. It takes more heat energy to warm water than to heat a land surface. Likewise, water takes more time to lose heat energy in winter than does land. As physicists explain, earthen materials such as rock have much lower specific heat values than does water, meaning that it takes less solar heating to increase their temperature than to warm water. In addition, all solar heating is absorbed by earth materials at their surface while sunlight can penetrate many feet into ocean or lake waters. Of course, once a lake is covered with ice, it behaves more like snow-covered land (Figure 3.2). Thus, given its continental position, Wisconsin's winter temperatures are much colder than those observed at the same latitudes in western Oregon and southwestern Washington. Conversely, during the summer Wisconsin's temperatures are considerably hotter than those observed in coastal Oregon and Washington.

FIGURE 3.2. Snow atop ice covers the surface of Wisconsin's lakes during the winter. This December 31, 2017, aerial photograph shows ice covering the southern part of the Bay of Green Bay, the city of Green Bay, and extending up the Fox River and across Lake Winnebago in the distance. Lakes remain ice covered well into March in typical years.

Prevailing winds, those that occur most commonly, in Wisconsin are the westerlies, sometimes from the southwest (particularly in summer) and sometimes from the northwest (more common in winter), or simply from the west. In Milwaukee, for example, the prevailing winds in November through March come from the west-northwest, while the prevailing winds in July and August come from the southwest, and those in September and October blow from the south-southwest. In contrast, Milwaukee's prevailing winds in April through June are from the north-northeast (WSCO 2017). For any given community, its prevailing winds are influenced by both global and local circumstances. Furthermore, the prevailing winds are the most common of a large variety of wind directions that shift both during and over the days of a month.

Considerable distance separates Wisconsin from the moderating influence of the Pacific Ocean. While winds do blow from the southwest, tapping maritime tropical air masses moving from the Gulf of Mexico, air masses from the Atlantic Ocean are far rarer, even though winds do blow at times from the east. Both continental polar and continental arctic air masses come from the northwest, moving into Wisconsin from the Canadian Prairie Provinces or arctic areas to their north. Maritime polar air masses, coming from the Pacific Ocean, are highly modified by the distance they have traveled across the continent and their transit over high mountain ranges, altering their temperature characteristics and reducing their moisture content. Because of the prevailing westerlies, most storm systems that track across Wisconsin come from the west (varying between the southwest and the northwest). Such midlatitude cyclones develop along the boundary of unlike air masses, which are separated by fronts. Thus, a cold front marks the location where a warm moist maritime tropical air mass is abruptly replaced by a cold dry continental polar air mass.

Given constraints imposed by Wisconsin's location, its winters are long and cold. Average temperature in January, the coldest month of the year, between 1981 and 2010 ranged statewide from 9.8°F in Minong and 9.9°F in Gordon to 22.6°F in Racine and 23.4°F in Kenosha (Figure 3.3). Yet, Wisconsin's climate has definitely warmed, particularly in the winter (Kucharik et al. 2010). This change is also illustrated by climate statistics reported in the *1941 Yearbook of Agriculture*, which showed thirteen Wisconsin stations with average January temperatures below 10°F, including Solon Springs in Douglas County, which averaged 8.3°F, based upon a 30-year record (Coleman 1941). During the 1981–2010 period, only Minong and Gordon were below 10°F, while Solon Springs averaged 11.7°F (NCDC 2015).

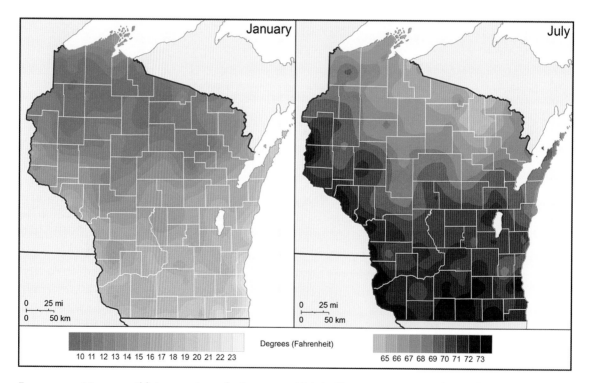

FIGURE 3.3. Mean monthly temperatures for January and July in Wisconsin. Patterns plotted from 1981–2010 climate statistics for 204 weather stations in Wisconsin and 350 in neighboring states. Map created from data in NCDC (2015).

Madison has an average mean daily January temperature of 18.8°F, with its daily high temperature averaging 26.4°F and the low temperature averaging 11.1°F (1981–2010 Normals: NCEI 2015a). All Wisconsin locations have average temperatures below 26.6°F during their coldest month, resulting in the state's humid continental or microthermal climate classification. The dividing line between those climates classified as having a "warm summer" versus "cool summer" is the 71.6°F average temperature isotherm for the warmest month. July is the warmest month in Wisconsin. In Madison the mean July temperature is 71.3°F, with the average daily high temperature being 81.6°F, while the average daily low temperature is 61.0°F (1981–2010 Normals: NCEI 2015a). Mean July temperatures in Wisconsin statewide vary from 64.3°F in Gurney, Iron County, to 74.1°F at Lynxville, Crawford County. Thus, spatial differences in temperatures are more pronounced in the winter than in the summer across Wisconsin.

### Regional Temperature Patterns

Temperatures vary regionally across Wisconsin, shown by both average annual temperatures and monthly temperatures, as well as diurnal (24-hour period) and annual temperature ranges. These differences are related to latitude, elevation, topography, proximity to large water bodies—in particular Lakes Superior and Michigan—land cover, and urbanization.

Latitude is one of the most conspicuous causes of temperature variation, but its influence is complicated by the proximity of northern Wisconsin to Lake Superior. Looking at the map displaying January mean temperatures, one sees that the coldest temperatures are not along the shore of Lake Superior, but farther inland. Parts of southeast Douglas, Vilas, and Forest Counties—all at approximately the same latitude—experience winter mean temperatures around 10°F. So do parts of Rusk and Sawyer Counties. In contrast, mean winter temperatures experienced in southern Wisconsin, along the Illinois state line, run from around 20°F to over 23°F. Between 1981 and 2010 Madison averaged 15.3 days annually with minimum daily temperatures of 0°F or colder, while Rhinelander experienced over twice as many zero or subzero days, 34.4 days (NCEI 2015a, NCEI 2015b).

Elevation and proximity to Lake Superior are responsible for the northern parts of Douglas through Iron Counties having winter temperatures that are not as cold as those experienced in the southern parts of those counties. Climatologists have long recognized that for every 1,000 feet increase of elevation, temperatures decrease an average of 3.6°F. The altitude of the Penokee or Gogebic Range is 1,100 to 1,200 feet above Lake Superior, which could easily account for temperatures averaging 4°F lower than those experienced in communities such as Superior, Bayfield, and Ashland. In addition, these cities are even milder in the winter given their proximity to Lake Superior, whose center remains ice-free during most winters.

Lake Michigan also moderates winter temperatures in lakeshore counties. However, because winds blow from a westerly direction far more often than from the east, the effects are not felt as strongly in Wisconsin as along the eastern shore of Lake Michigan in Michigan. Nevertheless, winds sometimes blow from an easterly direction, particularly with the approach of a midlatitude cyclone (low-pressure system), ahead of the arrival of a warm front or the passage of the low. Wind roses (showing the direction of winds) indicate these easterly winds occur occasionally during the winter months, particularly in February (P. N. Knox 1996). Thus, January mean temperatures exceed 21°F along the lakeshore areas of Sheboygan County south through Kenosha County, while temperatures average 2°F to 4°F colder in parts of western Sheboygan County, much of Washington County, and northwestern Waukesha County. For example, the 1981–2010 mean January temperatures were 21.4°F in Port Washington, 18.8°F in Germantown, and 17.7°F in Hartford (MRCC 2019).

The coldest extreme minimum temperature recorded in Wisconsin was -55°F, observed at Couderay in Sawyer County in 1996. Local topographic, sometimes microclimatic, conditions often create areas that are particularly prone to cold temperatures. Cold air, being denser than warmer air, often accumulates in valleys where cold-air drainage concentrates it. Even broad shallow depressions concentrate cold air, such that a farmer may notice that frost pockets develop in only part of a field. At Couderay, its recording "station is situated in a low-lying marsh that is subject to cold-air drainage," giving it a temperature that was about 4°F colder than "at other stations in northern Wisconsin" (Moran and Hopkins 2002, 155).

The growing season—the length of time between the last freezing temperature in the spring and the first day with a freezing temperature in the autumn—varies across Wisconsin, just as do the mean summer and winter temperatures. The shortest "frost free" growing season is less than 100 days (Figure 3.4), found in northern Forest and western Florence Counties. Large areas of Washburn, Bayfield, Sawyer, Ashland, and Iron Counties have growing seasons of fewer than 125 days, as well as those regions within 25 to 50 miles of the under-100-day zone in Forest and Florence Counties. In contrast, the growing season exceeds 180 days near Lake Michigan in Milwaukee, Racine, and Kenosha Counties, considering the 1981–2010 period. Those areas with at least 170-day growing seasons are found along the lakeshore as far north as Manitowoc County, and within the Mississippi River Valley in western Vernon, Crawford, and Grant Counties (WSCO, n.d.). As we will see in chapters 4 and 7, the growing season has had a profound impact on Wisconsin's crop production.

Just as cold-air drainage influenced Wisconsin's all-time minimum temperature record, land within shallow depressions more often experiences frost than nearby, slightly higher lands. This is particularly true if peat soils, such as those forming under marshes, are present, rather than sandy or loamy soils. (See discussion of soils later in this chapter.) Differences can be pronounced even within a single county. For example, the marshy area around Coddington has a growing season that is 30 to 45 days shorter than what is experienced at several other locations elsewhere within Portage County (Moran and Hopkins 2002). Even on a single farm, crops in low-lying patches in a field may suffer frost damage while emerging plants elsewhere in the field are unaffected.

Summer temperatures are highest within the Mississippi River Valley and across the southernmost tier of counties. Mean temperatures in July, typically the warmest summer month, exceed 73°F in western La Crosse, Crawford, and Grant Counties. In southeastern Wisconsin, mean July temperatures of 71°F are also found in Walworth, Waukesha, and western Milwaukee Counties (see Figure 3.3). Average high temperatures in July exceed 80°F within much of southern and central Wisconsin. The greatest daily high temperatures are found in southern Wisconsin, where several cities typically experience 16 to 18 days of 90°F or higher temperatures annually. In contrast, across a broad area

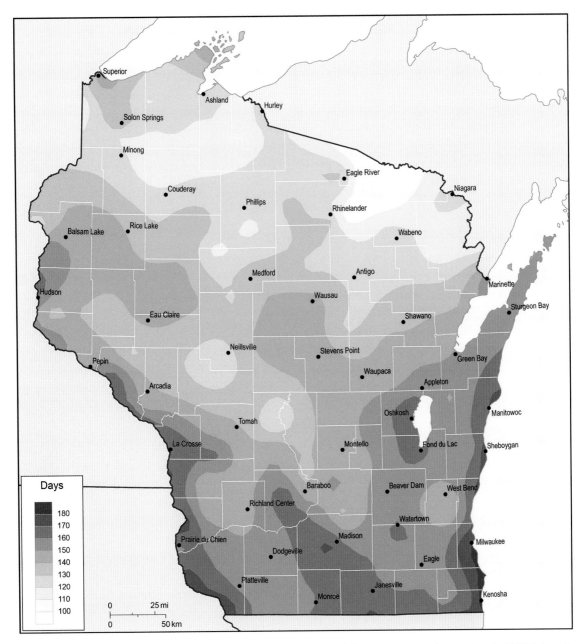

FIGURE 3.4. Frost-free growing season in Wisconsin. Map created from 1981–2010 climate data in NCDC (2015) for 160 weather stations in Wisconsin and 284 in neighboring states.

of north-central Wisconsin, south of Michigan's Upper Peninsula, such maximum temperatures are experienced less than twice a year.

Proximity to Lake Michigan moderates the summer temperatures of locations closest to the lake, given that the water remains colder than nearby lands. During the day lake breezes often develop, which are typically strongest in mid-afternoon, when temperature differences between the land and lake waters are the greatest, bringing cool winds from the east. These breezes occur on one-third to one-half of summer days. When strongest, they may bring cooler air up to 25 miles from the lake, yet at other times only several blocks from the lakeshore. They develop most commonly on those days when winds would otherwise be light to calm, such as when a high pressure system is positioned over the region. Thus, neighborhoods in Milwaukee and other lakeshore communities may have daily high summer temperatures that are 20 degrees cooler than western suburbs on days of pronounced lake breezes (Moran and Hopkins 2002).

### Regional Precipitation Patterns

Average annual precipitation totals in Wisconsin (over the 1981 to 2010 time span) range from a high of 38.08 inches in Dodgeville in Iowa County to a low of 28.32 inches on Door County's Washington Island. The southernmost third of Wisconsin receives at least 34 inches of precipitation annually, with northeastern and east-central areas receiving the least, typically under 32 inches (NWS-M, n.d.a). Such statistics indicate that all of Wisconsin experiences a humid climate, yet there are local climatic variations. For example, colder waters in both Green Bay and Lake Michigan undoubtedly slightly decrease convective activity over Door Peninsula and Washington Island, diminishing their precipitation. Thunderstorms bring the greatest share of precipitation to most of Wisconsin, and these storms cause the greatest amounts of precipitation that occur over the summer. In general, southwestern Wisconsin experiences about 45 days with thunderstorms annually, about 10 days more than are experienced in northeastern Wisconsin (Moran and Hopkins 2002).

Snow dominates the winter precipitation in Wisconsin, with most of the state receiving between 30 and 50 inches of total annual accumulation (Figure 3.5). Although much greater amounts of lake-effect snow occur near Hurley in Iron County, in most of Wisconsin snowfall is largely associated with the passage of the midlatitude cyclone, sometimes called extratropical cyclones, wave cyclones, or low-pressure systems. The tracks taken by these storms determine not only which areas receive the greatest snowfall but also the amount of snow. The band of greatest accumulation is typically between 100 and 150 miles north of the location of the surface low-pressure center associated with the midlatitude cyclone.

The low is often located where a warm front, which marks where warm moist air from the south or southwest reaches and overruns colder air, connects with a cold front. The cold front marks the location where cold dryer air advancing from the northwest bulges against the warm moist air coming from the southwest. At times the cold front catches up with the westernmost portion of the warm front, forcing the warmer moist air aloft, where it cools and its moisture condenses. The resulting front, called an occluded front, extends from the low to where the cold and warm fronts separate. The extra uplifting of the moist air ahead of where the occluded front is located on the ground results in

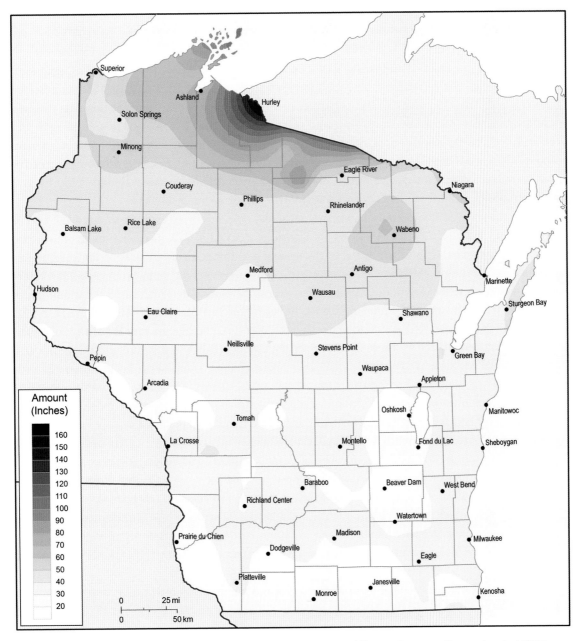

Amount
(Inches)

160
150
140
130
120
110
100
90
80
70
60
50
40
30
20

0        25 mi

0              50 km

FIGURE 3.5. Average annual snowfall across Wisconsin. Map created from 1981–2010 climate data in NCDC (2015) for 153 weather stations in Wisconsin and 276 in neighboring states.

particularly heavy precipitation 100 or so miles north of the low. Depending on the temperature of the rising air and the temperature of the air through which the precipitation falls, winter precipitation may be rain, freezing rain, sleet, or snow.

The greatest snowfalls (or, if the temperatures are warmer, winter rains) are typically associated with midlatitude cyclones whose tracks resemble the Panhandle hook. These storms often develop in the Panhandle of north Texas and swing across Oklahoma before heading for Michigan's Upper Peninsula. While the "average" path of these Panhandle storms typically passes just west of Green Bay, in reality they may track anywhere within 100 miles or so east or west of the idealized track. Developing so far south, these Panhandle storms tap into moisture from the Gulf of Mexico and bring it into Wisconsin, where the storms are often beginning their occlusion stage. Thus, Panhandle storms are responsible for most of the bigger snowfall events over Wisconsin, often bringing 6 or more inches of snow. In most of Wisconsin, mean annual maximum daily snowfall is between 9 and 15 inches, and snowstorms with 6 or more inches occur between two and three times a winter.

Midlatitude cyclones that track from west to east across northern Wisconsin are often called Alberta Clippers. Forming on the leeward slope of the Canadian Rockies in Alberta, these storms move rapidly across Wisconsin. Given their speed of movement and that their tracks are too far north to tap into much moisture from the Gulf of Mexico, Alberta clippers are not big snow producers. They rarely bring even four inches of snow, unless lake-effect snows develop. Other extratropical cyclone tracks that can affect Wisconsin in the winter are the North Pacific and the North Rocky Mountain tracks. The North Pacific track storm typically crosses the Pacific Coast near the Oregon and Washington border, loses most of its moisture as it successively crosses the Cascade Range and the Rocky Mountains, and travels across far southern Wisconsin. Midlatitude cyclones following the North Rocky Mountain track typically develop in central Wyoming and move across central Illinois before turning toward the northeast, crossing the lower peninsula of Ontario. Even if the storm is on the northern side of its average track, the path of heaviest snowfall only extends across southern Wisconsin.

For most of Wisconsin snowfall (of at least .1 inch) occurs between 30 and 45 days a year, yet snow remains on the ground for a greater number of days. As Moran and Hopkins (2002, 137) explain, "The average number of days with significant snow cover (1 inch . . . or more) varies from about 65 days along the Wisconsin-Illinois border to more than 140 in extreme northwestern Wisconsin." The deepest snow cover in Wisconsin is found near Hurley, where the lead author of this text once had to remove snow from his windshield in early June. Hurley, which received an average of 167.5 inches of snow annually between 1981 and 2010, provides a classic example of lake-effect snowfall (NWS-M, n.d.b).

Lake-effect snows develop when cold air travels across a large body of warmer, ice-free water. As the air moves across the lake, it gains water vapor evaporated from the lake, yet as that water vapor condenses, it releases heat, causing clouds to develop. When the air is cold, ice crystals—the constituent of snowflakes—form, bringing snows to the downwind shores. Although lake-effect snow rarely extends more than 25 miles downwind from the lakeshore, its impact is enhanced when the winds from the lake are forced to rise by topographic barriers. Hurley is located about 900 feet above the level of Lake Superior, and its annual snowfall is nearly 110 inches greater than that of Superior, which is located at the southwestern corner of Lake Superior. Nevertheless, lake-effect snows are experienced from eastern Douglas County across Bayfield and Ashland Counties, and into Vilas County. Even greater amounts of lake-effect snow are experienced in Michigan's Upper Peninsula, where locations on its Keweenaw Peninsula not only provide orographic uplift but benefit from winds arriving having traveled long distances across Lake Superior from multiple directions. In contrast, only north winds bringing lake-effect snows to Hurley would have such a long fetch.

Lake-effect snows also occur in eastern Wisconsin, but their total accumulations are far more moderate, given that easterly winds blowing from Lake Michigan are much less common. For example, Milwaukee receives just over 10 inches of snow more than what is experienced annually in Waukesha (NWS-M, n.d.b). Nevertheless, heavier snows can occasionally occur.

## VEGETATION OF WISCONSIN

The type and character of vegetation found at any locale is a function of several environmental conditions. Climate is undoubtedly the most important factor, but topography and soils play a huge role in determining whether a given amount of precipitation is appropriate for the optimal growth of many species. It is abundantly clear that plants growing anywhere in Wisconsin must be able to survive long cold winters, winters that typically have subzero temperatures (the statewide average is three days with maximum temperatures below 0°F every two years). Length of time between the last killing frost of the spring and the first such frost of the fall is too short in much of northern Wisconsin for many plants, including many commercial crops, to produce viable seeds. Thus, without the ability to successfully reproduce, such species cannot be part of the vegetative cover, even if they can survive the cold winter weather.

### Moisture Needs

Moisture availability is a key determinant as to whether grasses or trees dominate a landscape. Not only does total precipitation influence moisture availability, but so does the form of precipitation—whether rain or snow that melts months after falling—and the seasonal distribution of the precipitation. Throughout Wisconsin summer is the season during which plants need the most moisture for growth, and summer is also the season of maximum precipitation. Nevertheless, topography concentrates moisture along streams and river valleys, while sloping lands shed more of their precipitation. Slopes, particularly longer and steeper ones as in the Driftless Area, may face the sun during the warmest time of the day, also enhancing evaporation, if facing toward the south and the west (Figure 3.6). In contrast slopes may provide shade if they face north, reducing both their temperature and moisture loss.

In addition, various soils have different moisture-holding capacities. Soils rich in organic matter, such as those that consist of peat, a poorly consolidated accumulation of partly decayed organic matter, have an ability to store large amounts of moisture in reach of shallow roots. Likewise, soils with high clay content, such as those formed from weathered shale or that developed on lake beds, can store considerable moisture. In contrast, sandy soils have little moisture-storing ability. Precipitation moves rapidly through sand, such that sandy soils may be quite dry just a few days following a heavy

FIGURE 3.6. Flowering plants and grasses characterize Dewey Heights Prairie State Natural Area within Nelson Dewey State Park, near Cassville in Grant County. Located atop a 300-foot-high limestone bluff facing the Mississippi River and the afternoon sun, these prairie plants are well adapted for warm dry conditions in the summer.

rain. Thus, drought-resistant trees, such as pines, often dominate the forest cover of sandy soils, even when rainfall is otherwise adequate, while poplars and birch may be abundant on peat soils that receive no more precipitation than locations where pines dominate.

## Periodic Fire

The presence of periodic fire also tilts the balance between conifers (softwoods) and deciduous trees (hardwoods), given that pines with their thick bark are more fire tolerant than hardwoods, particularly young hardwoods. In addition, certain conifers, such as the jack pine, have serotinal cones that are tightly closed until such time that heat and dryness provided by a forest fire "causes their scales to open and scatter the seeds on the burned ground" (Curtis 1971, 206). Given that such burned ground is an ideal seedbed for the jack pine and that the pines' scattered seeds face little competition from other species that lack this survival mechanism, jack pine forests may dominate a locale where periodic fires occur at intervals greater than a decade. Even without the serotinal cones, other pines, such as the iconic white pine that so dominated the quest of the lumbermen who cleared northern Wisconsin's forests, are much more resistant to fires than deciduous forests, particularly when ground fires do not spread into the forest crowns.

More frequent fires, however, place the survival of both the pine and deciduous forests in jeopardy. Lands that are burned regularly, at least once every five years or so, are dominated by prairie grasses. Grasses can tolerate fires given that much of their living structure is at or below the ground surface. Indeed, new blades of grass are often conspicuous less than two weeks following a fire that raced across the prairie. Even swamp white oak, which typically grows in river floodplains or other swampy locations in southeastern Wisconsin, is somewhat fire tolerant, and "when burned it can regenerate from the stump" (Curtis 1971, 137).

## VEGETATION DISTRIBUTION ACROSS WISCONSIN

At the time the first European settlers arrived in Wisconsin, much of southern Wisconsin was covered with prairies (Figure 3.7), marsh grasses, as well as oak savanna, a mixture of prairie grasses and open woods of burr oak and white oaks. Yet even in the savannas considerable woods existed, such that in the early nineteenth century about 85 percent of Wisconsin was covered by forests (NRCSA 1956). Because the surveyors of the public lands recorded details regarding the vegetation of each section of

FIGURE 3.7. Prairie grasses on a dry prairie preserved in Thomson Memorial Prairie, located along the Dane-Iowa County line south of Blue Mound, which is visible in the distance.

land they surveyed, much is known about the pre-European vegetation of Wisconsin (Finley 1976). The subsequent clearing of the trees from the land, the burning of the debris on the logged-off lands, and the drainage of swamps for cultivation dramatically changed the composition of the vegetation that one sees today in much of southern Wisconsin. Now the most commonly seen plant species are maize (corn), soybeans, and alfalfa—all grown to feed livestock or humans. Because the cut-over land was not replanted but reforested naturally, there are fewer conifers and more deciduous trees in northern Wisconsin (Rhemtulla, Mladenoff, and Clayton 2009). Only three-tenths of one percent of Wisconsin's forests are remnant old-growth, and one of the best places to find forests that resemble Wisconsin's original virgin forests is within Menominee County (Figure 3.8).

### Wisconsin's Tension Zone

The distribution of vegetation in Wisconsin two centuries ago clearly showed what has been described as a "tension zone," which, as explained by John T. Curtis (1971, 16), separates the ranges "of a large number of species . . . [such that the] band includes the southern limits of many northern species and the northern limits of many southern species." Thus, the tension zone separates those flora containing prairie plants in southwestern Wisconsin from those influenced by boreal plants in northeastern Wisconsin. Therefore, to the southwest, prairie grasses and hardwood forest dominate, while to the

FIGURE 3.8. Pine forest along Wolf River north of Five Islands within Menominee County.

north forests are composed of hardwoods and conifers. As the vegetation map (Figure 3.9) indicates, the location of the tension zone corresponds with the "southern limit of pine as a common forest tree." To the south of the tension line running from about St. Croix Falls to near Sheboygan, hardwoods such as oaks, beech, sugar maple, and basswood dominated the forests and savannas of most of southern Wisconsin, including much of the Driftless Area. To the north, although various hardwoods accounted for about half of the forest area in the middle of the nineteenth century, nearly a tenth of the forest was hemlock and white pine, another tenth was occupied by red pine, jack pine, fir, and spruce, and another tenth was cedar and tamarack. Aspen and birch were more widespread in northern Wisconsin than in southern Wisconsin. Given its transitional character between the hardwood forests of the south and the boreal forest dominated by conifers to the north, particularly north of Lake Superior, the northern ecological province is termed the West Laurentian Mixed Forest (Perry et al. 2008).

The tension zone, while best illustrated by considering forest species, orchids, and many other plant genera, also marks the ranges of certain animals, such as the bobwhite and mourning dove. Climatic factors that correspond to the tension zone have been meticulously described by Curtis (1971), who cites summer temperature and evaporation amounts, winter snowfall, precipitation variability, thunderstorm and hail frequency, and summer maximum temperatures among other meteorological variables that differ between the two sides of the tension zone. Not only do these meteorological variables correspond to the limits of many floristic ranges, but there are also phenological differences, demonstrated by the lags in the time of flowering and fruiting of various plants.

In comparing changes in Wisconsin's forests between the middle of the nineteenth century and the 1930s and with the beginning of the current century, distinct differences have been noted between what has happened in northern versus southern Wisconsin. Within northern Wisconsin the forest composition of today more closely resembles that of the mid-1800s than did the forests in the 1930s. In contrast to this partial recovery in northern Wisconsin, the forest species composition in southern Wisconsin is even more different from that of a century and a half ago than it was during the Great Depression (Rhemtulla et al. 2009).

Yet, even in northern Wisconsin, the forests of today are dominated by smaller trees, and hardwoods account for a larger share of trees than they did a century and a half ago. Among conifers, while white pine has begun a comeback on drier soils, both hemlock and jack pine have declined. As Rhemtulla and colleagues (2009, 1073) explain, "fire suppression implemented since the 1930s, jack pine budworm . . . outbreaks, and more recent logging have slowly led to the loss of jack pine and increase in hardwoods, particularly oak and aspen." Yet more recently, sugar maples are increasing their dominance in the northern forests (Figure 3.10).

In southern Wisconsin, where the oak-hickory savannas were cleared for agriculture, most remaining woodlots have been protected from periodic fire, resulting in denser forest growth and a greater dominance of red maple and sugar maples among the hardwoods. The forests of southern Wisconsin overall are increasingly "less fire resistant and more shade tolerant" (Rhemtulla et al. 2009, 1074).

### Regional Vegetation Patterns of Wisconsin

Wisconsin's native vegetation was classified into 21 major communities and 13 minor communities, covering less areal extent, by Curtis (1971). Drawing upon his groundbreaking and authoritative work,

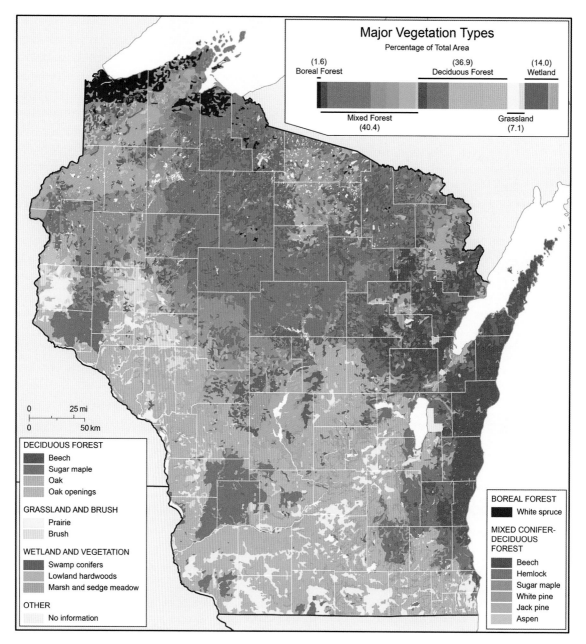

FIGURE 3.9. Original natural vegetation in Wisconsin, redrafted from data gathered by Finley (1976) and obtained from Wisconsin Department of Natural Resources Open Data website (https://data-wi-dnr.opendata .arcgis.com/datasets/3e952715b0d549c39cd8e26b4b274a0c_1). The L-shaped area east of Lake Winnebago lacks categorization, as it was originally designated an Indian Reservation whose survey did not record vegetation data.

FIGURE 3.10. Sugar maples are a significant contributor to fall color across much of Wisconsin. This view is toward Horseshoe Island and Nicolet Bay from Sky Line Road in Door County's Peninsula State Park.

we will focus upon the characteristics of some of the more prominent major communities, beginning in the prairie-forest floristic province, south and west of the tension zone.

Southwest Wisconsin includes grasslands, of which five communities have been identified; savannas, a mixture of prairie grasses and woods, of which four communities can be seen; and forests, of which five variations are noted depending upon their wetness. The forests south and west of the tension zone "occur on a full range of moisture sites, from very wet places along streams and lakes, through mesic sites with deep soils, to very dry places with thin soils of exposed hills and bluffs" (Curtis 1971, 87). Oaks dominate those drier forests. Sugar maples in particular, as well as red maples and basswood, are prominent in the mesic locations. Swamp white oak, silver maple, ash, and elm—before the spread of Dutch elm disease—characterize the wettest soils. In all of these locations a variety of other tree species supplement the dominant species.

Because moisture and soil drainage environments can change over very short distances, transitions in vegetative cover characterize much of Wisconsin south and west of the tension zone, particularly within the Driftless Area. Thus, mesic forests, dominated by sugar maples, are "scattered quite widely in separate islands" (Curtis 1971, 103), rather than occupying a large swath of land, even though they covered nearly a tenth of the state two centuries ago. Xeric forests, marked by oaks, are on sandy soils, on slopes facing the sun during the warmest time of the day, or atop ridges with thin soils, interspersed with either savanna prairie grasses or mesic species. Thus, these soils are often droughty. Today such xeric forests characterize most woodlots on farms in southern Wisconsin, which typically cover less than a quarter of a quarter section of land, thus less than 40 acres. Yet collectively they occupy more land area than was the case before European settlement of the land (Curtis 1971). Covering about a sixth of the area of the mesic forests, floodplain forests (Figure 3.11) are found not only along the Mississippi River and the glacial sluiceways that fed it, but also beside the Rock, Pecatonica, Sugar, and Fox Rivers.

The forest floor under the xeric forests is typically covered with shrubs, while that under mesic forests is more open. There, spring ephemeral plants sprout, blossom, and fruit all within a few weeks after the winter snows melt, but before the dense growth of the forest canopy shades the ground. Thus, the forest floor is often covered with flowers in late spring, while the only evidence of such growth in the mid- to late summer is the underground tuber or bulb of the plant. In addition to these ephemeral plants are others, such as wild ginger and trilliums that bloom in the spring but keep their leaves during summer like a shade-tolerant species.

FIGURE 3.11. Dense growth of hardwoods in standing floodwater near the Black River from the Great River State Trail in the Upper Mississippi River National Wildlife and Fish Refuge, southeast of Trempealeau.

Just as the forests south and west of the tension zone vary depending upon the available moisture, grasslands can also be distinguished into those that display xeric, mesic, and wetland characteristics. In all of these settings the prairies are dominated by grasses, particularly considering the number of individual plants, although in some xeric prairies the number of species of flowering plants rivals the variety of grasses. In contrast to the ground herbal layer in forests that typically concentrates its blossoming in the spring, the majority of prairie plants flower during the summer, with a quarter blossoming after mid-August (Curtis 1971). Shrubs and scattered young trees also characterize the prairies.

Savanna lands cover nearly three times the area of Wisconsin's grasslands, of which the oak savanna is most common (Figure 3.12), covering a fifth of the state's land surface (Curtis 1971). Savannas represent a transitional area, where trees are more abundant than widely scattered oaks amid prairie grasses, which because of fire frequency sometimes resemble brush in stature, and forests, where trees shade at least half of the ground. Within such oak openings burr oak is the most common variety, although areas dominated by black oak and white oak are known. Such oak barrens are highly fire dependent. Without periodic fires, trees and brush replace the prairie grasses, and with further growth, the dominant varieties of oak also change.

FIGURE 3.12. Oak savanna within the Buckhorn Barrens area of Buckhorn State Park, Juneau County.

Northern Wisconsin was once dominated by the northern mesic forest, "by far the largest community in Wisconsin" (Curtis 1971, 184), covering a third of the state's land area. Although the forest covered large parts of north-central Wisconsin, plus the Lake Michigan counties north of Sheboygan, its constituent species varied regionally. For example, beech was an important tree along Lake Michigan and Green Bay while hemlock grew west into Sawyer and Bayfield Counties (Curtis 1971). Even though conifers are typically associated with the northern forests (Figure 3.13), overall the sugar maple is the leading dominant species, with hemlock being the second most prominent species when numbers of seedlings and smaller trees are counted, although at specific sites hemlock may be more prominent than sugar maple. In contrast, white pines are often the tallest and most conspicuous trees. While they constituted a small minority of the standing trees in northern Wisconsin, because of their size and longevity—with the big pines cut in the late nineteenth century being about 400 years old, they were profitably logged from hardwood forests that contained only two or three white pines per acre (Curtis 1971). Thus, the white pine was distributed far more widely than just those conifer forests where it was the dominant species.

The northern forest contains several distinct areas of pine barrens, which are more open, savanna-like areas within the pine-hardwoods forest that include "continuous ranges of red pine . . . and jack

FIGURE 3.13. Deciduous trees dominate the lower level of the pine forest along Heart of Vilas Trail in Northern Highland–American Legion State Forest south of Boulder Junction, Vilas County.

pine" (Vogl 1970, 175). Their locations "correlate with the distributions of sandy soils, great forest fires, present fire hazard areas, [and] sites subject to local drought" (205), with the presence of relatively infertile sandy soils being of critical importance in their presence. Although such pine barrens are found along the Manitowish and Peshtigo Rivers, the most extensive barrens extend from the Brule River to the St. Croix River, approximately the route of the former glacial sluiceway leading southwest from Lake Superior.

Large areas of conifer swamps can be found in Wisconsin, accounting for about 6 percent of the land area. While Curtis (1971, 223) describes the presence of "tamarack forest with a groundlayer of herbs and shrubs" south of the tension zone as being "relic outliers of the northern forests," such as in the area of the former Glacial Lake Wisconsin and in the Sheboygan and Manitowoc swamps, they are more extensive in those counties south of Michigan's Upper Peninsula. Another particularly prominent area of conifer swamp occupies the glacial sluiceway once occupied by the St. Croix River, including the Bois Brule River Valley. In the wettest conifer swamps black spruce and tamarack are the dominant species—and at times may constitute nearly pure stands—while white cedar, balsam fir, and jack pine may also be present. Tamarack is a deciduous conifer, the only conifer in Wisconsin to shed its needles in the autumn, a more common characteristic of the boreal forest north of Lake Superior.

FIGURE 3.14. Muskeg with tamarack and black spruce along Heart of Vilas Trail in Northern Highland–American Legion State Forest west of Boulder Junction, Vilas County.

In some of the wettest forests, a savanna-like muskeg is found (Figure 3.14), where tamaracks and black spruces are surrounded by low-growing bog plants. On slightly drier areas, cedars and fir, as well as several hardwoods, including sugar maple, may be present.

True boreal forests were originally found along portions of Wisconsin's Lake Superior shore and in the northern portion of the Door Peninsula (Figure 3.15). Comprised of balsam fir and supplemented by white spruce, white pine, white cedar, and white birch, plus aspen and birch, this forest represents a conifer-hardwood mix, yet one in which conifers dominate the canopy layer. The boreal forest accounted for about 2 percent of Wisconsin's land area before European settlement (Curtis 1971).

## SOILS OF WISCONSIN

Beneath the vegetation lies the soil, which is inhabited by a multitude of microorganisms, worms, and a few burrowing animals. The soil itself is a product of both its parent material, determined by the local geology and often those glacial materials atop the bedrock, and the type of vegetation that grows upon it. Indeed, we often describe soils as having developed under prairie grasses or forming beneath a hardwood or coniferous forest. Soils can be defined as being either young or old, as the length of time that they have been forming influences both their depth and the nature of their development. Soils in

FIGURE 3.15. Conifers dominate within the boreal forest found along the rugged coast of Madeline Island within Big Bay State Park, Ashland County.

most of Wisconsin are relatively young, given that their development could only begin with the retreat of the last ice sheets 11,000 to 12,000 years ago. Many soils in the Driftless Area developed on thick deposits of wind-blown silt, or loess, that were deposited by winds blowing from glacial sluiceways, particularly the biggest one, the Mississippi River Valley.

Francis Hole, who spent his long and distinguished career as a professor of soil science at the University of Wisconsin–Madison, was an avid violinist who composed an ode to promote his favorite soil, the Antigo silt loam. That soil was designated by the Wisconsin legislature as the official Wisconsin state soil. Thus, we begin our discussion of soils by examining the Antigo silt loam. As its name indicates, its type location is the Antigo area of Langlade County, and the soil has a silt loam texture. Soils are typically defined based upon their texture, considering the percentages of their material that are sand, silt, and clay. Sands are the coarser particles, while clays are the finest, with a diameter of less than .002 mm. Sands have a diameter exceeding .05 mm or .02 inch, while silts comprise those particles sized between sand and clay. A soil that is predominately sand would be described as a sandy soil, one that is mainly of silt-sized particles is termed a silty soil, while a loam is a soil that contains a relatively even mixture of sands, silts, and clay.

The Antigo silt loam is composed of a mixture in which silt comprises at least 50 percent of the material. The U.S. Department of Agriculture's Official Series Description indicates that this type of soil "consists of very deep, well drained soils formed in 50 to 100 centimeters of loess or silty alluvium . . . [and that] these soils are on outwash plains, stream terraces, eskers, kames, glacial lake plains, and moraines" (USDA-NCSS 2009). The uppermost layer of this soil, ranging from 0 to 23 cm (approximately 10 inches) in depth, is a silt loam, and is a dark grayish brown in color. Immediately below this layer or horizon is a very pale brown zone, which is typically 23 to 30 cm below the soil surface. It overlies a dark yellowish-brown layer to brown layer to a depth of 84 cm or 33 inches, with a more sandy and gravelly layer, showing less weathering and influence of the vegetation above, extending to a depth of about 1.5 meters, or nearly 5 feet.

The Antigo silt loam is an example of an Alfisol, which develops under hardwood forests. Its uppermost layer contains more of the soil's organic matter, consisting of decayed leaves, twigs, and branches of the deciduous trees. As rainwater soaks through this top layer of the soil, it becomes slightly acidic, resulting in the bleaching of the lighter-colored zone. The darker zone, extending an average of 30 to 84 cm from the soil surface, is enriched by the deposition of organic matter and clay that is transported from the higher soil layers.

## Wisconsin's Soil Orders

Soil scientists have identified over 800 different types of soils within Wisconsin, which differ depending on the type of rock or geologic material that comprises their parent material, the type of vegetation that grows on them, the climate of the locale, the slope or drainage of the land that may result in anything from a dry, well-drained soil to a waterlogged peat, and the length of time that the soil has been forming. Nomenclature for soils has changed over time, and Wisconsin is a prime location to see these changes, given the long history of soil mapping in Wisconsin. Indeed, the first soil map published in the United States was published by the Wisconsin Geological Survey in 1882 (Hartemink,

Lowery, and Wacker 2012). It identified soils by their textural character or a combination of texture and vegetative cover, thus by such examples as "sandy loams," "prairie loams," and "forested, silty loams." Nearly a century later when he published his magisterial *Soils of Wisconsin*, Hole (1976) mapped 190 soil series in Wisconsin over five major physiographic regions of the state. Wisconsin's different soil series can also be grouped into soil orders (Figure 3.16), comprising soil types that share many similar characteristics. While 12 distinct soil orders are recognized nationally, within Wisconsin only 7 occur: Alfisols, Entisols, Histosols, Inceptisols, Mollisols, Spodosols, and Ultisols.

Alfisols, of which the Antigo silt loam is an example, are found beneath hardwood forests of much of Wisconsin. The most abundant soil order in Wisconsin, it covers 47 percent of the state's land area (Bockheim and Hartemink 2017). Alfisols are the dominant soil in a broad area extending from the Door Peninsula south to the Illinois state line and continue west to the Mississippi River. They are also common within the Driftless Area but are less dominant in portions of southwest and south-central Wisconsin where extensive prairies once existed. Described as "base-enriched forest soils with an argillic horizon" (Bockheim and Hartemink 2017, 3), Alfisols are relatively fertile even though moderately leached, with the presence of calcium, magnesium, and potassium and an accumulation of clays below their surface horizon. Although formed from a variety of parent materials, including clayey till, loess, and sandy outwash, Alfisols in Wisconsin are typically loamy in character. Three other soil orders are also strongly linked to their vegetative cover.

Spodosols, which cover 17 percent of Wisconsin's land area, typically develop under coniferous forests. They are the dominant soil in much of the northern quarter of Wisconsin, particularly that area from Florence and Forest Counties west through Washburn County. The decomposition of pine needles and other coniferous debris creates an acidic soil, one that is marked by a conspicuous layer of sand or sandy loam that is bleached almost white immediately below the uppermost soil horizon. The presence of this albic horizon, far more conspicuous than the light layer found within Alfisols, is a key diagnostic trait of this soil order. It overlies a reddish-brown spodic horizon. These soils typically contain less organic material in their upper horizons than do Alfisols and also have lower cation exchange capacities, reducing their utility for cultivation, even if their location's short growing season permitted it.

Mollisols develop under prairie grasslands, both tall-grass prairie as well as oak-savannas and marshes. They contain a dense network of plant roots within their top 2 feet, or 60 to 70 centimeters. Decomposition of the grass material causes the top foot or so of this soil, or the mollic epipedon, to be very dark in color. These soils are humus rich, as organic matter often comprises 4 to 5 percent of the soil material, and the organic carbon content of the soils exceeds 0.6 percent. These soils are often silt rich, formed upon loess, alluvium, and glacial outwash. Having a high cation exchange capacity, containing calcium and magnesium and being slightly alkaline in character, Mollisols are considered the most fertile of soils, particularly desired by farmers growing various grains. These soils are widely scattered across areas of the southern third of Wisconsin, as well as several areas of west-central, east-central, and southern-northeast Wisconsin. Overall, Mollisols cover 10 percent of Wisconsin's land area.

Histosols are typically described as peat or muck soils, and they form in wetlands, which are widely scattered across Wisconsin, covering patches throughout most of the state except the Driftless Area,

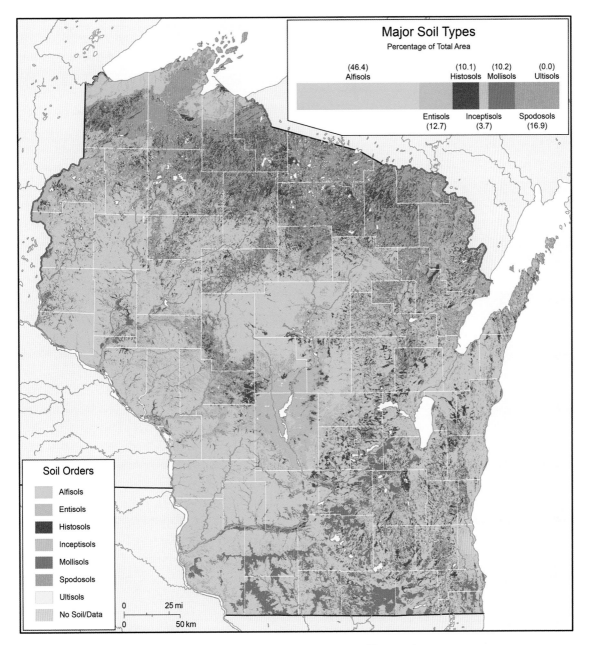

FIGURE 3.16. Soil orders of Wisconsin. Map data from Soil Survey Staff, Natural Resources Conservation Service, United States Department of Agriculture. Web Soil Survey at https://websoilsurvey.nrcs.usda.gov/.

where only a few small areas of Histosols exist. Nevertheless, they are most common within the colder areas of the state. Overall, they cover a tenth of the state's land area. Whether formed in a marsh, dominated by grasses, or a swamp, which includes trees that can withstand lengthy periods of inundation, these soils are largely composed of organic material—at least 20 to 30 percent by weight (Bockheim and Hartemink 2017)—whose decomposition is constrained by anaerobic conditions resulting when the soils are saturated with water. When such soils are drained, they are quite suitable for certain vegetable crops, such as carrots and onions, but they are quite vulnerable to oxidation and subsidence when exposed to the air. They are vulnerable to fire. Indeed, in parts of the world peat soils are cut and dried for use as fuel.

The other two soil orders widely found in Wisconsin are the least developed, showing little in the way of horizons below their topsoils. Entisols, which overlie 12 percent of Wisconsin, are typically considered of recent development, often characterized by having a topsoil atop their lower horizon, which is dominated by unconsolidated parent material, basically missing their middle or B horizon. Three-quarters of Wisconsin's Entisols display a sandy texture (Bockheim and Hartemink 2017). Inceptisols are slightly better developed and are often characterized by the accumulating sediments that comprise its parent material. Inceptisols cover only 4 percent of Wisconsin's land surface; the largest area of Inceptisols is the central part of the state, where they dominate the soils formed in the area of Glacial Lake Wisconsin and the outwash plains to its northeast. It is this same region where Wisconsin's Entisols are most commonly found, although they are also common in several areas of substantial glacial outwash deposits, including Burnett and southern Douglas Counties and much of Marinette County. Finally, a very small area, in total less than one square mile, is covered with Ultisols, which are highly leached, acidic forest soils that display well-developed horizons.

## WISCONSIN'S ENVIRONMENT

Wisconsin's physical environment displays the legacy of the Pleistocene Ice Age. This glaciation laid down the parent material upon which most soils of Wisconsin have developed, and the vegetation growing on these soils not only influenced the evolution of the soils but is also strongly shaped by the local climate. While the Ice Age has been over for 10 millennia, Wisconsin's climate includes a long winter, with frosts and freezing conditions experienced during half or more of the year. In such a climate forests dominated much of the landscape before the arrival of European explorers and settlers, and it was the natural resources that came from the ground, the soil, the forests, and the state's wild animals that attracted these groups, as we will see in the next several chapters.

*Chapter Four*

# Natural Hazards in Wisconsin

W ISCONSIN SUFFERED one of the nation's deadliest disasters when the deadliest wildfire in American history destroyed Peshtigo on October 8, 1871, killing approximately 1,200 residents (Gess and Lutz 2002). It occurred the same day as the Great Chicago Fire, which caused more property damage, but only a quarter as many deaths, and the Great Michigan Fire, which burned the area around Port Huron, Manistee, and Holland, the nation's second-deadliest wildfire. Besides the simultaneous fires in Illinois and Michigan, much of Wisconsin's Door Peninsula was ravaged by fire the same date. Including hurricanes, floods, and earthquakes, the Peshtigo Fire was the seventh- or eighth-deadliest natural disaster to occur in United States. Although wildfires are often considered a natural disaster, and the exceptionally hot, dry, and windy weather conditions definitely conspired to accentuate the wildfire hazard in Wisconsin, Illinois, and Michigan on that fateful day in 1871, human behavior in felling the forest and leaving large quantities of logging refuse on the ground in the vast cutover zone definitely contributed to the hazard.

The question that readily comes to mind is: "Can a disaster like this occur again?" The episode of clear-cutting that resulted in conditions that fed the Peshtigo Fire is unlikely to be repeated, and our response to fires is far different today. Wisconsin is located too far from tropical waters to receive anything other than heavy rains from the remnants of a hurricane. It is well known for its seismic safety. Yet, Wisconsin remains vulnerable to a variety of natural hazards. Whether they remain a hazard or spark a full-fledged disaster is largely dependent upon human behavior.

## HAZARDS VERSUS DISASTERS

Hazards represent the risks that society faces. When society has appropriately planned for these risks or threats, an event may prove to be an inconvenience, but it is unlikely to result in a disaster, where extensive property damage and loss of lives may occur. A disaster is a destructive manifestation of a hazard, while the hazard is the threat. The occurrence of any disaster provides strong evidence of the presence of natural hazards.

Besides the Peshtigo Fire, which burned 1.2 million acres, Wisconsin has experienced other major wildfires. While those fires experienced during the late nineteenth century are hardly indicative of the contemporary threat from wildfires, wildfires in the latter part of the twentieth century are more illustrative. Since 1977 Wisconsin has experienced nine wildfires that have burned more than 2,560 acres (4 square miles). The largest of these occurred in 1977, when the Brockway Fire burned 17,590 acres in Jackson County and the Five Mile Tower Fire charred 13,375 acres in Washburn and Douglas Counties. More recently, the Germann Road Fire incinerated 7,499 acres in Douglas and Bayfield Counties in 2013 (Garrett 2017). On average the Wisconsin Department of Natural Resources responds annually to 1,700 fires that burn a total of 6,000 acres (Ecklund and Kassulke 2012).

The deadliest tornado in Wisconsin hit New Richmond in St. Croix County on June 12, 1899, killing 117 persons, the ninth deadliest tornado in American history. While the annual loss of life from tornadoes in Wisconsin usually ranges between 0 and 5, Wisconsin experienced an average of 22 tornadoes annually from 1982 through 2015, with the annual number varying from 4 to 62. Of the 767 tornadoes during that period, only 6 had EF4 or EF5 status, while 85.4 percent were ranked either EF0 or EF1, the weakest categories (WDMA 2017b).

Most meteorological hazards rarely result in loss of life. Rather, they are best known for the property damage that they cause. For example, drought may cause major crop losses and often contributes to enhanced wildfire risk, but drought-induced famine is limited to the poorest nations of the world. While hail can cause considerable damage to windows and roofs, most often it is farmers who see hail destroy their crops. Meteorological extremes constitute hazards, while normal conditions can be viewed as a resource. Typical rains are critical to successful production of the state's bountiful crops, while too much rain causes flooding, and too little rain leaves crops wilting in the field. Normal snowfall moistens fields when it melts in the spring and assists operators of ski slopes, cross-country ski trails, and snowmobile sales. Too little snow may diminish the income of winter sports facilities, while excessive snowfall contributes to traffic accidents, transportation delays, school closures, and other disruptions.

Natural hazards are not defined solely as physical events, but by their interaction with human behavior. Only those events that exceed our normal capacity "to reflect, absorb, or buffer that lead to harmful effects, oftimes dramatic, . . . characterize our image of natural hazards" (Kates 1971, 438). The consequences of hazards are often influenced by the vulnerability of the population, which is determined not only by their location relative to the threat, but by their socioeconomic situation. We next describe the physical and spatial character of natural hazard vulnerability in Wisconsin.

## REGIONAL DISTRIBUTION OF NATURAL HAZARD VULNERABILITY

### Geologic Hazards

Wisconsin is located in one of the most seismically stable regions of North America. The few earthquakes that have occurred in Wisconsin have only been locally felt, yet the U.S. Geological Survey's seismicity map of Wisconsin (Stover, Reagor, and Algermissen 1980) indicates three earthquakes that had a Modified Mercalli magnitude of V, which would have been "felt indoors by practically all," but whose damage would be minimal. Those Wisconsin residents most likely to feel earthquakes, and then less than once every decade or so, live in those counties along the Illinois border. Earthquakes that occur in the Wabash River Valley of southern Illinois and Indiana are occasionally felt, with this most recently occurring in 2008. In addition, several minor quakes closer to the Wisconsin-Illinois border and in central Wisconsin have occurred in the last decade.

While earthquakes that occur in the Wabash Valley with Richter scale magnitudes exceeding 5 are more likely to be felt in Wisconsin, a far more dangerous seismic region lies along the Mississippi River Valley between St. Louis, Missouri and Memphis, Tennessee. The New Madrid Fault produced four or five catastrophic earthquakes in 1811 and 1812, with a local Modified Mercalli magnitude of XII—the strongest category—and estimated Richter readings of 7.5, resulting in lesser magnitudes of shaking over much of the central and eastern United States. In its *Hazard Analysis for the State of Wisconsin*, the Wisconsin Department of Military Affairs (WDMA 2002, 23) notes that should an earthquake like those of 1811 and 1812 reoccur "anywhere in the New Madrid Seismic Zone," maximum Mercalli scale magnitudes of V to VII would be experienced in Kenosha, Milwaukee, Racine, Rock, Walworth, and Waukesha Counties. Such shaking would likely result in damage to poorly built structures, particularly those of masonry construction, and to chimneys. However, the biggest consequences for Wisconsin would be the disruption of major transportation links, including natural gas and petroleum pipelines, that cross the central Mississippi River Valley en route to Wisconsin.

Landslides and other earth movements occur in Wisconsin (Figure 4.1). Occasionally these are deadly, such as the mudflow that killed a Vernon County resident in 2016. Characteristics of landslide hazards vary across Wisconsin. Within the Driftless Area of western Wisconsin, steep cliffs are sites of rockfalls, and flooding on steep slopes can create mudslides. A U.S. Geological Survey report indicates that "along the limestone and sandstone bluffs on the east side of the Mississippi River, [the] incidence of rock falls is moderate and susceptibility is high" (Radbruch-Hall et al. 1982, 11). Indeed, The Rock in the House in Fountain City features a dwelling with a 55-ton boulder that crashed into its bedroom in 1995. On the opposite side of Wisconsin, erosion of unconsolidated glacial deposits by wave action along Lake Michigan, particularly during periods of above-average lake levels, has led to slope failures and slumping, visible at places from Kenosha County north into the Door Peninsula.

Most of Wisconsin lacks steep slopes that facilitate rapid earth movement or earthflows, yet lacustrine clay deposits—such as those created under glacial lakes in many areas of central Wisconsin—result in a moderate vulnerability to earth flows and lateral spread. In places this is most conspicuously shown by creep, which slowly displaces retaining walls and weakens basement walls and building foundations. In addition, these clay-rich soils expand when wet and contract when dry, also cracking foundations and damaging roadbeds.

Other geologic hazards are associated with the local bedrock. Within areas underlain by limestone or dolomite, karst hazards may occur. While sinkholes are not a significant problem in Wisconsin, highly fractured dolomite bedrock within Door and Kewaunee Counties enables polluted waters to rapidly move through the ground. In the eastern third of Wisconsin, but particularly concentrated in areas of Outagamie and Winnebago Counties underlain by the St. Peter Sandstone, the presence of small quantities of arsenic in the bedrock poses health risks from well water, forcing communities to use surface water (Schmidt 2016; WDNR 2017a).

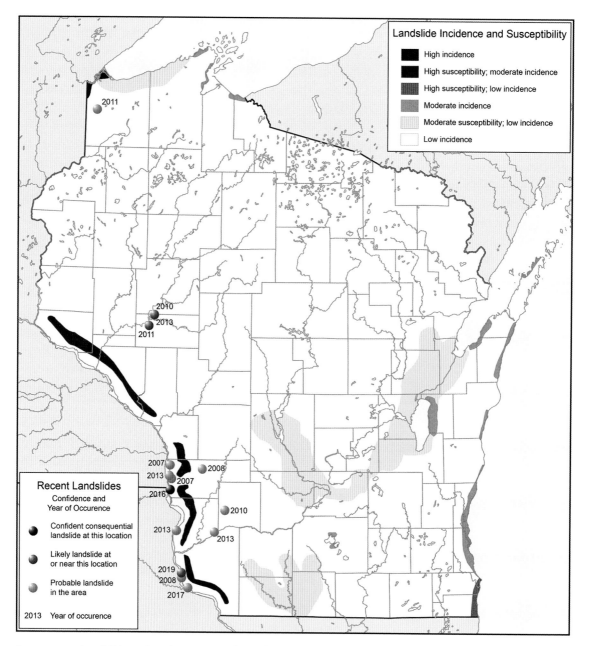

FIGURE 4.1. Landslide and earth movement hazards in Wisconsin, with dots indicating locations of landslides that occurred between 2008 and 2019. Revised from Radbruch-Hall et al. (1982) and updated using U.S. Geological Survey's web-based interactive U.S. Landslide Inventory Map (https://usgs.maps.arcgis.com/apps/webappviewer/index.html?id=ae120962f459434b8c904b456c82669d).

Radon, a radioactive gas within indoor air and dissolved in water, is a hazard in much of Wisconsin. Indoor radon gas levels exceed federal health standards locally in both northern Wisconsin, which has granitic bedrock, and farther south where sedimentary rock layers are atop granitic bedrock (Kochis and Leavitt 1997). While variations in radon exposure are partly related to local geology, they are also linked to a building's condition. Fractured or cracked basement floors and walls enhance the amount of radon that enters a building.

## River Flood Hazards

Flooding in Wisconsin has multiple causes. Major river flooding along the Mississippi and Wisconsin Rivers may occur with rapid melting of winter snowpack or from heavy region-wide rains that may occur when atmospheric circulation patterns stagnate over the area (Figure 4.2). Spring rains over still-frozen ground can also cause flooding. Flash floods are more likely to be local in extent and happen when heavy convectional precipitation results from a series of thunderstorms that train over an area. Rapid runoff, enhanced where steep slopes concentrate the runoff, often occurs within the coulees of

FIGURE 4.2. Sandbags protecting business buildings along State Highway 35 in Fountain City, Buffalo County, during the Mississippi River Flood of May 2001. The towering bluffs behind the community represent a rockfall hazard, and a 55-ton boulder crashed into a house in the city in 1995.

the Driftless Area. Indeed, Vernon, Crawford, and Grant Counties have the greatest number of recorded flood events among Wisconsin's counties (NWS-M 2019b). Flooding may result from dam failures, with the failure of the Lake Delton Dam in June 2008 being a prime example.

Flooding is not just confined to established stream and river channels. Heavy precipitation often results in sheet flow across fields and ponding, particularly where soils are rich in clay and relatively impermeable. Following heavy rains, one often sees low-lying portions of fields where crops are flooded, with flooded areas not connected to any stream channel. Many farmers have addressed this hazard by laying drainage tiles below the surface of the soil, or by burying perforated plastic drainage pipe, to convey excess water to ditches along field edges (Figure 4.3). Extensive areas that are subject to streamflow flooding have been mapped, both at the state level (WDA 1975) and at large scale by the Federal Emergency Management Agency (FEMA) in its nearly 7,000 Flood Insurance Rate Map panels issued for Wisconsin. These areas include not only the broad floodplains of the Mississippi River and those of its major tributaries, the Wisconsin, Black, and Chippewa Rivers, but also extensive flowages associated with the Fox and Wolf Rivers, plus the many tributaries of the Wisconsin River, including the Eau Pleine and the Yellow Rivers. Broad areas of flood-prone lowlands exist along the Rock River, the Fox River in southern Waukesha County, and the Sugar and Pecatonica Rivers that flow into Illinois.

FIGURE 4.3. Coils of drainage pipe awaiting placement into trenches cut in the soil in field south of Allenville in Winnebago County.

Significant flooding occurred in Wisconsin in 1905, 1912, 1922, 1938, 1941, 1959, 1960, 1965, 1967, 1973, 1978, 1986, 1990, 1993, 2008, 2010, 2016, 2017, 2018, and 2019, all of which locally exceeded the 100-year flood (Krug and Simon 1991; WDMA 2002; WDMA 2017b; NCEI 2018b; NWS-L 2018). The 1938 floods along the Wisconsin River breached the levees at Portage, spreading floodwaters into the Fox River. Flooding in 1993 was associated with widespread above-normal precipitation during the first half of the year. Not only was the Mississippi River over its banks, but significant flooding occurred along the Black River, which rose 2.5 feet above the 100-year flood level in Black River Falls. Significant flooding in 1993 also occurred in Darlington, where the Pecatonica River crested 7.6 feet above flood stage (WDMA 2002). Flooding there in 2019 was the worst since 1993, cresting just a foot lower.

Successive flooding occurred within counties along Lake Superior in 2016, 2017, and 2018. Flooding in 2016, which was particularly severe along the Bad River, exceeded the previous record by 5 feet and was calculated as having a recurrence of less than once in 500 years. Portions of that area received 10 to over 12 inches of rain, mostly over an 8-hour period, causing the Bad River's flow to increase 133 fold over a 15-hour period (Fitzpatrick, Dantoin et al. 2017). Flooding the following two summers resulted from the training of tropical moisture from remnants of Pacific hurricanes. These floods caused significant damage to roads, bridges, and resort facilities and prompted the abandonment of railroad service to Ashland.

The Driftless Area of Wisconsin is particularly vulnerable to flash flooding, given its steep slopes and many small watersheds, which concentrate runoff from strong thunderstorms. Significant flooding occurred multiple times this century. Floods in both 2007 and 2008 along the Kickapoo River at the Gays Mills area were estimated to be one in 500-year events (Fitzpatrick, Peppler et al. 2008), yet that area suffered major flooding again in 2010, and each year between 2016 and 2019. The Kickapoo River crested over 18 feet in 2016 and 2017, less than 2 feet below the all-time record set in 2008. Flooding during 2017 also inundated parts of Ontario, Soldiers Grove, Steuben, and Readstown. Cassville and Potosi, villages located in narrow valleys in Grant County, also suffered flooding in 2017 (Preusser 2017). Flooding in 2018 exceeded the previous all-time Gays Mills record by 3 feet. The Kickapoo River at Viola crested at 23.17 feet, a new record that was 9 feet over flood stage. Extensive flooding occurred throughout the area, including Coon Valley—where 75 percent of its businesses suffered damage, Ontario, La Farge, Elroy, and Norwalk, among others (NCEI 2018b). Extensive railroad and road closures, including an interstate highway, were necessitated given numerous washouts, mudflows, and submerged routes.

Several communities along the Kickapoo River, known much of the time for its tranquility and spectacular scenery that attracts canoeists, have suffered such repetitive flooding that they have either relocated out of the floodplain or have contemplated doing so. Soldiers Grove relocated its downtown business district to higher ground in 1980, following major flooding in 1978 (David and Mayer 1984). Following the 2008 flooding that inundated much of Gays Mills, fifty houses had been relocated before the 2016 floods.

Along the Mississippi River at Prairie du Chien, a U.S. Army Corps of Engineers flood control project acquired 121 properties between 1978 and 1984 on St. Feriole Island, the center of its historic settlement. The removal of these vulnerable structures was the corps' first nonstructural flood control project in the nation. Today, only Villa Louis State Historic Site, a few historic buildings relating to the early fur trading industry, and a nineteenth-century hotel remain on St. Feriole Island, other than structures related to a marina and river transportation. During the spring 2019 Mississippi River flooding, all roads leading to Villa Louis were submerged, even though the historic house was not.

FEMA's National Flood Insurance Program (NFIP) has identified and mapped Special Flood Hazard Areas, those areas within the 100-year (or 1 percent probability of flooding per year) floodplain, within 594 communities representing all of Wisconsin's 72 counties. Of these communities, 65 were not participating in the NFIP as of October 2016, thus making their residents ineligible to purchase flood insurance, acquire bank loans for mortgages for property purchases, or receive certain types of disaster assistance (WDMA 2017a, 6–182). Even in those communities participating in the National Flood Insurance Program, many residents have elected not to purchase flood insurance, some because of its cost and others because their properties are located outside of the 100-year flood zone and are thus perceived—often erroneously—by their occupants as being safe. A detailed study following widespread flooding across southern and central Wisconsin in 2008 showed that far more homeowners requested FEMA assistance than the number who had flood insurance policies. Indeed, 40,799 persons applied for assistance, yet only 1,400 received flood insurance payments (WDMA 2008).

Statewide, 241,356 improved structures, or 10.9 percent of the state's houses and other buildings, are located on parcels within Special Flood Hazard Areas (SFHA), those lands that have a 1 percent chance of flooding any year (WDMA 2017b). The *Wisconsin Threat and Hazard Identification and Risk Assessment* reports:

> Waukesha, Brown, Oneida, Dane, and Milwaukee Counties represent the areas with both the greatest number and greatest value of improved parcels in the SFHA. Waukesha County leads the state in both total number of parcels in the floodplain (11,093) and value of potentially-vulnerable improvements (over $3 billion). . . .
>
> The counties with the greatest proportion of improved parcels in the SFHA relative to the total number of improved parcels are Burnett (45% of improved parcels located in the SFHA), Washburn (38%), Oneida (28%), Forest (25%), and Rusk (24%). (WDMA 2017b, 3–88)

### Lakeshore Flooding and Erosion Hazards

Lakeshore areas are subject to floods and other hazards. Shoreline erosion has long been recognized as a hazard along the shores of Lakes Michigan and Superior. Lake Michigan water levels fluctuate not only seasonally, but also over time. The lowest water level in Lake Michigan since data collection began in 1860 was in December 2012 and January 2013, when the water's elevation was 576 feet. In contrast its level reached 581.92 feet by July 2019, less than an inch below its all-time record for that time of the year. It remains well above normal. During five months of 2019 (May through September) Lake Superior experienced its highest water levels in over a century of records (USACE 2020).

Fluctuations in lake levels pose hazards to shipping and influence shoreline flooding. The Bay of Green Bay and the East River area of the city of Green Bay are particularly vulnerable to flooding when strong northeast winds blow waters into the constricting mouth of the Fox River, raising water levels even higher, such as occurred in 2019 and 2020. During low lake levels the depths of harbors and shipping channels decrease, forcing cargo ships to "light load," reducing their freight by 5 to 8 percent, such as occurred in 2000. Marinas and harbor entrances may require additional dredging to keep ships and boats from going aground. Yet, higher than normal lake levels also pose problems.

During high water stands, lakeshore erosion occurs when waves break against sand dunes and bluffs of glacial deposits, which are composed of highly erodible unconsolidated sediments (E. A. Brown, Wu, and Mickelson 2005). High water levels inundate beaches and erode the toe of bluff slopes, leading to extensive slumping or massive slope failures, which undercut foundations of structures located too close to the bluffs (Figure 4.4). During the high-water stand of 1950–1951, when the lake level was less than a foot lower than during 2019, shoreline erosion from the Ozaukee–Sheboygan County line south to Port Washington Harbor averaged 10 feet, erosion south to Milwaukee Harbor ranged from 5 to 20 feet, erosion from Milwaukee to Racine Harbor averaged 15 feet, and erosion from Racine to the Illinois border ranged up to 75 feet (Hadley 1976, 15).

FIGURE 4.4. Eroding shoreline in St. Francis along Lake Michigan south of Milwaukee. At the time this aerial photograph was taken in July 1984, much of the area was experiencing bluff retreat.

Long-term net erosion along Lake Michigan, shown by bluff crest recession since detailed surveys of sections were made by the General Land Office in the 1830s to 1850s, was least where sand dunes or bluffs of dune sands dominated. There erosion averaged less than a half foot annually, given that sands accreted along the shore during low lake levels. In contrast, other Lake Michigan bluffs in Wisconsin averaged 1.61 feet annually, with the southernmost Kenosha County locations averaging 7.25 feet per year (Buckler and Winters 1983). Losses were greatest in bluffs of strata with variable permeability, where "groundwater seepage creates more instability" (98).

Lakes in Wisconsin typically freeze over every winter, even though the centers of Lake Superior and sometimes Lake Michigan may remain ice free. In the spring, during the presence of strong winds, the melting ice may be blown against the downwind shore, creating ice shoves (Figure 4.5). These ridges of heaped ice, which may reach 10 to 20 feet in height along the edge of large lakes such as Lake Winnebago and Lake Mendota, can damage structures located too close to the lake's edge (Moran 1995). Because of past destruction of docks and boathouses, many docks along Lake Winnebago are either protected behind breakwaters or are removable.

FIGURE 4.5. Ice shoves towering above rear of house along Lake Winnebago in Oshkosh on April 7, 2001. The threat from such ice shoves keeps homeowners from placing permanent docks and boat houses along the lakeshore.

## METEOROLOGICAL HAZARDS

Thunderstorms occur between 30 and 40 days a year, depending upon one's location in Wisconsin. Responsible for much of the annual precipitation, thunderstorms also bring hazards—too much heavy rainfall resulting in flash flooding, lightning, high winds, hail, and occasionally tornadoes. In her review of the severe thunderstorm hazard in Wisconsin, Waltraud A. R. Brinkmann (1985, 9) notes, "the migration pattern—which is very similar for severe thunderstorms in general, for tornadic storms, and for hailstorms—is from the southern portion of the state toward the *northwest* and back, which reflects seasonal changes in the degree of activity over the two main cyclone tracks affecting Wisconsin." Storms in spring, when tornadic activity is most likely, come from the Great Basin and Colorado, while summer storms are more likely to develop in the Plains between Alberta and Wyoming.

The southern third of Wisconsin, excluding the immediate lakeshore counties, has the greatest historical frequency of experiencing severe thunderstorm winds, while the northeastern and northwestern corners of Wisconsin have the lowest (NWS-M 2019f). Although their frequency of occurrence in northern counties, such as Bayfield, is less than a quarter that in southern counties, such as Dane, all areas of Wisconsin have experienced damaging thunderstorm winds. Damage from lightning is highly concentrated in southeastern and south-central Wisconsin (NWS-M 2019d), a function of the greater number of lightning events in the region and a greater concentration of structures and population vulnerable to damage or injury during lightning storms. Several hazards associated with thunderstorms deserve specific discussion.

### Tornadoes

Wisconsin experienced an average of 23 tornadoes annually between 1981 and 2010, two more yearly than recorded between 1971 and 2000. The most recorded was 62 in 2005, yet in 2012 only 4 were observed. While the average maximum width of the tornado track is 121 yards and the average tornado path length is 5.5 miles, based upon data from 1950 through 2007 (NWS-GB 2017), some tornadoes exceed these values. For example, the May 16, 2017, tornado, which tracked from eastern Polk County across Barron and Rusk Counties into western Price County, was on the ground for 83 miles (NWS-TC 2018).

The frequency of tornado occurrence varies regionally (Figure 4.6). Southern Wisconsin is considerably more likely to experience tornadoes than northern Wisconsin, with Dane County having the most recorded tornadoes (NWS-M 2019g). In that region five tornadoes are likely to occur annually within an area measuring 100 miles by 100 miles. In contrast, far northern Wisconsin—northern Bayfield County through northern Iron County—sees only one tornado within a 10,000 square mile block every other year (Changnon and Kunkel 2006, 29). Wisconsin's area with the greatest tornado frequency runs from Dane County through Fond du Lac County, including Green Lake, Columbia, Dodge, and Jefferson Counties (WDMA 2017b). An analysis of 1980 through 1999 occurrences showed that far southwestern Wisconsin experienced on average one tornado day per annum, defined as one tornado occurring within a box measuring 80 kilometers by 80 kilometers or within 25 nautical miles of a given location. That area roughly southwest of a line running from Kenosha to Hudson experienced

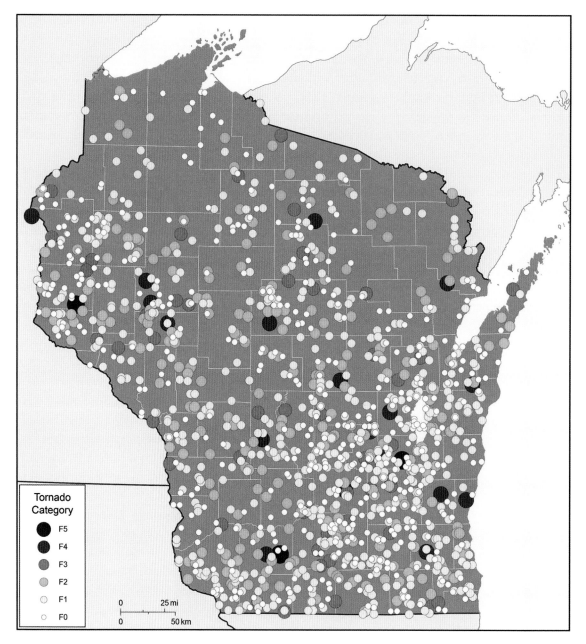

FIGURE 4.6. Tornado occurrences in Wisconsin between 1950 and 2017, indicating the maximum strength of the storm. Data from NOAA's Storm Prediction Center website (https://www.spc.noaa.gov/gis/svrgis/).

between .75 and one tornado day annually, while areas southwest of a line drawn from near Manito-woc to Grantsburg saw a tornado day on average every other year. Along the border between Wisconsin and Michigan's Upper Peninsula a tornado day occurred about one in every four years (Brooks, Doswell, and Kay 2003). Fewer tornadoes form near Lake Michigan because lake waters cool the air, "suppressing the formation of deep thunderstorm convection" (P. N. Knox and Norgord 2000, 2).

Recent research indicates an increased variability of tornado occurrence in the nation, as tornadoes appear to be occurring on fewer days per year, yet on tornado days a greater number occur (Brooks, Carbin, and Marsh 2014). Thus, the outbreaks of 10 tornadoes in northeast Wisconsin on both April 10, 2011, and June 14, 2017, and the 19 tornadoes on August 28, 2018, that occurred from Marquette through Sheboygan Counties represent record activity in the region (NWS-M 2018). On July 19–20, 2019, 15 tornadoes occurred from west-central to east-central Wisconsin.

The greatest tornado threat occurs in most of Wisconsin during the last half of June. Northeast of a line running from near Manitowoc to Bayfield County the threat is highest in July, while the south-eastern corner of Wisconsin is most at risk in the first half of June (Brooks, Doswell, and Kay 2003). It has been noticed that tornadoes that occur in August and September typically have longer track lengths (P. N. Knox and Norgord 2000), yet one of the longest on record occurred in May 2017.

Relatively few tornadoes, notwithstanding the deadly Barneveld tornado of 1984 that hit just before midnight, occur during the early morning hours through midday. The most frequent time for tornado occurrence in Wisconsin is between 4 and 7 PM standard time, with more occurring between 2 and 4 PM than between 7 and 9 PM (NWS-GB 2017). Although tornadoes have been observed coming from all compass directions, the most common path in Wisconsin is from the southwest.

Derechos are another type of severe wind in Wisconsin. These complexes of straight-line winds can produce hurricane force winds and "are most common from May through August and generally follow a northwest-southeast track" (P. N. Knox and Norgord 2000, 7). Extensive swaths of blown-down forest resulting from a derecho are distinguishable from tornado destruction because derecho-downed trees typically all face in the same direction, unlike the swirling pattern of destruction from tornadoes. In July 2011 extremely strong straight-line winds blew down 81,000 acres of forest within Burnett County, with some of the winds exceeding 100 miles per hour (WDMA 2017b). In July 2019 a macroburst associated with straight-line winds of 100 miles per hour downed hundreds of thousands of trees in a 10-mile wide swath extending 60 miles across Oneida, Langlade, and Oconto Counties (NWS-GB 2019).

## Snowstorms and Ice Storms

Snowstorms pose a hazard when they exceed residents' ability to cope with the accumulations of snow. Thus, a two or three inch snowfall in Wisconsin is usually a routine winter event. Those snows that constitute the biggest hazards in Wisconsin are those that bring abundant snowfall, sometimes a foot or more, and those snows that occur early in the fall or late in the spring, when trees are covered with foliage and most vulnerable to limb breakage.

Snows of 6 inches or more within 24 hours are viewed as being snowstorms in the Midwest (Changnon and Kunkel 2006), and there are "vast regional differences in what snowfall magnitudes equate to a snowstorm" (Changnon and Changnon 2006, 374). Further complicating the definition of snowstorm damage is that some of the destruction comes from high winds, flooding, icing, and freezing rains that often accompany a snowstorm. Using the 6 inches of snow definition, the northernmost tier of counties in Wisconsin, running from Douglas to Marinette Counties, receive three to four such snowstorms a year, while southwest Wisconsin and the southern part of south-central Wisconsin average fewer than one and a half snowstorms annually. At least once a decade snowstorms bring over 18 inches of snow to those northern Wisconsin counties between Douglas and Vilas, while the maximum snowfall expected once in a decade averages less than 14 inches in southwest and south-central Wisconsin plus the southern portions of west-central and southeast Wisconsin (Changnon and Kunkel 2006).

Blizzards, which combine sustained winds exceeding 35 miles per hour with blowing and drifting snow, may result in whiteout conditions, where visibility is severely limited. Not only is this a severe hazard to motorists who are blinded by the snow, but drifts of snow may bury roads and railroads under several feet of snow, blocking transportation. The risk of blizzards varies across Wisconsin. Areas of greatest annual snowfall are not those most likely to experience blizzard conditions. Indeed, National Weather Service records show no blizzard occurrences in Iron County or in a tier of counties extending from Price County west to Burnett County and south to St. Croix County from the winter of 1982–1983 through the winter of 2017–2018. Adjacent Vilas County only recorded one blizzard (NWS-M 2019a). In contrast, all counties along Lake Michigan experienced between five and nine blizzards during the same period.

The blizzard of April 13–15, 2018, was noteworthy not only for its severity, but its late-season occurrence. A large area of east-central Wisconsin extending from near Waupaca to the Door Peninsula received over two feet of snow, with several localities in Shawano, Oconto, and Door Counties receiving in excess of 30 inches, all-time snowfall records. The heavy snow resulted in collapsed roofs and barns. Not only were many roads impassable, but power failures and airport closures occurred (NCEI 2018a).

Snowstorms that are unseasonal pose particular problems. The latest major spring snowstorm to occur in Wisconsin happened on May 10, 1990, bringing up to eight inches of heavy wet snow from Marinette County south through Waukesha County (NCDC 1990). Because trees were covered with foliage, extensive damage occurred. It was estimated that 80 percent of the trees in Waukesha were damaged or destroyed (Foran 2016). The lead author of this book had to make multiple detours as he drove the normally three-mile journey between UW–Oshkosh and his home, given the number of fallen trees and power lines that blocked the route. His cherry trees, which had been covered with white blossoms the previous day, were barren of flowers the following day. That snowstorm not only broke many branches but kept the cherry trees from producing a crop that year.

Ice storms have historically occurred more often south of Wisconsin, where moist air from the Gulf of Mexico is sufficiently warm to begin falling as rain, freezing as it descends through colder air. Thus, the storm that brings freezing rain to central Illinois would more likely produce snow in Wisconsin if it received any precipitation at all. Wisconsin sees far fewer hours of freezing rain and less precipitation intensity during those events than locations in Missouri and Illinois (Houston and Changnon 2007). Within Wisconsin, ice storms are experienced more commonly within a swath of counties extending diagonally across the state from the southwest into those counties west of Green

Bay. In this region, many counties experienced between six and nine ice storm events over a 35-year period (NWS-M 2019c). In contrast, ice storms only occurred three or four times within the counties along Lake Michigan, and only three times along Lake Superior. Ice storms cause havoc for motorists and, if the ice is sufficiently thick, cause widespread damage to trees and downed power lines.

## Agroclimatic Hazards

Farmers are among those Wisconsin residents most vulnerable to vagaries of the state's weather and climate. Just as the lead author saw his cherry trees denuded of blossoms, cherry growers are also vulnerable to late frosts that occur after the trees in their orchards blossom. Most of Wisconsin's cherry and apple orchards are located in the Door Peninsula, where the cool waters in the Bay of Green Bay delay the trees' blossoming for a couple of weeks. This is typically sufficient time to provide protection against a late frost. Nevertheless, occasionally late-season frosts occur after the area's cherry and apple orchards are in blossom, such as occurred in 2015, which destroyed that year's harvest of sweet cherries and caused a 21 percent decline in tart cherry production (Peterson 2016). In 2008 nearly the entire crop of tart cherries, which is much larger than that of sweet cherries, was lost to a combination of a late-season drought followed by abnormally warm weather in January, which caused buds to emerge just before temperatures plunged below zero (M. L. Johnson 2008).

Fruit and vegetable crops are particularly vulnerable to a variety of weather hazards, but adverse weather conditions, such as drought, excessive heat, hail, and flooding, can result in large losses of many field crops, even those that are normally tolerant of frosts. Given the necessity of producing crops during a relatively short growing season, particularly in the northern half of Wisconsin, farmers face the challenge of delaying their crop planting until the threat of a killing frost is past, yet still having sufficient time for their crop to mature and produce grain, fruit, or vegetables before the first killing frost of the fall. If the farmer delays planting too long, there may be insufficient time to grow his or her crop to maturity. Yet, the timing of the last killing frost of the spring (and the first in the fall) varies yearly. For example, considering the 1980 to 2010 period for much of the Illinois border area, the earliest last 32°F freeze averaged between March 21 and 31, while the latest such freeze happened between May 11 and 20. In much of the northernmost quarter of Wisconsin, the range of dates for the earliest and latest last 32°F freeze runs from April 21 to May 10 and from June 11 to 20 (Moran and Hopkins 2002). Farmers who plant too early risk losing their crop.

### *Drought Hazards*

Farmers risk periodic droughts, which depending upon their length and severity may either reduce yields or result in crop failures. The most recent serious drought in Wisconsin occurred in 2012, with extreme drought conditions reported throughout the southern third of Wisconsin beginning in mid-July and persisting in the area closest to the Illinois state line into September. Severe drought conditions occurred throughout much of central Wisconsin. The consequences of this drought are well illustrated by crop yields (Figure 4.7). For example, corn for grain yields in Green, Lafayette, and Iowa Counties averaged between 75 and 100 bushels per acre in 2012, while two years later they ranged between 150 and 175 bushels per acre (NASS-CM 2013, 2015).

FIGURE 4.7. Drought-stricken cornfield west of Winneconne, Winnebago County, on June 25, 1988. At this time of the growing season the corn should have been a foot higher.

The drought of 1988 was even more severe and widespread, with extreme drought conditions occurring within six of the state's nine agricultural districts, with the remaining three (southeast, south-central, and central Wisconsin) experiencing severe drought. Crop losses were staggering. In Polk County, the 1988 corn harvest was just 17.8 percent of the 1987 harvest, while that in Marathon County was only 24.5 percent. Eighteen counties reported alfalfa hay crops that were less than half of those of 1987 (Cross 1992). These losses were particularly troublesome for the state's dairy farmers, who normally grow most of the feed for their cows. Their crop losses forced them to make large purchases at drought-inflated prices to maintain their herds. The cost of hay doubled or tripled in parts of Wisconsin. As a consequence of this drought, 87 percent of Wisconsin's dairy farmers experienced at least some decrease in net farm income; three-quarters sustained moderately to strongly felt shortages of hay or alfalfa, and a third reported a greater need for off-farm income (Cross 1992). Between March 1988 and March 1990, given that the drought persisted—but at a less severe level in parts of the state—through the spring of 1989, Wisconsin lost 3,119 dairy farms, a little less than 9 percent of its total, with the drought being a major factor in the loss.

Just as the 1988 drought was most severe in a different part of Wisconsin from that seen in 2012, Wisconsin droughts are highly variable in both their areal extent and their duration. Drought conditions that occurred between 1929 and 1934, roughly corresponding to when the Dust Bowl ravaged the Southern Great Plains, were "probably . . . the most significant in Wisconsin history, considering its duration as well as its severity" (Krug and Simon 1991, 571). The recurrence interval for a drought of that severity is estimated to be at least 75 years, with some areas of Wisconsin experiencing that level of drought severity less than once a century. While droughts are often associated with agricultural losses, droughts also affect municipal water supplies, navigation on rivers, hydroelectric generation, and fisheries. They also increase the likelihood of wildfires, such as those that occurred in 1977.

*Wet and Flooded Fields*

Wet fields in the spring are another threat that Wisconsin farmers face, even if fields are not flooded. A cool spring following a year with heavy snow can result in excessively muddy fields, just as frequent rainfall may leave the soils waterlogged. Because many Wisconsin soils are clay rich, they require sufficient time to shed any excessive moisture. Such was the case in 2019, when soils were excessively

FIGURE 4.8. Flooded hayfield with floating baled hay rolls in Outagamie County's Town of Center, located north of Appleton, on September 14, 2019.

moist following a cool and wet winter and spring. The *Wisconsin Crop Progress & Condition* report for the week ending June 2, 2019, noted that "Corn planting was 58 percent complete, 13 days behind last year and 17 days behind average" (USDA-NASS 2019a). The report issued two weeks later stated that corn planting was 87 percent complete and 18 days behind average. Only 70 percent of the soybeans had been planted, problematic given its longer growing season, and it was running 14 days behind average (USDA-NASS 2019b). Much of the acreage not yet planted would not be cultivated in 2019.

Excessive rainfall plagued the state's farmers throughout the 2019 growing season, with statewide precipitation totals exceeding normal means by 12 inches, with many stations in north-central and northeast Wisconsin breaking previous records. Green Bay broke its old record, set the previous year, by 9.42 inches, while Appleton's new record was 49.03 inches, nearly 5.5 inches above its prior year (NWS-GB 2020). The excessive rains led to flooded fields and destroyed crops (Figure 4.8) and caused further delays in harvesting. Indeed, the December 9, 2019, *Wisconsin Crop Progress & Condition* report indicated that statewide the corn for grain harvest was 74 percent complete, 24 days behind 2018. In north-central Wisconsin only 46 percent of the cornfields had been harvested (USDA-NASS 2019c).

### Hail Hazards

Hail is another hazard that is often associated with agriculture, although it can cause extensive damage to roofs, siding, and windows, both of structures and vehicles. Baseball-sized hail, accompanied by winds exceeding 60 miles per hour, fell over parts of Waushara, Winnebago, Outagamie, Calumet, and Brown counties on May 12, 2000, causing $100 million in property damage (Moran and Hopkins 2002). NOAA's *Storm Data* reported, "Wind-driven hail shredded west-facing sides of homes and businesses. . . . The worst damage in Calumet county was in and around Chilton, where 2 inch hail and hurricane force-winds gusts (estimated near 80 mph) knocked holes in siding and tore the walls and roofs from some buildings. Over 1,000 homes were damaged in the city of Chilton" (NCDC 2000, 307–308). In September 2016 hail of up to two inches in diameter pelted Madison, resulting in thousands of insurance claims. Individual insurance claims for hail damage averaged $6,700 in Wisconsin between 2000 and 2013, and Wisconsin had the seventh-highest increase nationally in average claim size over that time period (Lekas, Gannon, and Moghul 2014).

Severe hailstorms, defined as those with hail of at least one inch in diameter, pose an uneven hazard across Wisconsin (Figure 4.9). National Weather Service statistics from 1982 through 2018 indicate that Dane County experienced 277 such hail events, the most of any Wisconsin county (NWS-M 2019e). Because the typical hail swath includes scattered hail streaks where the hail is concentrated, a larger county presents a bigger target than a smaller county and thus would be expected to experience more hail. Yet, Dane County had more reported hail than Iowa, Green, and Rock Counties combined or Sauk and Columbia Counties together. While biases in reportage also influence historic records of hail, several spatial patterns of hail hazard are discernable in Wisconsin. In general, counties across the northern third of Wisconsin and in east-central Wisconsin, particularly along Lake Michigan, experience fewer severe hailstorms than counties of south-central and west-central Wisconsin.

Hail need not be large to harm crops. Even hail that is the size of a pea can destroy a field of small grain, such as wheat, oats, or barley, that is approaching harvest. Hail can knock the grains of wheat

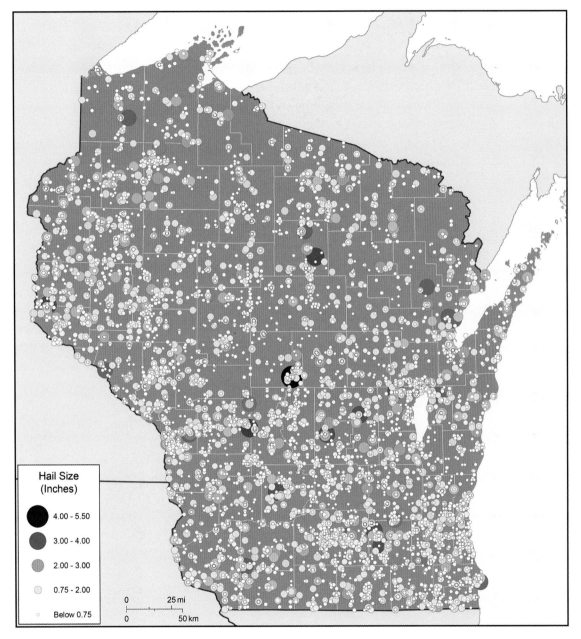

Hail Size
(Inches)

- ● 4.00 - 5.50
- ● 3.00 - 4.00
- ◉ 2.00 - 3.00
- ○ 0.75 - 2.00
- ∘ Below 0.75

0        25 mi
0             50 km

FIGURE 4.9. Hail reports in Wisconsin between 1955 and 2017, indicating the size of the hailstones. Data from NOAA's Storm Prediction Center website (https://www.spc.noaa.gov/gis/svrgis/).

or other cereals from their stalks, making harvest impossible. Hail is only one of a variety of hazards that threaten farmers. When surveyed several decades ago, over half of Wisconsin dairy farmers reported experiencing losses from drought, frost, and excessive cold over a three-year period. Nearly a third had losses from excessive rain, while at least a tenth had experienced losses from flooding, frost, hail, and windstorm. Ten percent reported losses of livestock from lightning (Cross 1994).

Given that farmers face many hazards, the U.S. Department of Agriculture's Risk Management Agency provides multihazard crop insurance that covers some, but not all, of the farmers' financial losses from a variety of agroclimatic hazards. While the indemnity to farmers obviously varies from year to year, given highly variable weather conditions, the 2012 crop year was particularly damaging to Wisconsin's crops. In that year payments exceeding $10 million per county were paid to farmers in eight counties, with the largest payment, of $83,963,997, going to farmers in Lafayette County, largely to compensate them for drought losses. In contrast, for the 2016 crop year, the largest payment within any Wisconsin county was $1,761,140 (USDA-RMA 2013, 2017). Reports for 2019 show that farmers in Outagamie County alone received $12,690,488 (USDA-RMA 2020).

## OTHER HAZARDS FACING WISCONSIN

Outdoors sports enthusiasts and hunters experience other hazards that threaten their activities. Persons working outdoors or engaged in many sports are exposed to flying insects and ticks. Mosquitoes in western Wisconsin, particularly in the Mississippi River Valley, occasionally spread La Crosse encephalitis, yet the total number of diagnosed cases in the entire Midwest is fewer than 100 a year. Since 2001 Wisconsin has experienced human cases of West Nile virus, which is spread from infected birds by mosquitoes. While at least four-fifths of persons bitten by infected mosquitoes do not develop any noticeable illness, it can be deadly. In 2002, basically the year of its arrival in the state, and in 2012 over 50 human cases of West Nile virus were reported in Wisconsin (WDHS 2017b).

Lyme disease is spread by the bite of a deer tick. Even when using insecticides it is often difficult for hikers and hunters to avoid encountering ticks. Wisconsin Department of Health Services (WDHS 2019b) statistics indicate that between 2009 and 2018 the number of annual cases reported statewide ranged from 2,500 to 4,300, with 3,105 cases estimated in 2018. While incidence rates fluctuated widely from year to year, infection rates are generally higher in much of northern and western Wisconsin, plus several counties in central Wisconsin. Incidence rates exceeded 100 per 100,000 persons within 15 counties during 2018. Four counties (Marquette, Oconto, Vernon, and Washburn) had rates exceeding 150 per 100,000. In contrast, few cases were reported anywhere in the eastern third of Wisconsin, with the exception of Door County.

Not only are deer hunters at risk of Lyme disease infection, but deer are increasingly likely to be infected by chronic wasting disease. Although for many years the state was relatively lax in routinely testing for the disease, infection rates, as shown by the percentage of tested deer that test positive, have escalated in parts of south-central and southwestern Wisconsin. Over 15 percent of tested deer in four Public Land Survey Townships, largely within Iowa County, were positive for chronic wasting disease between 2010 and 2013. By 2018 over half of the adult bucks tested in several towns of Iowa, Sauk, and Richland Counties tested positive for the disease (WDNR 2019a). Although human consequences of consuming venison from the animals are not fully understood, the presence of the disease threatens the sports hunting industry in much of central Wisconsin, just as it threatens the deer population.

A warming climate is likely to result in the spread of diseases into Wisconsin that were previously found only in warmer, more southerly locations. Indeed, during the summer of 2017 *Aedes albopictus* mosquitoes were discovered to have spread to Wisconsin, which could potentially spread the Zika virus, which had rapidly spread from South America into southern Florida just a year earlier (Shastri 2017). With expansion of ranges of a variety of mosquitoes, Wisconsin might become vulnerable to yellow fever and malaria in the future. Indeed, until nearly a century ago malaria—although fortunately not the most virulent varieties—occurred in much of the Mississippi River valley, including Wisconsin (Acknerkecht 1946). Warmer summers, combined with a lengthened growing season, may facilitate the expansion of fungal diseases and insect pests that threaten certain crops. Thus, exposure of Wisconsin to hazards is not static, as we must increasingly consider the risks of global warming.

## HAZARDS ASSOCIATED WITH GLOBAL WARMING

Abundant climatological data show that the climate of Wisconsin is warming. During three of the past ten years (2009 to 2018), the average temperatures within Wisconsin were among the top ten warmest years in 124 years of records (NCDC 2019). In its examination of temperature data from 1950 to 2006, the Wisconsin Initiative on Climate Change Impacts, a committee of over twenty university and government scientists, found that "winter temperatures increased significantly in northwestern Wisconsin, and these increases extended into the central part of the state. . . . During winter, nighttime minimum temperatures warmed at a faster rate than daytime maximum temperatures. . . . Northwestern and central Wisconsin experienced 14 to 21 fewer nights with temperatures below zero degrees Fahrenheit" (WICCI 2011, 6). Although average annual temperatures statewide increased by 1.1°F over the 56 year period, winter temperatures rose an average of 2.5°F. The greatest increases in winter minimum temperatures occurred "in the central regions through the entire western portion of the state" (Kucharik et al. 2010, 14). The diurnal (24 hour) range in temperature decreased across Wisconsin. Overall, spring has arrived earlier, while the average date of the first fall freeze is nearly a week later than in 1950 (WICCI 2011).

### Observed Changes in Climate

With the earlier occurrence of the last killing frost in the spring and a later date for the first killing frost in the fall being observed in much but not all of Wisconsin, the growing season has increased. This is particularly pronounced "in the northwest and central regions, where typically the growing season has been extended by two to three weeks" (Kucharik et al. 2010, 18). While a longer growing season enables farmers to grow crops that would otherwise have been grown in more southerly climates, it enhances the spread of plant disease and insects.

Warmer conditions threaten winter activities, including snowmobiling, skiing, and ice fishing. For example, "the American Birkebeiner cross-country ski race in Wisconsin was cancelled due to a lack of snow in February 2017" (Angel et al. 2018, 878). The average duration of ice cover on Lake Mendota has decreased by 29 days since 1852. It did not freeze over in winter 2018–2019 until January 10, and until January 12 in 2019–2020 (WSCO 2020). Within the next few years it will cease to reliably have complete ice coverage, just as is already occurring on Big Green Lake (Sharma, Blagrave, and Magnuson 2019). The length of the winter season during which ice fishermen and women can safety drive across Lake Winnebago decreased. Deteriorating ice conditions in 2020 resulted in sturgeon spearers setting out only 2,439 ice shanties, as counted on opening day, the fewest in eight years and under half the number placed in 2017 (WDNR 2020). Yet in 2019, Lake Winnebago remained ice covered into late March, courtesy of the polar vortex that brought subzero temperatures to Wisconsin in late January.

Annual precipitation increased by two to four inches over much of Wisconsin between 1950 and 2006. The largest changes were in the summer, with strong regional differences. Much of the southern two-thirds of Wisconsin gained one to two inches of summer precipitation, while the northernmost third, particularly near Michigan's Upper Peninsula, experienced that size of a decrease. In contrast, all of Wisconsin saw increased autumn precipitation, with the amounts varying regionally (Kucharik et al. 2010). Of greater concern from a hazards perspective "is evidence that heavy precipitation has become more common in the U.S., including Wisconsin, during the past several decades" (Kucharik et al. 2010, 10). For example, the number of days that Madison annually experienced two inches or more of precipitation doubled since the 1950s, and occurrences of three inches increased even more dramatically (WICCI 2011). Indeed, 3 of the last 10 years (2010–2019) have ranked among the 10 wettest springs in Wisconsin since records began 125 years ago.

## Future Climate Changes and Hazards

Accurately predicting future changes in climate is difficult. While climatologists almost universally agree that increases in greenhouse gasses, particularly carbon dioxide and methane, result in a warming climate, extending that prediction to issues of precipitation, heavy precipitation events, and severe weather (including tornado outbreaks) is more challenging. Yet, the Wisconsin Initiative on Climate Change Impacts suggest several changes. For example, winters would be about four weeks shorter, with nearly 14 inches less snow (WICCI 2011). By 2055, depending upon one's location in Wisconsin, there will be "about one to four more weeks each year with daily high temperatures topping 90°F" (WICCI 2011, 26). Later in the twenty-first century climatologists anticipate even more days of intense heavy rain separated by longer periods without rainfall (WICCI 2011), simultaneously exposing residents to enhanced risks of both floods and drought.

While Wisconsin's climate will undoubtedly display changes brought by global warming, these changes may be less challenging than what farmers may experience elsewhere in the country. Indeed, dairy farmers in California, Arizona, Texas, and Florida—all areas that compete with Wisconsin's milk producers—face far more serious threats to their sustainability (Cross 2015). Yet, the *Fourth U.S. National Climate Assessment* portrays many challenges for midwestern growers, even as we approach the middle of this century.

> Higher growing-season temperatures also shorten phenological stages in crops (for example, the grain fill period for corn) . . . [so] overall yield trends will be reduced because of periodic pollination failures and reduced grain fill during other years.

Increases in humidity in spring through mid-century are expected to increase rainfall, which will increase the potential for soil erosion and further reduce planting-season workdays due to waterlogged soil. . . .

Midwest surface soil moisture likely will transition from excessive levels in spring due to increased precipitation to insufficient levels in summer driven by higher temperatures. . . .

Projections of midcentury yields of commodity crops show declines of 5% to over 25% below extrapolated trends broadly across the region for corn (also known as maize) and more than 25% for soybeans in the southern half of the region, with possible increases in yield in the northern half of the region. (Angel et al. 2018, 881, 882)

While Wisconsin's soybean farmers may benefit from the expanded growing season, given that the state is currently near the northern limit for soybean cultivation, Wisconsin's forestry industry faces obstacles.

Changes in climate and other stressors are projected to result in changes in major forest types and changes in forest composition. . . .

In the Upper Midwest, the duration of frozen ground conditions suitable for winter harvest has been shortened by 2 to 3 weeks in the past 70 years. The contraction of winter snow cover and frozen ground conditions has increased seasonal restrictions on forest operations in these areas, with resulting economic impacts to both forestry industry and woodland landowners through reduced timber values. (Angel et al. 2018, 885, 886)

As the twenty-first century progresses, the environment of Wisconsin, whether viewed from the perspective of its climate, vegetation, or hazards, will increasingly deviate from that experienced in the twentieth century. These environmental changes will interact to have profound impacts upon the state's people and economy. Some changes may be advantageous, yet many others could produce unanticipated negative consequences.

*Chapter Five*

# Settlement of Wisconsin 1

## *Fur Trading and Mining*

NATIVE AMERICANS HAD LIVED IN WISCONSIN for over ten millennia before the first European explorers ventured into the area in the third and fourth decades of the seventeenth century. Sent forth from Quebec by the governor of New France, Samuel de Champlain, both Etienne Brule and Jean Nicolet were seeking new routes to the Pacific Ocean. Brule focused upon Lake Superior. Nicolet traveled along the shore of Green Bay and then up the Fox River to Lake Winnebago. From there he journeyed upriver to an easy portage that linked the Fox River to the Wisconsin River. He proceeded downriver to near the mouth of the Wisconsin River before retracing his route. French exploration during the seventeenth century had three goals: to search for river routes into the interior—and hopefully to the Pacific Ocean; to convert the indigenous population to Christianity, and to develop the fur trade (Kellogg 1925; Wyman 1998). Several French explorers were priests, including Father Claude Allouez, whose celebration of mass along Lake Winnebago in 1670 is celebrated by a historical marker in Oshkosh's Menominee Park, Father Jacques Marquette, and Father Louis Hennepin. Just as the French were drawn to Wisconsin by its natural resources, with fur traders engaged in a type of primary production, so were those miners who sought lead and iron ore.

At the time the French began exploring the land that is now Wisconsin and establishing trading posts along the waterways that their large canoes navigated, three Native American peoples dominated the landscape. These were the Menominee, the Winnebago (now named the Ho-Chunk), and the Dakota (Sioux). This was soon to change as other indigenous peoples moved into Wisconsin from the east, partly in response to European settlement along the Atlantic coast. Furthermore, Wisconsin then conspicuously displayed, and to a certain degree still shows, much evidence of prehistoric Indian inhabitation.

## LANDSCAPE LEGACY OF PREHISTORIC INDIAN POPULATION OF WISCONSIN

Wisconsin has a very rich archaeological history, with a phenomenal number of earthworks and mounds located along many lakes and rivers (Figure 5.1). At the time Europeans began arriving, Wisconsin had between 15,000 and 20,000 mounds scattered over 3,000 sites, more than in any other comparable

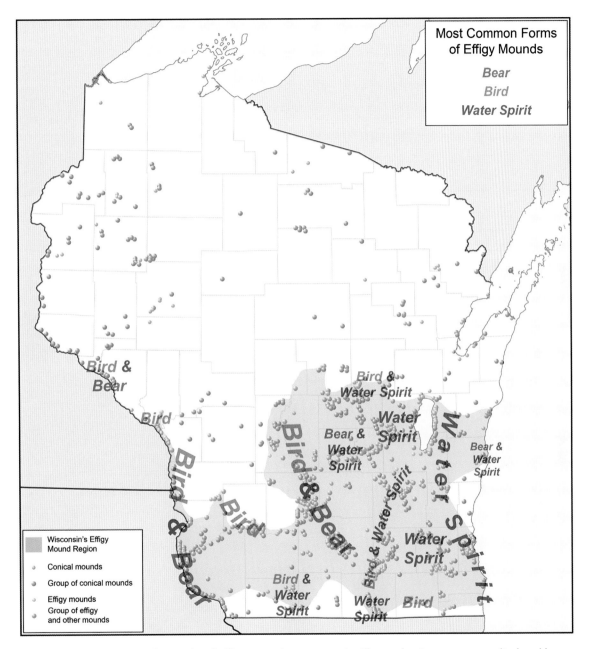

Most Common Forms
of Effigy Mounds

*Bear*
*Bird*
*Water Spirit*

Bird &
Bear

Bird

Bird &
Water Spirit

Water
Spirit

Bear &
Water
Spirit

Bird &
Bear

Bird
&
Bear

Bear &
Water
Spirit

Bird & Water Spirit

Water Spirit

Water
Spirit

Bird &
Water
Spirit

Water
Spirit

Bird

Wisconsin's Effigy
Mound Region

Conical mounds

Group of conical mounds

Effigy mounds

Group of effigy
and other mounds

FIGURE 5.1. Distribution of conical and effigy mounds in Wisconsin. The predominant creatures displayed by the effigies vary geographically, with birds most common in the west and water spirits to the east, while bears and combinations are found elsewhere. "Map Showing the Distribution of Indian Mounds in Wisconsin" used and revised with permission from the collection of the Wisconsin Historical Society, WHS-92138. Revised from C. E. Brown (1916) with effigy type distribution from Birmingham and Rosebrough (2017).

area of the Midwest (Birmingham and Eisenberg 2000). Largely constructed between 700 and 1100 AD (Birmingham 2010), many of the mounds have been destroyed by modern farm machinery and urban development, yet an estimated 4,000 remain. Thus, as Birmingham and Eisenberg (2000, 9) aptly note, "In Wisconsin, the Indian mound is the most visible legacy of a Native American past."

### Effigy Mounds

Effigy mounds shaped from soil and other earth materials were most frequently erected along river terraces and atop bluffs overlooking rivers and lakes in the southern half of Wisconsin. Thus, one sees many mounds within Wyalusing State Park (Figure 5.2), atop the bluff facing the confluence of the Wisconsin and Mississippi Rivers. Farther south along the Mississippi River, mounds are located atop bluffs in Nelson Dewey State Park, also in Grant County. Grant County and Dane County are both the sites of over 1,000 Indian mounds. Within Dane County numerous mounds are along the shores of Lakes Mendota, Monona, Waubesa, and Kegonsa, where over 1,200 mounds were constructed among 160 separate locations (Birmingham 2010). Wisconsin was the focus of the effigy mound culture among Late Woodland period peoples. Near Madison the largest such mound was created. On the grounds of Mendota State Hospital, an effigy mound of a bird displays a 624-foot wingspan.

FIGURE 5.2. Effigy mounds within area of mowed grass and several tall trees atop Sentinel Ridge, a bluff overlooking the Mississippi River in Wyalusing State Park, Grant County.

Just considering the Lake Mendota area, 370 mounds were scattered across over 50 locations, including the University of Wisconsin campus, where a partly defaced bird effigy mound can be seen atop Observatory Hill.

Birds are the animal most commonly depicted by effigy mounds in Wisconsin, accounting for about one-third of them, yet their prominence varies across the state. Bird effigies are lacking in eastern Wisconsin, but they are prominent between the Rock River and the Mississippi River. In eastern Wisconsin, the water spirit or panther with its long tail is the most frequently displayed animal. In the Driftless Area and along the shore of former Glacial Lake Wisconsin the bear, or often a series of bears in a line, is most commonly seen (Birmingham and Eisenberg 2000; Birmingham and Rosebrough 2017; Gartner 1999). Examples of lizards are common at some sites, yet the effigies preserved at Lizard Mound County Park near West Bend are now interpreted as being panthers. Turtles are also displayed, and one mound northeast of Baraboo features a human that originally extended 218 feet from the top of its head to its feet. Most effigy mounds are relatively low, not rising more than 3 to 5 feet above the surrounding ground. Some of them can be identified by the character of the sands or soils used in their construction, even after farming destroyed their elevations. Although far rarer than the raised mounds, in Fort Atkinson an intaglio "mound" is preserved, in which the effigy of a panther is displayed in a shallow excavation. Although it is the only survivor of nearly a dozen that were known in the mid-nineteenth century, when viewed following heavy rains, water marks the outline of a panther, which corresponds with it being a water spirit.

Scattered among the effigy mounds are many other low earthen mounds, many that are either conical in shape or linear, sometimes extending 100 or more feet. While burials did occur in some of the mounds, many effigy mounds lack burials. Although archaeologists attribute effigy mound creation to their ceremonial usage (Birmingham and Eisenberg 2000), the reasons they were constructed and their geographic display of specific animal effigies have prompted considerable scholarly inquiry. Specific images may correspond to contemporary clan groupings which relate to sky, earth, and water (Birmingham and Rosebrough 2017). It has been argued that their display, along with the extensive creation and use of ridged fields for growing their crops, "classify area, communicate boundary and enforce control" and provided "a means of maintaining local identity" (Gartner 1999). It has been suggested that "effigy mound culture . . . [developed] in opposition to the mounting threat of new peoples [Mississippians] who began to encroach upon their landscape" (Guéno 2017, 52). Thus, "mound construction ritually realized the existence, importance, and boundaries of a community" (67). Yet effigy mound builders erected "their mounds on sacred or hallowed ground well removed from their everyday domiciles" and "led a seasonally mobile existence" (Stoltman and Boszhardt 2019, 106). The building of effigy mounds in Wisconsin ended around the close of the twelfth century, yet erection of burial mounds continued into the time of contact with Europeans.

## Mississippian Temple Mounds

Temple mounds were erected at several Wisconsin sites by Mississippian peoples as the effigy mound period was waning. Trade developed between these Wisconsin sites and Cahokia, the center of Mississippian culture in the American Bottoms area of Illinois, across the Mississippi River from present-day

St. Louis. An area along the Crawfish River, near the West Branch of the Rock River close to Aztalan in Jefferson County, became a Mississippian outpost, complete with three temple mounds. The largest mound preserved in Aztalan State Park was constructed on two levels, the uppermost being the site of a large house or structure (Figure 5.3). Another platform mound contains the remains of a mausoleum or charnel house to accommodate the bones of the deceased. Surrounding the 15-acre site was a high wooden palisade, whose existence was discerned by archaeologists noting evidence of long-filled postholes in the soil. While it existed from around 700 to 1,000 years ago, the site had a population of about 500, yet there is little evidence of cultural links with nearby Indian groups (Kassulke 2009).

Recent archaeological research indicates that a shorter-lived Mississippian settlement, complete with three temple mounds, was erected overlooking the Mississippi River at Trempealeau about 1050 AD (Pauketat, Boszhardt, and Benden 2015). A short walk from its post office leads to the temple mounds on Little Bluff at the eastern edge of the much higher Trempealeau Bluffs (Figure 5.4). This mound complex, which extended downslope into part of today's village of Trempealeau, where a degraded platform mound remains along Third Street, was occupied for a couple of decades and then abandoned. Petrographic analysis of potsherds from both Trempealeau and the Stoddard Terrace site,

FIGURE 5.3. Temple mound displaying two platform levels in Aztalan State Park, west of Lake Mills in Jefferson County. A wooden building once stood atop the upper platform, and the entire community was surrounded by a wooden palisade.

including the Fisher Mounds complex, near where Coon Creek empties into the Mississippi River about 30 miles downriver, confirms their origins in Cahokia. Likewise, gaming stones and tri-notched arrow points found at Stoddard are of the Cahokia style (Pauketat, Boszhardt, and Benden 2015). These sites may have had significant spiritual significance to the Cahokians. "Trempealeau has all of the hallmarks of being an important shrine complex. . . . Cahokia's appropriation of the mysterious Drift-less Area and the geomorphically unique Trempealeau Bluffs, with their Effigy Mound spirits and other natural wonders, may have been integral to a cosmic claim" (284–285). Aztalan, in contrast, was occupied far longer and displays considerable archaeological evidence of domestic settlement.

The temple mounds at Aztalan and Trempealeau are the most impressive in Wisconsin. Yet other sites displaying Mississippian works are located along the Grant River near Burton in Grant County and at the Diamond Bluff site overlooking the Mississippi River in Pierce County, where both Wood-land period effigy mounds and Mississippian potsherds have been found. Immediately across the river in Goodhue County, Minnesota, two platform mounds exist, which together with those in Pierce County are referred to as the Red Wing Locality (Birmingham and Eisenberg 2000). A different artifact of Mississippian culture is displayed at the Gottschall Rockshelter site in the Wisconsin River Valley

FIGURE 5.4. Lower portion of platform mound overlooking Mississippi River and Lock and Dam Number 6 from Little Bluff Mounds Trail in Trempealeau. The upper part of the mound was destroyed by construction of the community's former water tank.

near Muscoda. There, as Birmingham and Eisenberg (2000, 149) explain, "paintings are rendered in an undeniable Mississippian art style that is completely foreign to Wisconsin." Other Native American pictographs exist in Wisconsin, but none display Mississippian styles.

### Ridged Fields and Other Landscape Features

Besides the thousands of mounds—both effigy and burial, early settlers in Wisconsin encountered other visible prehistoric landscape features. These include the previously mentioned ridged fields or garden beds, which ranged "in size from one to more than several hundred acres" (Gartner 1997, 229), spread over more than 175 known sites. These ridges, typically six to eight inches high and three to five feet wide, not only aided in the drainage of the soils, with the swale between the ridges acting as a moisture reservoir, but they also aided in cold air drainage, reducing the vulnerability of the fields to unseasonal frost (Gartner 1997; Doolittle 2000). A few sites still endure, such as at Kletzsch Park in Milwaukee, Mirror Lake State Park, and Rocky Arbor State Park, where ridges a foot high remain (Doolittle 2000). Although some are probably contemporaneous with the ridged fields, corn hill sites, or small circular accumulations of soil rising a foot or so above the surrounding field, were utilized to grow maize when Europeans first ventured into Wisconsin. While they are no longer visible, on the Sauk Prairie near Prairie du Sac prehistoric Indians had covered a 400-acre field with such small circular hillocks.

## WISCONSIN'S NATIVE AMERICAN POPULATION ENCOUNTERS THE FRENCH

Both the Woodland and Mississippian cultures, or the effigy mound and the temple mound builders, respectively, had vanished several centuries before Europeans first ventured into Wisconsin. From the Woodland Culture the Oneota evolved, and there is considerable evidence that both the Ho-Chunk (or Winnebago) and Menominee possessed belief and kinship systems that came from these earlier societies (Birmingham and Eisenberg 2000). By 1600 the Objibwe dominated northwestern and north-central Wisconsin, while much of northeastern Wisconsin, including the Lower Fox River Valley, was the home of the Menominee. The Potawatomi occupied much of east-central and all of southeastern Wisconsin. While Wisconsin's indigenous population is estimated to have been about 70,000 in 1492 (Gartner 1997), the spread of disease and warfare far in advance of the actual arrival of the Europeans had reduced their numbers before the first French explorers arrived in the western Great Lakes.

### French Fur Trading

The history of the French in Wisconsin is intricately interwoven with the history of Wisconsin's Native American population in the seventeenth and eighteenth centuries (Kellogg 1925). French missionaries played a major role in the early exploration of the Great Lakes region, and they gained a wealth of information that aided in the development of the fur industry. The first European settlement in Wisconsin was the La Pointe mission on Madeline Island along Chequamegon Bay, established in 1665 by Father Claude Allouez, a Jesuit missionary (Wyman 1998). Several trading posts were established in the area (Figure 5.5).

FIGURE 5.5. Reconstructed fur-trading post housing Madeline Island Museum, operated by Wisconsin Historical Society at La Pointe, Ashland County. The museum building consists of several historic structures that were joined together in 1955 to form the museum. One section was used by the American Fur Company in the 1800s.

Trading involved both licensed traders, who were officially approved by the colonial authorities and who paid for the privilege of trading within specific regions, and unauthorized individuals, referred to as *coureurs du bois*, who operated relatively freely given the few government officials and missionaries across the vast area. Native Americans embraced the trade in furs, which encouraged a small number of French traders to establish trading posts in Wisconsin, and also saw Indians transporting their furs directly to Montreal. Because of the demand for French trade goods, warfare ensued among indigenous populations, not only in Wisconsin but also along the routes the voyagers and their indigenous helpers took to Montreal. Such warfare altered the location of various Indian groups and affected the location of Indian settlements where the French could successfully situate their trading posts. Thus, beginning in the mid-1650s "the Sauk, Foxes, Kickapoo, and Mascouten . . . wandered hither and thither in Wisconsin forests, until in the next decade peace with the Iroquois permitted them to form settled villages on Green Bay, the Fox and Wolf rivers" (Kellogg 1925, 99). Even when French traders arrived, various groups, such as the Potawatomi, tried to monopolize the trade with them, discouraging trade with other Indians.

We often think of the role of the French in the fur trade, with the French organizing the shipments of furs to Montreal for transshipment across the Atlantic Ocean to the ultimate consumers, yet Native Americans numerically dominated the fur industry. They were the ones who trapped the animals, prepared the furs for shipment, and facilitated their transport. For example, the flotilla of canoes that departed for Montreal in 1670 included five French traders and 900 Indians, including a "delegation of Outagami" (Kellogg 1925, 136). Frenchmen were never numerous in the region, and given that many French traders took Native American wives, many traders were Métis, or persons of mixed race. Indeed, in summarizing their demographic impact, an early settlement geography of Wisconsin notes, "while very few of these French, and French and Indian half-breeds became permanent settlers there must have been a few hundred of them who spent a major portion of their lives in Wisconsin" (G-H. Smith 1929, 57).

French fur-trading posts were established at strategic locations along lakes and waterways that provided transportation links to both Sault Ste. Marie, along the St. Mary's River draining Lake Superior, and Fort Michilimackinac, along the Straits of Mackinac linking Lakes Michigan and Huron. The two most prominent fur-trading centers in Wisconsin during the French regime were along Lake Superior and near the mouth of the Fox River. Along Chequamegon Bay, a short-lived trading post was established near Ashland in 1659 and on Madeline Island, where the French operated Fort La Pointe from 1693 to 1698 and again from 1718 until 1762. Along the Lower Fox River, the French established St. Francois Xavier mission, at the northernmost rapids along the river. It functioned between 1671 and 1687 at De Pere, which was named after the Jesuit priests who served the mission. The French operated trading posts at Fort La Baye from 1684 until 1717 and at Fort St. Francois from 1717 to 1760 near the mouth of the Fox River. While the French had intermittently operated a fort and a trading post at Prairie du Chien since 1686, it reached its greatest importance after the French ceded their territory to the British, even though most of the traders remained French Canadian.

## Legacy of the French in Wisconsin

During the eighteenth century French trading activities in Wisconsin were affected by both conflict with the British and warfare with Native Americans related to the rivalry with the British over control of the fur trade. French colonial control of Wisconsin ended with the conclusion of the French and Indian War, or the Seven Years' War as it was termed in Europe, which was simultaneously fought in North America, the Caribbean, and Europe. Nevertheless, by that time the French had made a noticeable impact upon Wisconsin. As Louise Kellogg summarized in her history of the French effort,

> Wisconsin had been the keystone of the arch of the French empire in the New World. With one end resting upon Quebec and one on New Orleans, Wisconsin was the connecting link between the Great Lakes and the great river, and was central in all plans for French civilization in America. . . . Wisconsin had been made known to the world by French explorers, had been occupied by French officers and soldiers, had been peopled by French residents. . . .
>
> French Wisconsin has played an insignificant part in the development of our commonwealth. Its population has merged indistinguishably with the American. Its forts have fallen and disappeared until

hardly their sites are known. . . . Yet France has left traces upon our map that are ineradicable. Fourteen of our . . . counties bear French names. (Kellogg 1925, 440–441)

Besides the French names for a fifth of its counties, Wisconsin displays French toponyms for many rivers, lakes, and settlements. Many additional locations display Native American names as first recorded and written by the French. On the U.S. Geological Survey's topographic maps (Figure 5.6) around the French settlements of Prairie du Chien and De Pere, one notices that field patterns and road orientations deviate from those of the Public Land Survey, which characterize the rest of the state. French settlers around these two communities delineated their lands into long lots, which extended perpendicularly from the edge of the river toward higher lands, often over a mile distant. These French surveys were later recognized by American surveyors in the early nineteenth century, who numbered the plots but did not overlay these lands with the Public Land Survey's grid pattern. (See Patterson, Endrizzi, and Lippelt 1997 for a detailed map illustrating the survey of such lands along the Lower Fox River near Green Bay.) Even though the fields around De Pere have long been subdivided into

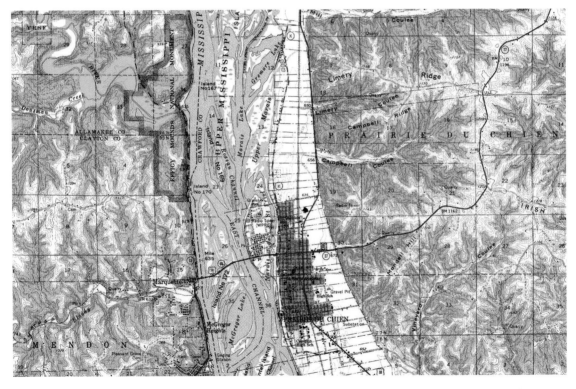

FIGURE 5.6. French-surveyed long lots, whose boundaries are shown in dashed red on the map, extend on the river terrace from the Mississippi River to the bluffs to the east. Fields whose boundaries extended perpendicularly from the river are conspicuously visible both north and south of Prairie du Chien on the U.S. Geological Survey's 15-minute Prairie du Chien quadrangle published in 1967. One-mile-square Public Land Survey sections cover those areas away from the Mississippi Valley.

residential neighborhoods, the street pattern displays the French survey, with streets running parallel and perpendicular to the river, rather than north-south and east-west.

Many French Canadian fur traders remained in Wisconsin after the French and Indian War. They continued to trade with the Indians to obtain furs, but they were sending the furs to British companies, such as the Northwest Company, established in Montreal in 1779. La Pointe on Madeline Island, Prairie du Chien, and Green Bay remained prominent centers of the fur trade. They were joined by small new posts established at Milwaukee, Sheboygan, Manitowoc, and Kewaunee along Lake Michigan; at Lac Vieux Desert, Lac du Flambeau, and Lac Court Oreilles in northern Wisconsin; at Kaukauna, Butte des Morts, and Portage up the Fox River, and by Hudson and Chippewa Falls in western Wisconsin (Wyman 1998).

Rendezvous took place annually at Prairie du Chien, where traders met to transfer their furs to voyageurs who transported them eastward. On St. Feriole Island there remains the stone masonry Brisbois Store, erected on a site facing the Mississippi River long associated with fur trade by French, British, and finally American companies. A mile to the north a two-century-old log cabin dates from the fur trade (Figure 5.7), displaying the classic pièce-sur-pièce French style of construction in which hewn logs are placed horizontally between vertical timbers (de Julio 1996). Another example, with

FIGURE 5.7. Vertefeuille House, a French fur trader's cabin erected about 1800 just north of Prairie du Chien, Crawford County, illustrating pièce-sur-pièce construction using hewn logs.

vertical corner timbers clearly displayed, is preserved at Heritage Hill State Park in Allouez. While fur production has dramatically changed in the past two centuries, Wisconsin remains the nation's leading producer of mink pelts. The French and their Native American associates also set the stage for the next major development in the state's historical geography, the establishment of the lead mines.

## LEAD AND ZINC MINING IN SOUTHWESTERN WISCONSIN

The presence of lead deposits in what is now southwestern Wisconsin, adjacent northwest Illinois, and near Dubuque, Iowa, was known to both the French and the Native Americans with whom they traded since the mid-1600s (Schockel 1916). Crude drift mines were utilized by Native Americans to obtain the ore, which was traded to the French. French fur trader Nicholas Perrot opened a lead mine near the present site of Potosi in the 1690s (Blanchard 1924). Nearly 20 small mines were reported in the Galena area by the French in the 1740s (J. C. Knox 2019b). Yet distance to market and political issues slowed the development of the lead mining industry until the British exited the region following the War of 1812. By 1818 it was reported that "U.S. factors at Prairie du Chien received almost 200,000 pounds of lead from Indians in the area" (Wyman 1998, 136). Shortly thereafter miners began to rapidly settle the region.

A federal program for leasing lands for lead mining in the area was initiated in 1822. The first mines in what is now Wisconsin opened near New Diggings, Lafayette County, in 1824 (Blanchard 1924). Unlike government programs that sold lands to the public, the lands in the Upper Mississippi Valley mining region were not sold initially as farmland but were leased to miners who paid 10 percent of their output as a royalty (Schafer 1932). The extensive nature of Wisconsin's lead deposits is shown by the fact that of the 1,940,000 acres reserved for leasing, over 73 percent were in Wisconsin (Heyl et al. 1959, 68).

Lead ore deposits in this region are found in sedimentary rocks that had been subjected to secondary placement of lead and zinc deposits in cracks, crevices, bedding plane and reverse faults, and solution voids in the soluble carbonate rocks long after they were deposited. Mineralized solutions, containing lead, zinc, and other metals, transported metal-rich brines that were leached from older, lower-lying rocks and then precipitated as galena, or lead sulfide, and zinc sulfide in the carbonate rocks, which became increasingly dolomitic and siliceous (Heyl et al. 1959; Millen, Zartman, and Heyl 1995).

Initially lead mines were small, simple, and crude. Miners sought lead ore that had become concentrated by weathering processes in unconsolidated residuum. It was simply extracted by digging a pit into the ground, leaving a pock-marked landscape (J. C. Knox 2019b). Later miners dug shafts to reach the sedimentary strata containing the ore. The greatest concentrations of the lead and zinc ores are found within the Galena formation, with lead primarily being produced from deposits above the water table and zinc coming from deeper mines (Blanchard 1924). Sphalerite, or zinc sulfide, became the primary zinc ore produced in Wisconsin, dominating production from mines in the Benton, Platteville, and Hazel Green mining districts, while farther north, in the Highland district, northwest of Linden and Dodgeville, zinc carbonate was the main ore (Schubring 1926).

## Lead Mining Expands

After the first lease for a lead mine in Wisconsin was granted in 1822, the extraction of lead ore grew in size, complexity, and prominence. Mines were rapidly established across the lead region of southwestern Wisconsin and adjacent northwestern Illinois (Figure 5.8), with the onrush of miners swelling the population "from 74 in 1823 to 10,000 by 1828" (Blanchard 1924, 57). In 1829, 13.3 million pounds of lead ore was extracted (Schubring 1926). The greatest concentration of mines in Wisconsin that year was located in the Dodgeville and Mineral Point area south to the state line, including the Gratiot, Shullsburg, New Diggings, and Hazel Green communities. To the east mines developed south of Blue Mounds, while to the west lead was extracted at Potosi. By 1830 the region had become the foremost area of lead mining in the nation, and it maintained that distinction until 1871 (Heyl et al. 1959). Although the mining operations were initially crude, often conducted by individual miners, who simply burrowed into the ground where lead outcropped, by the early 1840s most lead was produced from larger excavations and underground mines by "chartered companies."

The rapid expansion of lead mining in southwestern Wisconsin reshaped the region's economic, physical, and cultural landscape. Hundreds of mines, ranging from small badger holes dug into surface outcrops to larger shaft mines that tapped underground deposits, dotted the landscape. Many of the early miners came from Kentucky and Missouri, where lead was already being mined, and from Illinois and Tennessee. They reached the region by traveling up the Mississippi River. By 1830 the larger mines attracted skilled miners from Cornwall, the southwesternmost county in England, a region known for its tin and arsenic miners and skilled stone masons. Two decades later, shortly after the lead mining had peaked, it was estimated that "there were about 6,000 native Cornish in Wisconsin, making up about one-sixth of the total population of Grant, Iowa, and Lafayette counties" (Blanchard 1924, 31).

The mining of lead ore prompted the building of smelters (Figure 5.9) to extract the metal from the ore. By 1840 southwestern Wisconsin had 49 stone furnaces, where "the oven was placed beneath an immense chimney, thirty or thirty-five feet high" (Blanchard 1924, 58) that converted the galena ore into "pigs" of lead. In addition, a shot tower was erected atop the bluff overlooking the Wisconsin River at Helena, from where droplets of molten lead fell 180 feet into a pool of water at the bottom of the shaft, forming into spherical gunshot as it dropped. This site is now preserved in Tower Hill State Park.

Initially the Mississippi River provided the easiest means to ship the lead. While some lead moved down the Fever River, now called the Galena River, to Galena, Illinois, river ports developed in Wisconsin at Cassville, Potosi, Paris, and Muscoda, with the Wisconsin River only accessible by shallow draft boats. Overland transport of lead was by wagons pulled by a team of eight oxen, with Milwaukee receiving "three or four teams" daily in 1841 (Blanchard 1924, 45).

The steep decline in lead production within the region by the middle of the nineteenth century was largely a consequence of two events: the exhaustion of the easily mined lead ore and the discovery of gold in California. It is asserted that "in some places over one-half of the population" departed for California (Blanchard 1924, 58), while a history of Mineral Point reports that 1,000 miners left the vicinity of that community for California (WPA 1979). Yet lead continued to be mined for well over a century, often as a byproduct of zinc mining, whose production began in 1860.

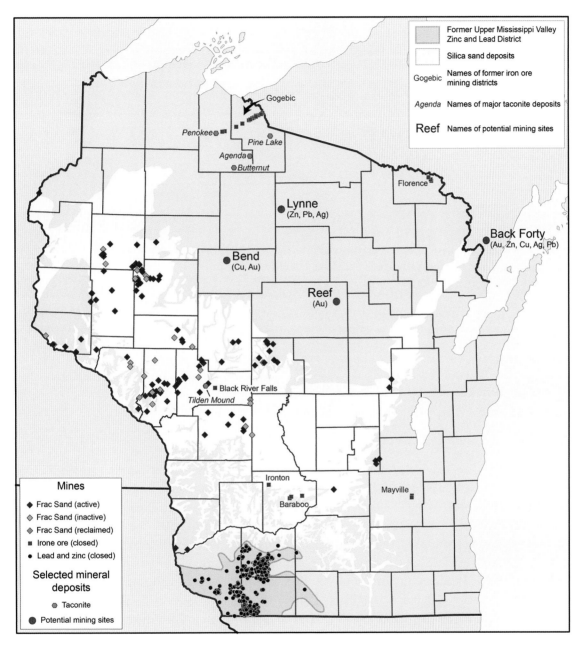

FIGURE 5.8. Historic and contemporary mining in Wisconsin. Map shows areas of the earliest lead and zinc mining, locations where iron ore was mined, and recent frac sand mine developments. Locations of significant unexploited metallic ore deposits are also shown. Drafted from Heyl et al. (1959), WDNR (2014, 2017b), U.S. Geological Survey's Active Mines and Mineral Plants in the US website (https://mrdata.usgs.gov/mineplant/), and FracTracker Alliance website (https://www.fractracker.org/2017/01/new-frac-sand-resources/).

FIGURE 5.9. Open-hearth lead furnace, operated by Richard Straw and Company, on Roundtree Branch of the Little Platte River in Grant County, photographed in 1872. Image used with permission from the collection of the Wisconsin Historical Society, WHS-08997.

## Zinc Mining Gains Prominence

Zinc production was relatively minor until it rapidly gained in prominence in the decade leading up to World War I. While zinc ore was initially considered a waste material, requiring a more complicated smelting process than what was necessary to extract lead, it was often obtained in mining lead. As Blanchard's (1924, 64) geography of southwestern Wisconsin notes, "Most of the zinc mines are further developments of old lead mine shafts." Because zinc ores were more often found below the water table than were lead ores, their extraction required deeper mines and more drainage than the earlier lead mines. In addition, Wisconsin's zinc ores were of low grade, requiring investments of well over a million dollars (and that is in dollars of over a century ago, when gold sold for under $21 an ounce) in concentrating plants and smelters. The Mineral Point Zinc Company's plant, just south of Mineral Point's old railroad depot, involved an investment of $6 million (Blanchard 1924). In its heyday, from World War I until 1925, 5,000 workers were employed in southwestern Wisconsin's 50 to 60 zinc mines, which produced over 10 percent of the nation's zinc. Lead was produced as a byproduct. Half of the zinc produced came from the Benton–New Diggings district.

Mining of both zinc and lead plummeted during the Great Depression. Only the Vinegar Hill Zinc Company, which operated an underground mine near Shullsburg with over seven miles of passages and ran a mill at Cuba City, remained open throughout the 1930s. Production of zinc again increased during World War II and immediately afterward, when several large new ore bodies began to be mined (Heyl et al. 1959). The last mine for lead and zinc in southwestern Wisconsin ceased operations in 1979 with the closing of the Calumet and Hecla Mine near Shullsburg. A U.S. Bureau of Mines report explained that this mine ceased operations "because of depressed market prices and U.S. Environmental Protection Agency Rules regarding quality requirements for water pumped out of the mines. The requirements put an excessive burden . . . because ambient water quality in the area already violates standards due to the extensive presence of metals in the surrounding terrain" (Hill and Evans 1981, 574). Competition from larger producers elsewhere had already resulted in the closing of Wisconsin's other mines.

## Landscape Legacy of the Lead Mining

Although Wisconsin's lead mining days are over, several mines, including ones in Platteville and Shullsburg, offer tours of their historic facilities. Furthermore, the mining legacy of the region is conspicuously displayed on the landscape. When looking at topographic maps of the area, locations of numerous "inactive mines" and "mine dumps" are seen. These are also visible on the landscape, particularly in the region between Shullsburg and Hazel Green, which includes the New Diggings area, where large piles of rock can be seen by the site of former mine shafts (Figure 5.10), and rock

FIGURE 5.10. Tailings around abandoned mine shaft east of New Diggings in Lafayette County.

cuts and waste from the furnaces that smelted the ore are visible in some of the valleys (Figure 5.11). A few furnaces remain, either in ruins or with the chimney still standing, such as in Dodgeville (Figure 5.12). Unfortunately, a lasting legacy of these sites is widespread contamination of soils and water from the mine tailings and processing residues throughout four counties of southwestern Wisconsin (Pepp, Siemering, and Ventura 2019).

As with many mining regions that suffered from a boom-and-bust economy, when the mines and smelters were fully operating, the population swelled and communities grew rapidly to house and supply the miners and their families. The 1830 census enumerated 3,635 persons in the three counties of Michigan Territory that were to become Wisconsin. Forty-four percent lived in Iowa County, which at that time included all of southwestern Wisconsin south of the Wisconsin River. A decade later, the population of Wisconsin Territory was 8.5 times what it was in 1830, with the population of Iowa County (which then included what is now Lafayette County) and Grant County totaling 7,904, just over 25 percent of Wisconsin's population. By 1850 the total population of Grant, Iowa, and Lafayette Counties had reached 37,225 (Forstall 1996).

Mineral Point was considered the de facto capital of Wisconsin during the 1830s. As a writer for the Depression-era Work Projects Administration explained, "Although Madison was legally the

FIGURE 5.11. Lead mine tailings remain visible along Diggings Creek east of New Diggings and southwest of Shullsburg in Lafayette County.

capital, until a settlement grew there most of the government business was transacted at [Mineral] Point and the town became the political as well as the mining center of the new territory" (WPA 1979, 61). Indeed, Henry Dodge took the oath of office as the first governor of Wisconsin Territory in Mineral Point in 1836. In 1837 Mineral Point was described as having a population of 1,200 to 1,500 residents, who lived in a community of 250 houses, 7 dry goods stores, 4 grocery stores, 2 tailor shops, 4 taverns, and a brewery. Two years later it boasted 3 hotels, a bank, a printing shop, and a newspaper. By 1850 three of the eight largest communities in Wisconsin were located in southwestern Wisconsin: Mineral Point, Platteville, and Dodgeville. Many buildings from this era remain.

The populations of both Iowa and Lafayette Counties were greater in the nineteenth century than they are today. Lafayette County reached its maximum in 1870, yet it had fewer residents in 2010 than it had in 1860. While Iowa County has grown over the last couple of censuses, it is still below its 1870 peak. Because so much of their construction was with stone masonry—and the Cornish are known for their adeptness at it, both in the Old Country and Wisconsin—impressive historic downtowns remain in several of the region's small cities. Pendarvis State Historic Site in Mineral Point preserves several stone masonry houses erected by Cornish miners in Mineral Point (Figure 5.13), and many

FIGURE 5.12. Bennett and Hoskins Slag Furnace erected in 1876 to process lead ore in Dodgeville, Iowa County. It is now preserved at the edge of a city park.

FIGURE 5.13. Pendarvis House (left) and Trelawny House are both examples of stone masonry structures built by Cornish immigrants. Pendarvis House, now a Wisconsin State Historic Site, was erected between 1842 and 1845 in Mineral Point, Iowa County.

other examples of Cornish stonework and miners' cabins remain in Mineral Point and in Dodgeville (Perrin 1981). The Iowa County Courthouse was built in Dodgeville in 1859 by Cornish immigrants. It is the oldest operating courthouse in Wisconsin.

On a hillside several miles east of Platteville a giant *M* has been formed of white painted rocks, similar to what one sees in many communities in the American West. This denotes the Wisconsin Mining Trade School that was established in 1907 in Platteville to educate mining engineers. Later renamed the Wisconsin Institute of Technology, this institution of higher education was merged in 1959 with Platteville's Normal School, the state's oldest, dating from 1866. It is now the University of Wisconsin–Platteville, which continues to offer engineering degrees.

## SMALL-SCALE IRON ORE MINING IN SOUTHERN WISCONSIN

Mining of iron ore was taking place in several locations in southern Wisconsin well before far larger ore bodies were discovered in northern Wisconsin (see Figure 5.8). It took place nearly a half century before the miners' attention focused upon the Lake Superior region.

### Early Mining in Dodge County

The earliest iron ore mines in Wisconsin were established in 1849 at Mayville and Iron Ridge in Dodge County, where a rich deposit of oolitic hematite outcropped in the Niagara Escarpment. The granular nature of this iron ore and the fact that the ore beds were several inches thick made extraction easy. The furnaces to process the ore were nearby, using dolomite as a flux and charcoal as fuel. While the iron ore was high grade, assaying around 45 percent iron, as it was at the bottom of the dolomite from which dissolved iron had been concentrated immediately above a layer of shale, it was relatively small in extent. By 1928 the mines had closed, "after the largest accessible deposits were used" (Schultz 2004, 135).

Although mining near Mayville and Iron Ridge ceased nearly a century ago, the legacy of this activity remains visible. Ruins of the old charcoal kilns were long visible from the main highway north of the city, and although the blast furnace has long been silent, metal working remains a key part of the local economy. Metalcraft of Mayville, Inc. has major operations in Mayville as well as in West Bend and Beaver Dam, involving precision metal crafting and custom manufacturing for a wide range of industries. Mayville Engineering Company, with factories in the Mayville area and four other states, manufacturers components used in a variety of products. Other industries include Mayville Die and Tool. The impressive brick buildings of Mayville's downtown historic district, plus its large limestone masonry school house dating from 1857, are a lasting legacy of the importance of this early mining center.

### Early Mining in the Baraboo Range

Iron mines were located along the Baraboo Range at Baraboo, North Freedom, and Ironton. Small quantities of iron ore were extracted near Ironton in Sauk County between 1850 and 1880, which "furnished sufficient ore to operate a small furnace for a number of years" (Weidman 1904, 159). Although some iron-bearing slate was mined in the late 1880s for use in mineral paint, higher grade iron ore of sufficient concentration for mining was discovered in 1900 (Weidman 1904). By 1910 three iron ore mines were operating in the Baraboo District, but mining stopped around 1930. There are no economically viable deposits remaining.

## IRON ORE MINING IN THE GOGEBIC-PENOKEE RANGE OF NORTHERN WISCONSIN

Iron ore deposits in northern Wisconsin are far richer and much more extensive, and their discovery initiated the settlement of the region. Mining within a 21-mile stretch of the Gogebic Range between Wakefield, Michigan, and Upson, Wisconsin, between 1877 and 1967 produced 325 million tons of high-grade iron ore (Cannon et al. 2008). This area, where 40 mines once operated, was the nation's third most productive iron ore mining district, trailing only Minnesota's Mesabi Range and Michigan's Marquette Iron Range, both of which remain centers of active mining. Within Wisconsin's section of the Gogebic Range, often called the Penokee Range, mining was underway by 1884, initially focused upon the twin cities of Hurley, Wisconsin, and Ironwood, Michigan, separated by the Montreal River.

## Montreal and Hurley

Montreal, just three miles southwest of Hurley, was a major center of mining within the Penokee Range. The Montreal Mining Company excavated high-grade iron ore of the Ironwood Formation from over a half mile underground. These ores of hematite are described "as irregular masses of soft, earthy iron oxides and hydroxides . . . [and] in some ores, relict bedding was preserved. The ore bodies are mainly localized along the keels of plunging structural troughs" (Cannon et al. 2008, 32). Overlying a diabase dike cutting across steeply dipping rock layers, the full vertical depth of the iron ore deposit is unknown, but it extends at least 1.5 kilometers, or nearly a mile, below the surface. Reached by huge hoists mounted within its headframe erected over the mine shaft, the Montreal Mine became the world's deepest iron ore mine, reaching a depth of 4,335 feet below the surface by the time it closed in 1962. Shipments of stockpiled ore continued through 1963. During the late 1930s this mine alone produced over a million tons of ore a year, employing 600 men working two shifts (WPA 1954). A mile closer to Hurley and just a mile from Montreal was the Carey Mine, another underground mine, which, like the Montreal Mine, began operations in 1886. It ceased operations in January 1965, having mined iron ore to a depth of 3,300 feet. The Plummer Mine, about three miles southwest of the

FIGURE 5.14. Headframe from Plummer iron ore mine near Pence, located two miles southwest of Montreal in Iron County.

Montreal Mine, operated between 1905 and 1932, yet it is the only mine in Wisconsin whose head-frame remains standing (Figure 5.14). Several other underground mines were located a few miles farther southwest.

During the three-quarters of a century that the iron mines operated, they had a profound impact upon the economy, culture, and population of the region. While a company town, with its houses owned by the mining company, developed at Montreal, Hurley was thronged by 7,000 miners and lumbermen. It gained a rather unsavory reputation. While Ironwood became the business and commercial center of the twin cities, Hurley developed "a reputation for bawdiness and crime . . . [and became] a rendezvous for booze runners, gunmen, and criminals" (WPA 1954, 375). It was once noted for its many brothels, and "in 1938, 80 of its 115 business establishments were taverns" (375). While the mines operated, Hurley functioned as a major shipping center for ore coming from both Michigan and Wisconsin.

## Ashland Served the Iron Mining Industry

Ashland grew rapidly to serve the iron mining industry. In 1890 it had a population of 9,956, and that number grew to 13,074 in 1900. Besides being the site of a mining stock exchange, Ashland became

FIGURE 5.15. Abandoned iron ore loading dock in Ashland extending into Chequamegon Bay from which railroad ore cars dumped their loads into ore ships. Photograph taken in 2007, before the dock's superstructure was removed in 2014.

a major transshipment point, where iron ore was brought to the city by four railroads that had "terminals . . . with facilities for dumping directly from flatcars to ship decks" (WPA 1954, 391). Not only were millions of tons of ore shipped annually from its 1,000-foot-long ore docks (Figure 5.15) to steel mills along Lake Michigan and Lake Erie, but quarries near Ashland shipped brownstone for building construction across the Midwest. Examples can be seen today on the older commercial and public buildings in Ashland.

The closing of the iron mines over a half century ago undermined the local economy, resulting in steady losses of population. Ashland's population now is a little over half of the 14,519 inhabitants counted by the 1905 state census (Beck 1907). The iron ore loading docks were removed early this century, with one of them being converted into a long fishing and recreational pier after the removal of the loading chutes and the railroad trestle leading to it. Even railroad service to Ashland has been discontinued, including both the east-west connections to Michigan and Minnesota and routes leading south to the iron mines and the rest of Wisconsin. Both Ashland County and Iron County, which includes Hurley and Montreal, reached their maximum populations in 1920 and have steadily declined since then.

### RENEWED MINING PROSPECTS IN THE PENOKEE RANGE

Early this century renewed interest in lower-grade iron ore deposits in the Penokee Range resulted in Gogebic Taconite LLC, a mining company, conducting exploratory geologic work and drilling of sample cores within the Penokee Range between Hurley and Mellon. The state legislature passed a bill in 2013 to regulate iron mining differently from other metallic mines and to streamline the permitting process. Taconite, which locally contains 25 to 30 percent iron, requires concentration before it is fed into blast furnaces and is the type of ore that is mined today in Minnesota's Mesabi Range. The taconite deposit that extends 20 miles between Upson and Mineral Lake, Wisconsin, is estimated to contain "3,711,000,000 tons of taconite [that] could be mined profitably" and "constitutes one of the largest undeveloped iron resources of the Lake Superior region" (Cannon et al. 2008, 33). A large open-pit mine was proposed for development.

Following many environmental objections, the mining company withdrew its application to mine in 2015. While concerns were raised by many environmentalists, who were worried about the impacts that the open-pit mine would have upon the aesthetics of the landscape and upon the quality of the waters draining the region, concerns of the residents of the Bad River Indian Reservation may have had the greatest impact. Waters draining the proposed mine site flow into the Tyler Forks River, which empties into the Bad River near where it flows through a gap in the Penokee Range. It then flows through the center of the Indian reservation to Lake Superior. Given the extensive wild rice ecosystem in that area, any degradation of the water would impact the Indians' economy and culture (Fitz 2012). Any litigation would need to be resolved in federal court, not state courts that might follow the legislature's recently weakened environmental regulations regarding mining. In addition, even if the mining was approved, its economic extraction would also face another hurtle. Ore docks no longer remain in Ashland, nor is the area served by any railroad that could transport the ore to either a smelter or shipping port.

## MINING ELSEWHERE IN NORTHERN WISCONSIN

Historically, mining also took place in several other locations in northern Wisconsin, both for iron ore and for copper, silver, and gold.

### Iron Ore near Florence and Black River Falls

Across the Marinette River northwest from Iron Mountain, Michigan, a small extension of the Menominee Range was mined near Florence. The Florence Mine operated from 1880 until 1931, producing 3,700,000 tons of iron ore. Nearby at Commonwealth, just south of Florence, several mines operated between 1880 and 1916 and then shipped ore that had been stockpiled between 1937 and 1943, with small quantities shipped until 1960 (Dutton 1971).

Much farther southwest, the Black River Falls area was also a center of iron ore mining. Small quantities of iron ore were mined and smelted at Black River Falls in 1856–1857 and from 1886 until 1892. A larger surface mine for taconite operated five miles east of the city from 1968 until 1982, when it proved uneconomical to deepen the mine. The site has now been reclaimed and forms the county-operated Wazee Lake Recreation Area.

### Copper Mining and Sulfide Deposits

The last metallic ore mine to operate in Wisconsin was the Flambeau open-pit mine, from which copper, silver, and gold were mined between 1993 and 1997. Located along the Flambeau River just south of Ladysmith, during its four years of operation the mine produced "1,720,000 metric tons of ore averaging 8.9% copper and 3.43 grams per ton gold" (USGS 1998). Although 334,000 ounces of gold and 3.3 million ounces of silver came from the mine, copper dominated production. Mining ceased when the open-pit mine became too deep and too close to the Flambeau River to expand, even though less than half of the ore body had been extracted.

A vastly larger sulfide ore deposit of copper and zinc was discovered in 1975 three miles south of Crandon, in Forest County. At the time the Flambeau Mine was closing and undergoing reclamation, the permitting process for the Crandon Mine was well underway. Proposed as an underground mine that would extract nearly 5,000 tons a day from an ore body estimated to include 50 or 60 million tons, environmental concerns focused upon pollution of both nearby waters and the Wolf River watershed (Riemer 2003). Fears were raised that sulfides in the crushed ore could acidify the waters, causing heavy metal pollution. Following environmental protests and state legislation that placed additional restrictions upon sulfide mining, the land and mineral deposit were purchased in 2003 by the Sokaogon Ojibwe of the Mole Lake Ojibwe Reservation and the Forest County Potawatomi, who had lands near the prospective mine site, effectively ending all efforts to open the mine.

There are several other known sulfide deposits in northern Wisconsin, including the Bend Deposit in Taylor County, the Lynne Deposit in Oneida County, and the Reef Deposit in eastern Marathon County, yet at the time of writing none of them are undergoing the permitting process to begin mining. However, the state legislature repealed its stricter permitting process for sulfide mining effective mid-2018, so such metallic mining may return to Wisconsin in the future.

## QUARRYING OF DIMENSION STONE

The only mining that is presently occurring in Wisconsin is of the nonmetallic category, the quarrying of dimension stone and gravel and the digging of sands and gravel. Earlier mention was made about quarrying of brownstone near Ashland in the 1890s, and several small communities in central Wisconsin based their economies upon the quarrying of granite in the early twentieth century.

### Granite

Red granite was quarried from the center of the village of Redgranite from the 1890s until the Great Depression. Initially it was used primarily for paving streets, since its ability to withstand wear greatly exceeded paving bricks that were widely used at the time. Later the granite was mainly used for building construction and gravel. Nearby, granite was also quarried at Lohrville, West Point, and Glenrock during the first two decades of the twentieth century. The quarry at Berlin operated over the same time period, also producing paving stone, while the quarry at Montello, which once employed 200 workers, continued until 1977. Granite from the Montello quarry was used in Grant's Tomb, and the quarry site today is a city park, with artificial waterfalls tumbling over the exposed granite (Figure 5.16). Granite was also quarried near Marquette and Utley in Green Lake County.

FIGURE 5.16. Abandoned granite quarry, now featuring artificial waterfalls, in Daggett Memorial Park in Montello, Marquette County. Granite was quarried in Montello from 1880 until 1976.

Farther north, over a century ago granite quarries operated in Waupaca County, near the Wisconsin River near Wausau, and near Amberg in Marinette County. Brownstone, the brown-colored Potsdam Sandstone, was excavated along Lake Superior, not only near Ashland—where much came from the Washburn, Houghton, and Bayfield communities in Bayfield County, but also from several of the Apostle Islands and along the lakeshore of western Bayfield County. A similar sandstone was quarried along the Red Cedar River at Dunnville, south of Menomonie, and near Ablemans, southeast of Reedsburg. Stone construction is not nearly as common today as it was over a century ago, and quarries once employed far more workers than today. Many of the state's prominent buildings of the nineteenth century utilized their stone, as Buckley (1898) documents in great detail.

### Dolomite

Dolomite was quarried extensively along the Niagara Escarpment, and from the Niagara formation to its east, for use in building construction, mostly locally, given that "soft spots [of the stone] weather out, leaving cavities which look much like worm borings" (Buckley 1898, 280). Large amounts of dolomite, locally called "Lannon Stone" and sought for its color and hardness, were excavated from 14 quarries in Waukesha County. During the late nineteenth century the village of Lannon was considered a leading center of building stone in Wisconsin. Quarries remain in the area today.

Much of the dolomite quarried throughout eastern Wisconsin was also used for gravel, and at one time kilns were operated at various locations to manufacture lime. While many abandoned kilns that once dotted the landscape in east-central Wisconsin were removed over the past two to three decades, examples still remain at High Cliff State Park, near the foot of cliffs that were quarried for their dolomite (Figure 5.17). Farther south, examples remain at Lime Kiln Park in Grafton and Lime Kiln Park at Menomonee Falls.

Many of Wisconsin's largest quarries, such as those near Wauwatosa and Menomonee Falls in the Milwaukee area or near Appleton and Oshkosh, have primarily produced gravel and aggregate for concrete that is used locally in construction of buildings and highways. Similarly, quartzite extracted from the Baraboo Range and several other sites was primarily crushed for use as gravel.

## CONTEMPORARY MINING IN WISCONSIN

It is only appropriate to have considered mining from a historical geography perspective in this chapter. Metallic mining provided the foundation for the settlement of both southwestern Wisconsin and Wisconsin's Lake Superior region. Yet, the last mine extracting a metallic ore closed over two decades ago. Quarrying of dimension stone for constructing buildings, bridges, and monuments was far more important a century ago than now, and it no longer underpins the economy of any community. Yet, Wisconsin still quarries granite and dolomite used in gravestones, monuments, and as dimension stone in buildings, with granite coming from near Wausau and dolomite quarried in Calumet, Fond du Lac, and Manitowoc Counties. In 2016 Wisconsin was the fourth-largest source of such stone in the nation, and it was ranked third in 2017, yet it only yielded $41.2 million in sales in 2015, the last year for which data is available, less than a third of the value of crushed stone, and far less than for sands and gravel (Dolley 2017b).

FIGURE 5.17. Lime kiln ruins at foot of Niagara Escarpment in High Cliff State Park, Calumet County. Quick lime was produced in these kilns between 1856 and 1956.

## Frac Sand Mining

Mining of industrial sands now dominates the mineral industry of Wisconsin (Figure 5.18). Industrial sands have long been sought for use in glass making, metal casting, and various chemical and metallurgical processes. The development of hydraulic fracturing or hydrofracking to enhance production of oil and natural gas dramatically increased the demand for sands. These sands are injected with chemical solutions into oil and gas wells, dramatically increasing their output. Fracking a single well requires 40 to 50 railroad carloads of sand. As the U.S. Geological Survey's *Minerals Yearbook* reports, "Wisconsin's industrial sand is widely regarded as ideal for use in hydraulic fracturing because the State has several formations containing sand that has a high silica content, high crush resistance, and ideal sphericity, including the St. Peter Sandstone on which industry standards are based" (USGS 2017, 52.1). A geological survey report notes that "The upper Midwest . . . has the principal supply of the ideal frac sand . . . in near-surface exposures that make it economic to mine" (Benson and Wilson 2015, 8). These sandstone deposits are poorly cemented, being quite friable and easily mined.

Given the quality and desirability of its quartz sands that are 99.8 percent pure silica, Wisconsin produced $1.21 billion in industrial sands in 2013, the most of any state. Indeed, "Wisconsin accounts

FIGURE 5.18. Aerial view of large open-pit frac sand mines and railroad loading facilities south of Chetek, Barron County.

for nearly one-half of all the frac sand capacity in the United States owing to its premium sand deposits, railway infrastructure, and long-term presence in the industry" (Benson and Wilson 2015, 53). Given the demand, the number of industrial sand mines in Wisconsin grew from 28 in 2011 to 105 in 2013 (Figure 5.19). While demand fell with falling oil prices in 2015, Wisconsin sold 38.3 million metric tons of industrial sands and gravel in 2014, valued at $3.15 billion, over twice the $1.39 billion received for 32.2 million metric tons in 2015 (Dolley 2017a). Output in 2016 was 43.9 percent of that in 2014, with sales totaling $637 million.

Wisconsin had 128 industrial sand facilities, of which 92 were reported as being active in 2016 (Figure 5.20). Of the operating facilities, 69 were mines—of which 38 had on-site processing operations, 19 were off-site processing facilities, and 4 were railroad loading services (WDNR 2017b). Barron County, with 17, had the most operating industrial sand facilities, followed by Trempealeau County with 12, and Jackson County with 11 (see Figure 5.8). In total, industrial sand mining was taking place in 21 counties. The largest concentration of these mines extended from Monroe and Wood Counties on the east toward Pierce, Dunn, and Barron Counties to the west-northwest (WGNHS 2014). There are several outliers, including mines in Crawford County to the south; Columbia, Green Lake, and Fond

FIGURE 5.19. Taylor Frac Sand Mine with storage facility operated by Badger Mining Corporation northwest of Taylor in Jackson County.

FIGURE 5.20. U.S. Silica's frac sand loading facility along Canadian Pacific Railroad west of Sparta in Monroe County.

du Lac Counties to the southeast; Waupaca and Outagamie County to the east; and Burnett County to the northwest.

Wisconsin's industrial sand mines are open-pit excavations with the exception of two underground mines in Pierce County, near Maiden Rock and Bay City. Estimates of the number of workers employed directly in industrial sand mining ranged from 2,300 to 4,300, not including those employed in transporting the sands (WDNR 2017b). Yet the expansion is over. Transportation costs and stricter environmental enforcement with changes in political leadership have closed some of the mines (I. Holmes 2020). Low oil prices have discouraged new drilling, and demand for sand has plummeted.

The rapid expansion of frac sand mining prompted environmental concerns and objections to be voiced by local residents (Pearson 2017). Frac sand mines differ from the more familiar quarries, which are smaller and provide construction materials used locally. "Frac sand mines bring a much bigger and permanent presence. In contrast to local quarries, frac sand mines range from two hundred to over a thousand acres, some operating twenty-four hours a day all year long" (Pearson 2017, 9). Objections focused upon dust from the mining that could cause silicosis, noise and increased traffic on rural roads, pollution of local waters from the chemical treatment and washing of the sands during processing, and the destruction of the landscape aesthetics—threats to the local tourism industry and property values. Spills into waterways and blowing sands from closed mines have also been reported (I. Holmes 2020).

## Frac Sand Mining and Regulation

In their essay regarding frac sand mining, two geographers frame the issue as involving a "sacrifice zone" linking "landscape change and the risk of negative impacts on human health, economic livelihood, and ways of life" (Holifield and Day 2017, 270). Yet, in sharp contrast with environmental concerns that halted development of taconite mining in the Penokee Range and sulfide mining near Crandon, legislative acts in favor of sand mining and its associated job creation squelched local opposition. While local oversight has long been a hallmark of Wisconsin's political geography, state legislators have sought to transfer the regulatory authority from local towns to other jurisdictions. In counties without zoning at the town level, local authorities and residents have few tools to effectively stop the permitting of an industrial sand mine.

As geographers studying the "frac sand mining controversy" have observed, it has highlighted "conflicts between local land use ordinances and statewide industrial regulations, and the associated tension between local and state interests and rights" (Holifield and Day 2017, 276). Early in 2010s the cost of production of the sands was only half of their selling price. Thus, landowners were being offered hundreds of thousands of dollars for their mineral rights, "plus royalties of $1.50 [to $3.00] per ton for their frac sand" (Pearson 2013, 32), providing individuals strong financial motivations to sell to mining companies, often to the disgust of their neighbors. While many factors influence the spatial distribution of frac sand mines, including accessible sand exposures with minimal overburden and railroad connections for efficient transportation, it has been found that "frac sand operations tend to cluster in unzoned townships and counties" (Pearson 2017, 24).

## PRÉCIS

This chapter is the first of three that explore the roles of the earliest industries in Wisconsin in shaping its historical geography of settlement, which helps the reader understand the resulting cultural landscape. The chapter began with an overview of Wisconsin's prehistory, displayed by its many effigy mounds and other archaeological features. Native American groups at the time of European contact helped shape the exploitation of Wisconsin's natural resources, in particular the furs and deposits of lead.

While the fur industry and the mining of metals in Wisconsin are now history, they provided the foundation for the early settlement of parts of the state. Descendants of the miners of lead, zinc, iron ore, and copper, who no longer turn to mining for employment, constitute important segments—at least regionally—of the population of parts of Wisconsin. Yet some mining still does occur, and the demand for industrial sands has put the national spotlight on Wisconsin.

The next two chapters examine other groups who came to Wisconsin to exploit its bountiful natural resources. These include those who were attracted to the forests that covered over two-thirds of the state's land area and those who came to cultivate the soils on the prairies and savannas and then to remake cutover forest lands into farmland. The efforts of these farmers and lumbermen reshaped Wisconsin's landscape and provided the foundation of today's agricultural industry.

*Chapter Six*

# Settlement of Wisconsin 2

## *Forestry*

THE FORESTS OF WISCONSIN beckoned to loggers and lumbermen in the same way furs attracted trappers and traders and lead and iron ore attracted miners. Just as many of those men sought to exploit the resources for their gain, many loggers rapidly cut the forest and moved onto other virgin lands, often leaving abandoned settlements in their wake. Yet, not all of them left once the trees had been cut. Some, akin to certain miners, remained in the rural settlements, or as we will see in chapter 7, turned their attention to establishing farms. In this chapter we first focus upon the clear-cutting of Wisconsin's forests during the last decades of the nineteenth century and the first decade of the twentieth century.

### CLEAR-CUTTING OF WISCONSIN'S FORESTS

Logging of Wisconsin's white pines was already underway by the 1850s, but it reached its greatest intensity between 1880 and 1910. During the 1890s, Wisconsin produced more timber than any other state. We review this industry from a historical geography perspective, inasmuch as the circumstances that created, shaped, and ultimately constrained the logging and sawmilling of the timber were unique to that time. Today's forest is quite different from what lumbermen encountered a century and a half ago. Demand for wood products has changed. Neither the way to transport the lumber nor the labor are the same. The historical geography of the activity was controlled by location and abundance of desired forest species, demand for lumber, availability and character of transportation to both sawmills and market, recruitment of labor to cut the forests, and characteristics of the various business groups that financed, organized, and facilitated the logging industry.

#### Forest Species That Loggers Sought

The forests that covered Wisconsin in the middle of the nineteenth century were not only more extensive and taller than the forests seen today, but conifers, particularly pines, were found within a far greater share of the forest. As we saw in chapter 3, within much of the northern two-thirds of Wisconsin, pines dominated the forest, either numerically or visually, as in many areas scattered pines towered over the other forest species. The white pine, "the largest and most valuable tree of the region" (Whitbeck 1913, 20), was intermixed with hardwoods on "heavy soils, predominant on lighter sandy and gravelly loams, and as pinery proper on the extensive loamy sand areas" (20). Two other species of pines, the red pine or Norway pine and the jack pine, were also found in the region. Both varieties grew on less fertile sandy soils, with jack pine sometimes found within pure stands.

A detailed reconnaissance of northern Wisconsin's forests in the mid-1890s, after much of the white pine had already been cut, determined that there remained 15,038 million board feet of white pine sawtimber and 2,317 million board feet of Norway pine sawtimber (Roth 1898, 9). While a combined 17,355 million board feet of these two species remained in 1898, 66,000 million board feet had been cut between 1873 and 1898, with approximately another 20,000 million board feet logged off between 1840 and 1873. In addition, "about 26 billion [w]as probably wasted; chiefly destroyed by fire" (16).

Given the diameter and stature of the white pine, which could reach 120 feet above the ground, yields were stupendous during the cut-over period. In portions of Oneida and Vilas Counties loggers encountered "mature stands of pure growth" that yielded as much as 2 million board feet per 40-acre parcel and at times cut 100,000 board feet from a single acre, although a "cut of one million feet per 40 acres, or 25 [thousand] feet per acre was and is considered a very good yield and generally the cut is less than half this amount" (Roth 1898, 15).

While 3,475 million board feet of jack pine remained standing in 1898, it was not considered sawtimber. A contemporary evaluation of the northern forest noted that jack pine "was not used to any extent, neither stumpage nor logs having real commercial value except in parts of the jack pine and oak openings, where it is used as fuel and for farm purposes" (Roth 1898, 21). Loggers primarily sought the white pine, which in 1899 comprised 71 percent of the total board feet of wood cut (Finley 1965, 69), followed by red pine. Jack pine was initially ignored, but later it was cut as pulpwood.

Loggers in the nineteenth century also largely overlooked hemlock, which grew on wetter soils, particularly gravelly loams and clay soils. Although often growing intermixed with hardwoods in northern Wisconsin, pure stands did exist. Yet, in 1898 it was reported that the "present supply of hemlock is much underestimated . . . [and it] was ordinarily not estimated at all or only the largest and best trees were considered" (Roth 1898, 23). At the time, while hemlock was harvested for its bark, which was used in tanning—for which Milwaukee led the world by the end of the nineteenth century (Bates 2018)—and for dimension lumber, most hemlock was "not yet appreciated, so that neither stumpage nor logs can readily be sold and millions of feet are wasting in the woods" (Roth 1898, 23). Yet there was great demand for hemlock bark, even though the debarked logs were often left to rot. This was all to change shortly. With the disappearance of readily cut white pine, lumbermen turned to harvesting hemlock, which floated downstream nearly as well as the pines. Sawmill operators discovered that its lumber was marketable, even though it had a greater tendency to splinter than did white pine.

## Demand for Lumber

Rapid growth of settlements in the Midwest, and shortly thereafter in the relatively treeless Great Plains, fueled a huge demand for lumber. Large urban centers, such as Chicago and Milwaukee, provided important markets. The necessity of rebuilding Chicago following the Great Chicago Fire of 1871 and then providing lumber to expand it to a city of 1.7 million residents by 1900 played a critical role in spurring the development of the sawmill cities. These developed at locations such as Oshkosh, strategically located downstream from the Wolf River watershed, which floated logs into Lakes Poygan and Butte des Morts, which provided "unequalled boomage capacity" (Whitbeck 1915, 48). From there the logs were easily rafted to the 30 sawmills in Oshkosh or the 18 lumber or shingle mills in Fond du Lac. The lumber from both cities could be transported by rail to points south. Other sawmill centers along Green Bay, such as Oconto and Marinette, had the advantage of shipping their lumber by schooner directly to Chicago and Milwaukee. Logs from the Chippewa River watershed could be floated to the Mississippi River and then downriver to cities as distant as St. Louis for sawmilling and sale. The market for Wisconsin lumber was huge, and Wisconsin led the nation in its production during much of the 1890s.

### Seasonal Work: Logging in Autumn and Winter, Rafting Logs in Spring

Work in the woods was highly seasonal. Much of the tree cutting took place in the fall and winter, while the logs were sent downriver in the spring. Thus, the census report on occupations simply lists

FIGURE 6.1. Phoenix log hauler, a steam tractor, pulling skids loaded with logs on a frozen path near Rice Lake in Barron County in 1914. Image used with permission from the collection of the Wisconsin Historical Society, WHS-05820.

the pursuit as "lumbermen and raftsmen." Felling trees in the winter, when the ground was frozen and snow covered, provided better access and enabled the lumbermen to drag logs or transport sleds of logs to the banks of nearby rivers. At other locations small locomotives on treads, or steam tractors, transported logs on iced roads (Figure 6.1). In the spring dammed rivers, which had accumulated the snowmelt, were often released in a torrent, transporting the logs downriver. Where lakes or slack water areas along rivers permitted, large boomages were created where the jumbled logs sent by many operators could be identified by their markings and sorted.

Several locations were well known for their boomage capacity. Boom Bay, where the Wolf River flowed into Lake Poygan, was one of the best known. Beef Slough, where the southernmost channel of the Chippewa River entered the Mississippi River just north of Alma, was another important boomage, where rafts of sorted logs were arranged for towing by steamboats in the Mississippi River. It was there that conflict between Mississippi River timber interests and those whose sawmills were located in the Chippewa River Valley, primarily at Chippewa Falls and Eau Claire, was resolved when large lumber interests, including Weyerhauser, imposed regulations upon the log transport. The industry came under control of the sawmill operators, who either directly or indirectly controlled the logging camps, where the trees were being cut, and the logs' transportation to their mills.

Lumber camps were established in close proximity to the forest being cut, with the men housed in bunkhouses and fed by the camp cook. When the nearby forest was cut, the lumber camp was simply relocated to another accessible site closer to the virgin forest. Some, but not all, of the lumber camps were near sawmills, but transportation was critical. Logging practices were described by a geographer over a century ago:

> Naturally the trees near the river were cut first, for the principal method of transporting the logs to the mills was by floating them down the rivers. At first only the best parts of the choice trees were taken. Each season the logging camps pushed farther up the streams and farther back from the banks as the timber was cut away. The logs were hauled by horses or by logging railroads to the rivers, and by thousands were piled on the ice and along the banks, awaiting the spring break-up, when the melting snow turned the river into a torrent. (Whitbeck 1913, 21)

Much of northern Wisconsin was viewed as favorable for floating logs (Figure 6.2). An 1898 report describing northern Wisconsin's forest industry noted that

> the basins of the Chippewa, Wisconsin, St. Croix, and Black rivers, which drain 70 per cent of the entire area, are covered with the most perfect network imaginable of small streams especially suited for purposes of driving timber. The rivers emptying into Green Bay also "drive" quite well, but have required more improvements, while those running into Lake Superior are in great part unfit for driving. (Roth 1898, 6)

Thus, the means of log transport varied among the watersheds. This influenced which areas were logged first and which locations developed into major sawmilling centers.

FIGURE 6.2. Log rafts delivered logs to sawmills, such as this example floating on the Chippewa River at Chippewa Falls in 1870. Image used with permission from the collection of the Wisconsin Historical Society, WHS-01894.

## Transportation to Sawmills and Market

Transportation of logs took considerable skill and organization (Figure 6.3). For example, along the Chippewa River upriver from Jim Falls, in 1868 there were "87 camps employing 1,103 men, using 339 teams of horses or oxen" (Wyman 1998, 256). Five years later the Eau Claire Lumber Company alone had "400 men, 150 horses, and eighty oxen in its fifteen camps" (256). Logs had to be moved from where they were cut to where they could be transported downriver.

### Driving the Logs Downriver

Floating logs downriver was not a simple or easy task. The Wisconsin River, with its numerous rapids, was considered "the most difficult of the major rivers to drive" (Wyman 1998, 259). Floating logs downriver was dependent upon water level, which was vulnerable to poor winter snow cover or spring drought. Low river levels often contributed to the formation of colossal logjams. For example, a 4-mile-long logjam formed near Wausau along the Wisconsin River, while one on the Jump River extended 13 miles (Wyman 1998). River transport was often slow, and a contemporary newspaper reported "that it took 'all summer to move pine logs from Post Lake (near Elcho [in Langlade County]) to Oshkosh'" (Connor 1978, 34–35). In years of low water, it could take two seasons to float the logs to the mills.

FIGURE 6.3. Statues of the legendary Paul Bunyan and Babe the Blue Ox with a load of logs are depicted outside the Vilas County Historical Society Museum in Sayner.

Improvement companies and cooperative efforts among the logging outfits worked to eliminate physical obstacles. They removed snags and brush along the river and sand bars from the river channel. They constructed dams to raise the water levels, which was particularly important around rapids. Given that many streams were "too shallow and choked with brush, rocks, or fallen trees" (Higgens and Reinecke 2015, 2) to be suitable for driving logs, channels were deepened and canals dug across bends to create drivable channels. By such efforts, in "the St. Croix Valley alone, the length of rivers navigable for log driving was increased from 338 miles in 1849 to 820 miles by 1869" (2). Upriver dams were erected to flood the channel when their waters were suddenly released, driving the logs downriver with a torrent of water. Along the Chippewa River 148 dams were erected, 24 in the Wolf River watershed, and about 65 along the St. Croix River.

The Wisconsin River provided several dangerous challenges, including Big Bull Falls and Little Bull Falls at Wausau and Mosinee, respectively; rapids at Wisconsin Rapids, called Grand Rapids in the 1800s; a very deep, narrow, and swift-flowing channel at Wisconsin Dells; and a channel clogged with shallows and shifting sandbars extending downstream to the river's mouth. In the northern part of the Wisconsin River watershed, from Merrill northward, over 50 dams were erected (Wylie 1992).

Along the shorter Menominee River 41 dams were built, with smaller numbers erected along the Peshtigo and Oconto Rivers. Such structures ranged from permanent dams to others that were crudely built of logs, brush, stone, and earth, functioning as "splash" dams when broken, rapidly raising the level of the water surging downstream, lifting and floating the logs as they went downstream. Needless to say, these dams radically altered the configuration and ecology of the streams.

Logs were also transported on Lake Michigan and Lake Superior. On Lake Superior huge rafts were operated at times, with single rafts commonly transporting "3 and 4 million board feet of logs" (Peddle 1980, 31), with the largest nearly twice that size. These rafts "consisted of a mass of free floating logs enclosed by a 'necklace' of three-foot diameter white pine 'boom sticks'" (Peddle 1980, 32). Some rafts traveled from the mouth of the Bad River to the Duluth-Superior harbor. Others floated shorter distances from the Apostle Islands to Ashland. Hardwoods, because of their greater density and vulnerability of sinking, were often loaded on scows and barges, which, like the rafts, were moved by tugboats. Rafting of logs on Lake Superior continued until the mid-1920s. Log rafts were also towed by steamboats on Lake Winnebago and Lake Michigan. Rafts of sawn lumber were also floated on Lake Michigan.

### Railroads and Logging

The arrival of railroads facilitated changes in the locus of logging, in the trees that could be feasibly cut and transported, and in the seasonal rhythm of the work. Forests too far from navigable rivers or creeks suitable for flooding could then be logged. Hardwood species, not suitable for floating, could move by rail to sawmills and thus began to draw the attention of lumbermen. Because railroads operated year-round, loggers could continue cutting throughout the year. By 1871 the Chicago and Northwestern Railroad reached Marinette. The Wisconsin Central served Stevens Point in 1871 and Medford and southern Price County by 1873. The Wisconsin Valley Railroad arrived at Wausau through Wisconsin Rapids and Mosinee in 1874, just to mention several early railroads. While the expansion of the railroad network into northern Wisconsin connected the region to more distant markets, many other locomotives ran in the woods on private tracks whose design differed greatly from that of the common carriers.

Logging railroads linked areas of active cutting both to nearby sawmills and to locations where the logs could be floated downriver. The first logging railroad began operations in 1881 near Shell Lake in Washburn County, although common carriers had earlier transported logs. The last logging railroad stopped running in 1948 near Rib Lake in Taylor County (Kaysen 1978). Wisconsin's northernmost logging railroads were located in the Apostle Islands, on both Michigan and Outer Islands, to move logs to the Lake Superior shore. To the east logging railroads operated in Marinette and Oconto Counties, while to the west logging railroads transported logs to the St. Croix River. The southernmost logging railroads operated in eastern Jackson County. Thousands of miles of logging railroad track were laid, with Ashland County alone accounting for 717 miles (Kaysen 1978).

Logging railroads were either standard gauge or three-foot narrow gauge. Many were crudely constructed, as "a spur intended for only one or two years' service would be built as cheaply as possible with no ballast or ditching and with steep grades and sharp curvature. Tracks were sometimes laid over frozen swamps . . . [and] removed before the spring thaw" (Kaysen 1978, 42). Given the uneven track, steep grades, and sharp curves, unsuitable for most locomotives, Lima Shay steam locomotives with their right-side-mounted pistons, bevel gears, longitudinal drive shafts, and small drive wheels were widely used (Kaysen 1978). Because it was only practical to drag logs using sleighs, big wheels, or steel cables attached to a drum or winch a relatively short distance from where they were cut to the logging railroad, logging spur lines were often located roughly 1,200 feet apart. Other logging railroads were built to higher standards for longer service, such as those connecting the spur lines to the sawmill, using standard-gauge track and more typical railroad grades and curvature. Furthermore, many short-line railroads linked sawmill centers to the trunk railroads. The Laona and Northern's Lumberjack Stream Train, which now operates as a heritage tourist attraction every summer, is a reminder of the critical role that railroads once played in northern Wisconsin's lumber industry.

By the outbreak of World War I clear-cutting had largely ended. While Wisconsin produced 2.84 billion board feet of softwood lumber in 1899, that amount plummeted to 864 million by 1914, falling to 594 million in 1919, only 20.9 percent of its output two decades earlier (NRCSA 1956). River rafting ended earlier, with the last logs being floated down the Chippewa River in 1905. The St. Croix River was "the last of the state's major rivers to run logs" (Wyman 1998, 277), and its final run occurred in 1914.

### Settlement Resulting from the Logging Period: Ghost Towns

Settlements developed around sawmills, including housing and commercial facilities. Most settlements had scheduled railroad service by common carriers, which both shipped out lumber and brought in logging and sawmill equipment, plus cargo and passengers to meet the community's needs. Logging provided the foundation for the urban centers that developed, and often died, in the region. The author of a historical geography of lumber settlements that became ghost towns explains, "Most of the cities north of a line from Fond du Lac to La Crosse began as lumber towns . . . [yet] the end of lumbering often meant [their] end" (Rohe 2002, 1). For many of these communities, "hardly a trace remains to show where a town once stood" ( 4).

The Chippewa River State Trail passes the site of Porterville, about five miles downriver from Eau Claire. A historical marker along the trail commemorates the settlement established in 1863. A decade after its founding, the sawmill and lath mills in Porter's Mill employed 120 workers, who "produced 100,000 feet of lumber, 30,000 feet of lath, and 50,000 shingles daily" (Rohe 2002, 39). With another mill and improvements, output grew. The 1890 census counted 1,194 residents, who were served by several churches and a school. Yet, before the end of the decade the log supply had waned. The mills closed, with their equipment being dismantled and sold in 1899. As residents departed, buildings were demolished for their lumber, and others were moved. By 1902 no buildings remained other than a few cabins, with only overgrown foundations and basements remaining today (Rohe 2002).

Unlike Porterville, which had a nearby city, many ghost towns mark remotely located lumber settlements, far from any community. Examples include Goodyear, McKenna, and Zeda, which were located in eastern Taylor County. When the mill closed at McKenna, it was disassembled and moved up a newly extended railroad line to a new site. It became Star Lake in Vilas County, illustrating the railroad's role in the shifting location of both the logging industry and the settlements that it spawned. By 1903 Star Lake reached its maximum population of nearly 700, but the mill closed within five years. The sawmill owner then relocated to Washington State, creating a new settlement named McKenna there, with most of Star Lake's buildings disassembled for their lumber. The hotel, however, remains and is now Hintz's North Star Resort, an early entrant into the region's tourist business, a topic that is addressed more fully in chapter 14. Other northern Wisconsin settlements were razed by wildfires, such as Heineman, a sawmill community of nearly 400 residents, whose mill was rebuilt following the 1910 fire, but at Merrill, nearly 10 miles away.

## Settlements Resulting from the Logging Period: Today's Cities

At the beginning of the twentieth century, when Wisconsin led the world in lumber production, over 1,000 sawmills were operating. Statewide, one-quarter of all manufacturing wage earners "were engaged in the manufacture of lumber and timber products" (Whitbeck 1913, 24). These wood products accounted for 18 percent of the value of all goods manufactured in Wisconsin. Sawmill and lumber mill cities rapidly grew to prominence at locations where logs were floated in abundance and where there also was good steamboat or railroad access to markets.

Those lumber centers that developed along the Fox, Wisconsin, Chippewa, and Mississippi Rivers included some of Wisconsin's largest cities. The 1885 Wisconsin state census (Timme 1887) showed that three of Wisconsin's four largest cities, Oshkosh, La Crosse, and Eau Claire, were major sawmill and wood-processing centers. Even Milwaukee, the state's largest city, was engaged in sawmilling and woodworking, but at a lesser scale. As explained by a geographer a century ago,

> Not less than forty places in Northern Wisconsin have been important saw-mill centers—Marinette, Oconto, Green Bay, Wausau, Stevens Point, Grand Rapids [now called Wisconsin Rapids], Merrill, Black River Falls, Eau Claire, Chippewa Falls, La Crosse, Ashland, and others—yet the situation of Oshkosh made it one of the foremost cities of the group. La Crosse, with its highly favorable position on the Mississippi, is the only one of the old lumber centers which was and still remains a close rival of Oshkosh in size and industries. (Whitbeck 1915, 48)

La Crosse was well situated to receive logs floated down the Black River, which entered the Mississippi River at the city, plus logs transported down the Chippewa River and then towed in rafts by steamboats plying the Mississippi River. Yet, many of the logs from the Chippewa River were sawn into lumber in Eau Claire, one of Wisconsin's two "Sawdust Cities." Lumber, in cribs measuring 16 by 32 feet, was also rafted down the Chippewa River from both Chippewa Falls and Eau Claire (R. C. Brown 1980). In contrast, there were no major sawmill centers along the Wolf River upstream from Oshkosh, the other "Sawdust City," and Oshkosh lumber barons had a tight control upon the logging within the Wolf River basin, which likely had the most extensive white pine forest in the state. In addition, Lake

FIGURE 6.4. Buckstaff Company's furniture factory was established in Oshkosh in 1850 and remained in business until 2011, the last of the city's once dominant woodworking employers to close. This photo was taken in 2004. The buildings were razed in 2016, and the site is now occupied by the Menominee Nation Arena.

Poygan provided "unequalled boomage capacity [and] Lake Butte des Morts and the broad, sluggish Fox provided good sites for mill ponds and the sawmills" (Rohe 1997, 227).

The presence of sawmills spawned many other industries in the rapidly urbanizing sawmill centers. Some cut logs into shingles, pallets, and finely finished lumber, while others utilized lumber to create steamboats, wagons, furniture, doors, windows, ladders, and matches. Early in the twentieth century Oshkosh housed the world's largest manufacturer of doors, the Paine Lumber Company, whose facilities stretched nearly a mile along the Fox River. It also contained the nation's largest match factory, operated by the Diamond Match Company, plus nearly 40 other woodworking shops (Figure 6.4). Using lumber shipped on Lake Michigan, Sheboygan was known as "Chair City," with nearly two-thirds of the city's workers employed in the furniture industry (Wangemann 2005). Loggers and sawmill operators also needed tools and equipment, and Oshkosh, Eau Claire, and Marinette attracted these manufactures. The manufacturing output of these urban centers has obviously changed over the past century, as discussed in chapters 11 and 12, but the legacy of the sawmill era remains in these cities' downtown business buildings and lumber baron mansions (Pifer 1980; Rohe 1999).

## Recruitment of Labor to Cut the Forests

Immigrant labor played a major role in the logging industry, and the affinity of Swedes and Finns for forest occupations has long been noted (Hale 2002; Knipping 2008). The 1900 census report on occupations noted that 5,572 Wisconsin men were employed as lumbermen and raftsmen. Of these workers, 47.4 percent were foreign-born whites and 28.8 percent were native whites with foreign parents. Of those lumbermen with foreign-born parents, 30.5 percent were from Norway, Sweden, and Denmark, with only Canadians being more numerous than these three Scandinavian ethnics. In 1900 Wisconsin had 12,114 sawmill and planing mill employees, of whom 54.5 percent were foreign-born whites and 28.9 were native whites with foreign parents. Overall, 28.0 percent of Wisconsin's sawmill workers were of Norwegian, Swedish, or Danish parentage (USBC 1904). Even though Finns were not enumerated separately or included as Scandinavian, they were heavily involved in lumbering in Wisconsin's northernmost counties. Indeed, as reported in *Finns in Wisconsin*, "Most of the Wisconsin Finns tried logging at some time or other, and many who later turned to farming continued to work in the woods to provide wintertime income" (Knipping 2008, 13).

A lasting legacy of the involvement of Finns and Swedes in Wisconsin's logging is that their ethnicities still dominate the population in parts of northern Wisconsin. At the time of the 2000 census, the most recent one that reported ancestry, Finns were the dominant ancestry group in 12 towns scattered across all four counties along Lake Superior. Swedes comprised the dominant ancestry group within two locations in Douglas County, plus single towns in Burnett and Polk Counties. Although Finnish farmsteads often included "many small structures, each for its own particular function, rather than . . . a few large multipurpose buildings" (Knipping 2008, 27), such as the *navetta* (a small cowbarn), the *lato* (the haybarn in the hayfield), and the *talli* (horse stable), very few of these ethnic structures remain today. Instead, there are a few examples of Finnish farm buildings, including a house, barn, and sauna preserved at Old World Wisconsin, plus a Finnish windmill, constructed to grind grain into flour, near Lakeside in Douglas County. The saunas that Finns erected near their homes are the best indicator of their ethnic identity (Mather and Kaups 1963). Even though far fewer remain today than a half century ago, their presence defines the Finnish ethnic area.

## THE FOREST INDUSTRY AT MID-TWENTIETH CENTURY

Following the clear-cutting of the forests, lumber production within Wisconsin steadily fell. Its harvest of 594 million board feet of softwoods in 1919 was nearly 80 percent less than just two decades earlier, and it was to continue falling. While the harvest of hardwoods remained relatively steady during the first two decades of the twentieth century, by 1924, when 563 million board feet of hardwoods were cut, it was greater than the state's diminishing harvest of softwoods. Then the harvest of hardwoods began declining, such that by 1950 it totaled just 249 million board feet (NRCSA 1956). Equipment that had been erected at large mills early in the century was becoming obsolete (Figure 6.5).

### Change in Forest Composition

In contrast to the sharp decline in the cutting of forests for lumber, reflecting the fact that so few large trees remained, cutting of smaller trees for pulpwood was reaching record levels. Furthermore, a

FIGURE 6.5. Contemporary view of historic pulpwood log stacker at Cornell in Chippewa County. It sorted and stacked cut logs between 1912 and 1971. The millyard in which it is located was between the Chippewa River and a nearby railroad.

dramatic increase in the prevalence and utilization of aspen occurred. Of the 709,943 cords of pulp-wood cut in 1955, aspen comprised 51.3 percent of the total. Pines accounted for 17.1 percent, hemlock for 8.2 percent, and balsam fir for 6.8 percent (NRCSA 1956, 89). A contemporary state study reported, "As a result of logging and fires, there has been a great change in the composition of forests. Aspen has captured most of the good pine lands and some of the northern hardwood sites, so that it has become the most extensive forest type, predominating on 5,900,000 acres or one-third of the forest land. Being intolerant of shade, the natural succession will be toward more shade enduring species, while effective fire control precludes increase of aspen" (89). The stage was being set for additional changes in the composition, utilization, and management of the state's forests.

## Public Forest Ownership

Substantial changes took place in the ownership and management of Wisconsin's forests following their clear-cutting. Initially, the cutover zone was often touted to prospective farmers, whose frequent failures led to an abandonment of many of these lands by the late 1920s and 1930s. In addition, the uniformity provision of the Wisconsin Constitution, which made it necessary to tax forest lands on the same basis as other lands, prevented a taxation policy favorable to the management of forest lands on a sustained yield basis until 1927. The timber on the land, as well as the land, was taxed annually, even though there was no income from it. A policy of 'cut out and get out' of land ownership was inevitable. Clear cutting followed by tax delinquency became usual. As these lands became increasingly tax delinquent, the situation of the 'cutover' counties became desperate. (NRCSA 1956, 94)

Although the land taxation process was changed, so that the timber value is only taxed when the trees are cut, vast acreages of forest land and forest land that were temporarily worked as farmland were tax delinquent in the late 1920s and early 1930s. Beginning in 1929, when "tax certificates on tax delinquent forest lands could not be sold, the counties took the lands on tax deeds" (NRCSA 1956, 95). By mid-1955, 27 counties held 2,175,748 acres of land in county forests.

### State Forests

State forests were first established early in the twentieth century for environmental protection, such as "stabilization of stream flow and public recreation" (NRCSA 1956, 96). Early state forest lands were obtained by grants from lumber companies, with the state legislature authorizing the purchase of lands for inclusion in state forests beginning in 1937. By 1955 Wisconsin had 277,601 acres of state forests, including Northern Highlands State Forest, Brule River State Forest, Flambeau River State Forest, and Kettle Moraine State Forest. In addition, the Central Wisconsin Conservation Area, largely within Clark, eastern Jackson, and northern Juneau Counties, comprised over 100,000 acres, of which 61,000 acres in Jackson County became the Black River State Forest in 1957.

### National Forests

National forests in most western states were created by withdrawing lands from those available for homesteading. In Wisconsin, there were no lands in the federal domain that could be withdrawn; however, there were vast acreages for which there was little demand following their clear-cutting. Thus, beginning in 1925, the federal government was authorized to purchase lands, subject to state approval, to be included within national forests. Two national forests were established in 1933: Chequamegon National Forest, the larger unit, in north-central Wisconsin and Nicolet National Forest in northeast Wisconsin. Because much of this land had been denuded by clear-cutting, followed by forest fires, reforestation was necessary. Besides extensive replanting of tree plantations, which covered 169,947 acres of the 1,464,027 acres of national forest that existed in 1955, between 1933 and 1942 the Civilian Conservation Corps made major contributions to the restoration and improvement of these forest lands. CCC Camps were established in over 100 locations across Wisconsin, employing 92,000 men (Apps 2019). Twenty-one camps were in Chequamegon National Forest, where the workers built roads, trails, bridges, camp shelters, and fire lookout towers, as well as contributing to reforestation efforts. Their efforts were not limited to federal lands, as many Wisconsin state parks still display roads, trails, and shelters built by CCC workers.

## WISCONSIN'S FOREST INDUSTRY TODAY

Today's Wisconsin forests are the legacy of both their abuse in the nineteenth century and the many efforts to remediate their ravages during the twentieth century. The quality and size of the standing forest has dramatically improved over the past century, the use of the forest has become more multi-purposed, and management of the forests has become more scientific and enlightened. Wisconsin had 17,025,000 acres of forest land in 2017, with its estimated annual harvest of growing-stock trees totaling 287.7 million cubic feet (Kurtz 2018).

### Contemporary Forest Management

Over a quarter of Wisconsin's forested acreage is in public ownership. The largest share, 2,400,916 acres representing 14 percent of all forest land, is within county forests (Haines, Roberts, and Blaha 2019; WCFA 2013). Wisconsin's county forests are scattered across 29 counties and comprise the second-greatest acreage of county forests in the nation (Figure 6.6). Only Minnesota has a greater acreage, and Wisconsin and Minnesota together account for nearly 95 percent of county forest acreage in the United States (S. M. Davis 2013). The greatest acreages of county forests are found within Douglas and Marinette Counties, which had 280,144 and 229,939 acres, respectively. In addition, nine other counties across northern Wisconsin each had over 100,000 acres within their county forests. The two counties with the smallest area of county forests, Vernon and Monroe Counties, together had 9,328 acres (WCFA 2013) and were the southernmost counties with such forests. Wisconsin's county forests are managed not only to produce forest products, employ local residents, and return revenues to their counties, but to provide recreation opportunities to residents and visitors.

### *State and National Forests*

Wisconsin's state forests have nearly doubled in size since the 1950s, now including 526,947 acres, roughly 3 percent of the state's forested area (Haines, Roberts, and Blaha 2019). The Northern Highland–American Legion State Forest, which sprawls across 234,366 acres primarily in Vilas and Oneida Counties, is well over twice the size of Flambeau River State Forest, the second largest. The ten state forest properties are managed for forest production, recreation, and tourism, a topic covered in greater detail in chapter 14. Both the state and county forests are managed from a multiple-use perspective, yet their timber harvest per acre far exceeds that of Wisconsin's national forests (S. M. Davis 2013).

Today the Chequamegon-Nicolet National Forest is managed as a single national forest. It currently covers 1,519,800 acres, which is separated into not only the former Chequamegon and Nicolet forests, but several noncontiguous units. As with state and county forests in Wisconsin, there are significant inholdings of private lands within the national forest. Such privately owned lands range from family-owned recreational parcels to small farms and corporate forest tracts. Additional forest land is owned by the U.S. Fish and Wildlife Service, the National Park Service, and the U.S. Department of Defense, among others, bringing total federal forest land ownership in Wisconsin to 1,623,334 acres. Native American tribes own 395,369 acres of Wisconsin forest land (Perry 2014).

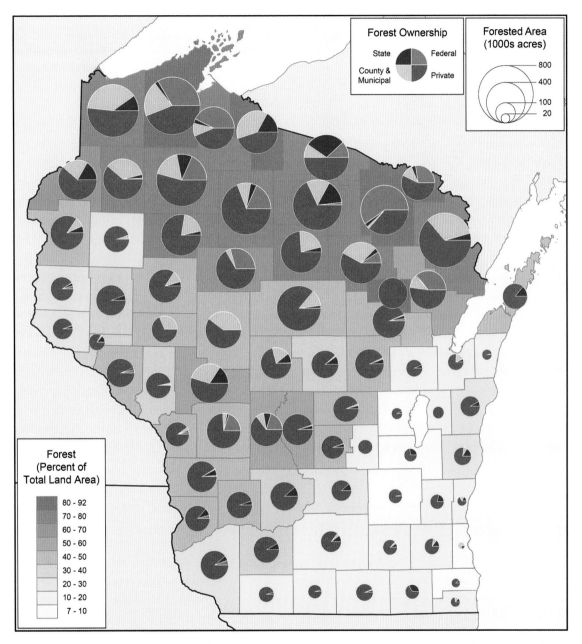

FIGURE 6.6. Proportion and acreage of land in forests by county in Wisconsin, with the pie chart showing the split in ownership of forest land among county, state, federal, and private holders. Map generated from data provided by USFS (2019).

*Private Ownership*

Seventy percent of Wisconsin's forest land is under private ownership, with 9,786,928 acres held by individuals in 2013; 1,049,358 acres are owned by corporations that lack a wood-processing facility; and 412,541 acres are under forest industry ownership, where the forest owner also owns a sawmill or other wood-processing facility (Perry 2014, Kurtz et al. 2017). Collectively, 183,000 family forest ownerships, each holding at least 10 acres, control 9 million acres of the individually owned forest land. A recent U.S. Forest Service report notes: "The primary reasons for [individuals] owning forest land are related to wildlife, aesthetics, hunting and nature. The most common activities on their land are personal recreation, such as hunting and hiking, and cutting trees for personal use, such as firewood" (Kurtz et al. 2017, 28). Nevertheless, nearly a third of the acreage owned by individuals is subject to cutting trees for sale.

Over 3.1 million acres of privately owned forest land in Wisconsin is managed under two state programs that promote sustainable production and public access to the lands. The Forest Crop Law, enacted in 1927 with new enrollments stopped at the beginning of 1986, and the Managed Forest Law, for which enrollments continue, reduce the assessed property tax rates. Instead, owners pay a 5 percent yield tax upon their land's commercial harvest and follow certain forest management practices, including "Harvesting timber according to sound forestry standards; Thinning plantations and natural stands for merchantable products; Releasing trees from competing vegetation; Tree planting to maintain necessary forest density; Treating before and after harvest to ensure adequate forest regeneration; [and] Controlling soil erosion" (WDNR 2017c, 3). Most lands enrolled in the Managed Forest Law program are open for public access, which provides their owners with additional tax savings. These lands are open to public access "only for the purposes of hunting, fishing, hiking, sightseeing and cross country skiing" (WDNR 2017c, 3) and access is only by foot without the landowner's permission. Thus, such lands expand the forest acreage in Wisconsin that is open for public recreation, while assuring that forest lands are sustainably harvested.

*Industrial Forests*

Management and ownership of industrial forests, those covering thousands of acres and whose harvest was traditionally linked to a specific lumber company or paper manufacturer, have dramatically changed. In 1976, 14 industrial owners had 1,351,535 acres of tree farms in Wisconsin. The three largest owners, Owens-Illinois, Consolidated Papers, and Great Northern-Nekoosa, held nearly 720,000 acres (Fixmer 1977). These vertically integrated industrial forest operators controlled the entire process of producing the logs needed by their forest industries, including the planting, managing, and harvesting of trees from their lands. This began to change rapidly by the 1990s, as timber companies sold their holdings to real estate investment trusts and timber investment management organizations. In 1999, Consolidated Paper still owned 322,979 acres of Wisconsin forest, Nekoosa Paper held 228,098, and vertically integrated forest product companies held about 930,000 acres. By 2015, such vertically integrated paper and lumber companies held less than 61,000 acres (L'Roe 2017). Real estate investment trusts, such as Plum Creek Timber Company, which by 2002 had acquired

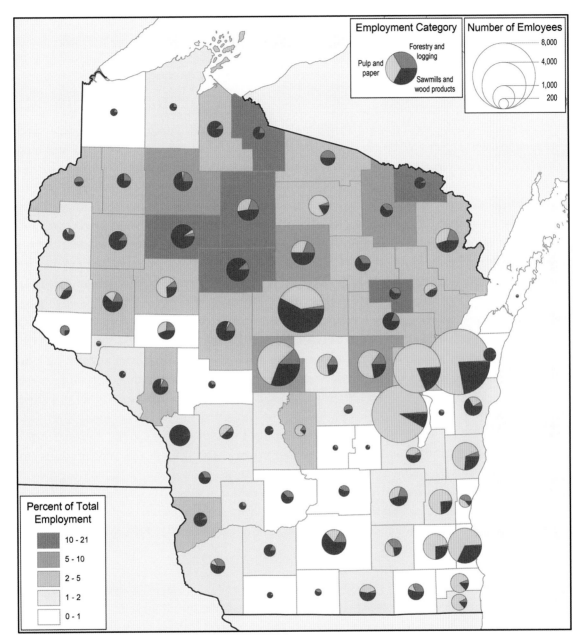

FIGURE 6.7. Total primary and secondary forestry employment in 2016. Map shows the percentage of total employment that comes from forestry in each county, and the proportional circle indicates the share of forestry employment that comes from forestry and logging, sawmills and wood products, and paper manufacturing. Data from WDNR (2016) and Menominee Tribal Enterprises website.

551,000 acres of forest in Wisconsin, looked for ways to maximize the return upon their investments, selling some lands for residential and recreation purposes, while seeking to increase harvests upon other lands. However, the link between particular forests and specific major forest industries no longer exists.

## Contemporary Forest Production and Employment

Forestry and logging directly employed 5,853 workers in Wisconsin in 2016, more than those counted in 1900. Processing the logs provided an additional 24,013 jobs at various sawmills and wood products manufacturers, plus 30,614 jobs in the state's pulp and paper industry. Indirect employment engaged 104,425 workers (WDNR 2016). While paper and wood product manufacturing are discussed in chapter 11, many forestry and sawmill jobs are located close to where the trees are harvested.

Forestry jobs directly employ 1.8 percent of Wisconsin's workers—a number that rises to nearly 5 percent when indirect jobs are added (Figure 6.7). Rusk County has the highest percentage of its work force employed in forestry, 14.2 percent; followed by Florence County with 14.0 percent; Price County, 10.9 percent; and Iron County, 10.7 percent (WDNR 2016). Although missing from the state statistics, Menominee County should be in this top group, given the employment at Menominee Tribal Enterprises in forestry, logging, and sawmill operations. Forestry employment in 13 additional counties was over 4.0 percent: Ashland, Brown, Forest, Langlade, Lincoln, Marathon, Marinette, Sawyer, Taylor, Washburn, Waupaca, Winnebago, and Wood (WDNR 2016).

More than 100 persons worked directly in forestry and logging in 23 counties in 2016, with the largest numbers in Wood County with 433, Waupaca with 231, Marinette with 218, Lincoln with 210, Vilas with 205, and Jefferson with 203 (WDNR 2016). Counties with the largest number of employees working in sawmills and wood products (excluding pulp and paper) are Marathon County, with 2,412 employees in 2016; Brown County, 1,762; Wood County, 1,109 jobs; Rusk County, 957; and Taylor County, 907 (WDNR 2016). An additional 15 counties had over 400 sawmill and wood products employees. While 9 of these counties were in northern Wisconsin, wood products manufacturing continues within several of Wisconsin's most metropolitan counties, as well as several counties whose engagement with sawmilling and wood products dates from the nineteenth century. For example, Outagamie County had 842 wood products workers, La Crosse County had 802, Milwaukee County had 737, and Dane County counted 734 (WDNR 2016).

## Contemporary Forest Cutting and Species Harvesting

In 2003 Wisconsin loggers harvested 561.5 million board feet, better than double that of a half century earlier. Although the harvest fell to 441.3 million board feet in 2008 with the onset of the Great Recession, it rebounded to 503.5 million board feet in 2013 (Haugen 2013, 2017, 2020). Of this total, hardwoods comprised 367.0 million board feet, while softwoods were 136.5 million board feet. In the U.S. Forest Service report for 2013, "red oak, red pine, and hard maple accounted for 56 percent of the total harvest of saw logs" (Haugen 2017, 2). Aspen was the leading species harvested by volume, largely used as pulpwood. Of Wisconsin's total industrial roundwood harvest, 53 percent went to pulp mills,

FIGURE 6.8. Stacks of lumber outside of MacDonald and Owen Lumber Company's sawmill along the Gandy Dancer State Trail in Luck, Polk County.

FIGURE 6.9. Large sawmill that converts logs into preservative-treated railroad ties along Canadian Pacific Railroad west of Rockland in La Crosse County.

29 percent was cut into lumber at sawmills, and 12 percent went to composite panel or plywood mills (Kurtz et al. 2017).

Logs were processed at 205 Wisconsin sawmills in 2013 (Figures 6.8 and 6.9), a decrease of 22 over the previous decade. Of the sawmills, 38 produced over 5 million board feet of lumber a year, of which 15 exceeded 10 million board feet (Haugen 2017, 2020). While the total number of sawmills had fallen since 2003, these losses occurred among the smaller sawmills, as those cutting under a million board feet fell from 125 to 107, and those handling between 1 and 4.9 million board feet decreased from 69 to 60. The number cutting over 5 million board feet increased by 5 (Reading and Whipple 2007; Haugen 2017, 2020). Twenty-nine percent of the industrial roundwood harvested in Wisconsin in 2013 went to sawmills (Figure 6.10), 2 percent went to its four veneer mills, and 12 percent was used by the four composite product or particleboard mills. By far the biggest consumers of the harvested wood (Figure 6.11) were Wisconsin's nine pulp mills, which accounted for 53 percent of the industrial roundwood production (Haugen 2017). Residues, including sawdust and wood chips, from many primary processing facilities, particularly sawmills, are raw materials used in manufacturing other products, including fiberboard and paper.

The type of trees harvested influences their utilization for lumber. Hardwoods comprised 72.9 percent of the output of Wisconsin's sawmills in 2013, with red oak accounting for 21.2 percent of the total; hard maple, 14.5 percent; and aspen or balsam poplar, 11.5 percent. Softwoods constituted 27.1 percent of the lumber production, with red pine being the leader, comprising 17.6 percent of the state's total lumber production (Haugen 2017, 2020). White pine ranks second in softwood lumber output, making 4.6 percent of Wisconsin's saw lumber.

Red oak led lumber output in 2013 was 106.8 million board feet, followed by red pine, which totaled 88.8 million board feet. Over the previous five years the cutting of red pine for lumber decreased, while the utilization of red oak grew. White pine, which produced 23.0 million board feet in 2013, had increased by a third (Haugen 2017, 2020). Pulp mills took 60.3 percent of softwood harvested from Wisconsin forests in 2013 and 51.0 percent of the hardwood.

### Regional Patterns of Forest Production

Industrial roundwood production, including the cutting of trees for sawlogs, pulp, poles, posts, bolts, and chips, focuses upon different species within the various regions of Wisconsin. Not only do hardwoods comprise a much greater share of the harvest in southern and western Wisconsin, but the types of softwoods vary across northern Wisconsin. For example, in only one county did the 2013 harvest of white pine and hemlock exceed that of jack pine and red pine, and that was in Menominee County. It was responsible for 11.5 percent of the entire state's white pine output and 66.1 percent of the hemlock harvest. In contrast, within most other counties of northern Wisconsin the red pine harvest was 5 to 10 times that of the white pine. In Douglas County, it was over 30 times greater. Hard maple dominated hardwood roundwood production in northeastern Wisconsin, while the harvest of hard maple and soft maple were relatively similar in northwestern Wisconsin (Haugen 2020). In contrast, within central Wisconsin, red oaks and white oaks together constituted twice the amount of roundwood as

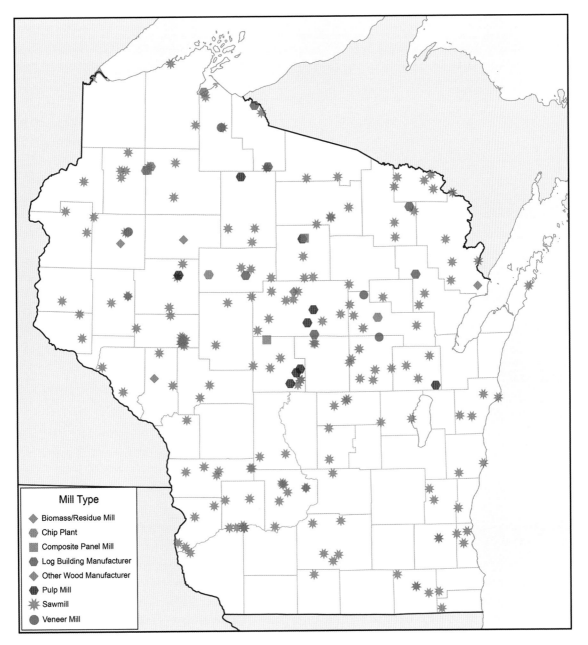

Figure 6.10. Location of sawmills, pulp mills, veneer mills, and other mills in Wisconsin in 2013. Map generated from data provided by Wisconsin Department of Natural Resources' Forest products industry listings webpage (https://dnr.wi.gov/topic/ForestBusinesses/industries.html).

FIGURE 6.11. Three types of railroad cars transport pulp logs in an Escanaba and Lake Superior Railroad train in Pembine in northern Marinette County.

did the two varieties of maple combined, with hardwoods providing twice the harvest of softwoods (Haugen 2017).

## Most Prominent Species of Sawlogs

The most prominent species cut for sawlogs varies regionally and temporally (Figure 6.12). In northeastern Wisconsin red pine led the sawlog harvest, accounting for 19.7 percent of that region's output in 2003, but falling to 14.9 percent in 2013. The most heavily harvested species group consisted of aspen or balsam poplar, 23.6 percent in 2013, followed by hard maple at 17.2 percent. In Wisconsin's northwestern forest unit, red oak led with 22.3 percent, followed closely by hard maple with 20.9 percent in 2003. By 2013 red oak had fallen to third place (11.4 percent), exceeded by both aspen (32.5 percent) and red pine (12.9 percent). In central Wisconsin red oak led with 27.1 percent, followed by red pine with 22.0 percent in 2003, yet their proportions were only slightly changed at 25.3 percent and 19.6 percent, respectively, in 2013 (Reading and Whipple 2007; Haugen 2017, 2020). In the southeastern Wisconsin forest unit, which extends from the tip of the Door Peninsula to the Illinois state line, and east of a line running from the western borders of Outagamie County to Dane and Green

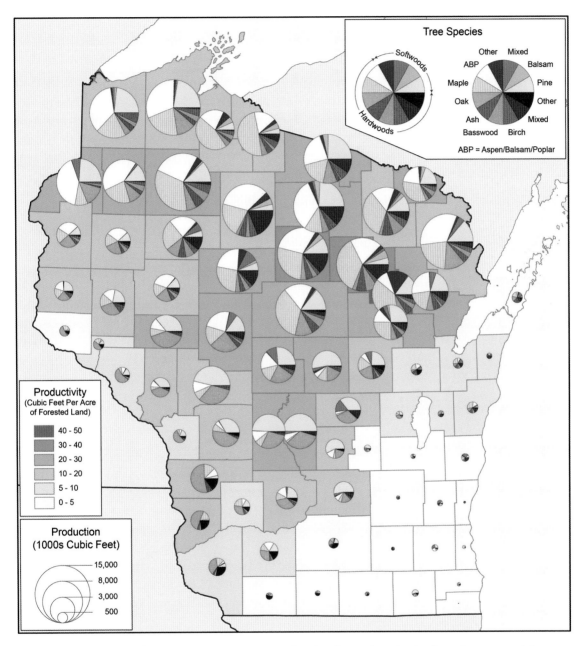

FIGURE 6.12. Forest production in Wisconsin in 2013, showing the total production in each county and the average production per acre. Forest species that dominate the forest production in each county are indicated by the pie charts. Map generated from data provided in Haugen (2017).

Counties, total sawlog harvest for all species is less than that of the leading species group in each other unit. Red oak, the most widely cut tree, accounted for 13.8 percent of the southeastern sawlog harvest (Haugen 2017, 2020). In the southwestern Wisconsin forestry unit, red oak comprised 46.1 percent of the sawlog harvest in 2003, followed by white oak with 14.9 percent (Reading and Whipple 2007), yet red oak fell to 30.9 percent in 2013 (Haugen 2017, 2020).

Red pine sawlog production exceeded 5 million board feet within five counties during 2003. Four of those counties were located in northeastern Wisconsin, led by Forest County, which cut 8,349,000 board feet of red pine (Reading and Whipple 2007). By 2013 over 5 million board feet of red pine sawtimber were cut in five counties, with the greatest amounts coming from Bayfield County (8,737,000 board feet) and Vilas County (8,324,000 board feet), followed by Douglas, Oneida, and Marinette Counties. Forest County's harvest had fallen below 3 million board feet (Haugen 2017, 2020).

Hard maple sawlog cutting exceeded 5 million board feet within five counties, spread among three forest units in 2003. Those counties included Langlade, which cut 7,115,000 board feet, followed by Sawyer, Shawano, Juneau, and Forest Counties. By 2013 Menominee County led with 11,867,000 board feet, followed by Forest County with 9,614,000 and Marathon County with 7,817,000 board feet. Both Langlade and Iron Counties also harvested over 5 million board feet of hard maple (Haugen 2020).

Red oak sawlog production exceeded 5 million board feet within six counties, headed by Sauk County with 7,628,000 board feet, and trailed by Buffalo, Richland, Jackson, Clark, and Sawyer Counties in 2003 (Reading and Whipple 2007). By 2013 it exceeded 5 million board feet within eight counties, but the leading county was Vernon with 7,435,000 board feet, while Sauk County harvested less than a third of what it had a decade earlier (Haugen 2020). Species of trees that dominate harvesting are highly variable over time.

*Tree Species Harvested for Pulp*

The primary species, and even whether it is a softwood or hardwood, utilized for pulpwood varies by region. Within the southwestern and southeastern Wisconsin forest units, softwoods comprise slightly over half of all pulpwood produced. Three of the four Wisconsin pulp mills operating today are within central Wisconsin, and given their small number, 2013 regional statistics are unavailable. Softwoods comprised 39.6 percent of pulpwood production within central Wisconsin in 2003 but only averaged 17 percent in the two northern Wisconsin forest units. Local conditions clearly influence production. For example, within Adams County jack pine and red pine comprised 61.9 percent of pulpwood production in 2003, while aspen comprised 44.2 percent of pulpwood in Lincoln County and hemlock accounted for 21.8 percent of pulpwood in Menominee County (Reading and Whipple 2007).

### Character and Intensity of Forest Harvesting

Logging companies in Wisconsin are typically small owner-operators without employees, but using subcontractors to help with the work. Although some are largely chain-saw based, the "cut-to-length" logging system dominates, in which "a harvester fells and processes (cuts to logs and removes limbs) and a forwarder moves the logs to a landing" (Geisler et al. 2016, 541). These loggers are highly

FIGURE 6.13. Piles of stacked logs at edge of forest east of Mount Morris, Waushara County. Loggers often cut the trees in the winter and move them across the frozen ground to await transfer to mills once the local roads can tolerate the weight of the loaded trucks.

mechanized, with a median capital investment of $610,000 (Geisler, Rittenhouse, and Rissman 2016). One-quarter of the trees are harvested on a contract basis, either for a mill or a landowner, yet three-quarters of the harvest involves a logging company buying standing timber and then selling the cut logs to a mill. Given the dramatic change in forest ownership over the past two decades, logging companies now play the key role of linking forest harvest to companies needing wood for their sawmills or pulping mills.

Logging remains highly seasonal, with winter the best time to move heavy logging equipment into the forest upon frozen ground. While over a century ago spring melt provided a mechanism to move the logs, today it presents more challenges to transport, given that thawed ground becomes too soft and muddy. Weight restrictions limit the transport of logs on rural roads when roadbeds are thawing. Thus, one often sees piles of logs along rural roads in the spring, awaiting transport after weight restrictions are lifted (Figure 6.13). Heavy rains in the summer and fall hinder access within the forest, prompting some loggers to prefer logging trees grown on better-drained sandy soils (Geisler, Rittenhouse, and Rissman. 2016).

FIGURE 6.14. Amish sawmill west of Hay Creek in Augusta Amish community in Eau Claire County. While many Amish households remain engaged in farming, as evidenced by the hay rolls, 43 Amish households in the community were either operating sawmills or were sawmill workers in 2018. Additional Amish households engaged in woodworking, pallet making, and furniture manufacturing.

Northern and central Wisconsin remain the focus for Wisconsin's loggers, yet the harvest intensity, measured by the volume of wood removals per acre of forest land, is highest within Eau Claire, Sauk, Menominee, and Langlade Counties (Haugen 2017). The explanation for the different intensities is multidimensional. Menominee County has an enviable record for managing its white pine forests, with these trees producing far greater volumes when cut than other tree species that are smaller in stature and younger. Many sawmills in Eau Claire County are clustered in that county's southeastern corner, the center of the Augusta Amish settlement (Figure 6.14), where 30 percent of its households are primarily engaged in sawmill work (Cross 2018), enhancing that county's harvest intensity.

## CHRISTMAS TREES AND MAPLE SYRUP

The forests of Wisconsin are noted for production of maple syrup and Christmas trees. While the former takes place within standing forests and produces a product without felling the trees, the later represents the growing of trees over a far shorter time horizon than within conventional tree farms.

FIGURE 6.15. Rows of conifers on Christmas tree farm southwest of Saxeville in Waushara County.

Both activities are highly seasonal, with sugar maples tapped in the spring when the sap runs, and Christmas trees cut in the late autumn so shipments of freshly cut trees arrive a few weeks before the holiday.

### Christmas Trees

Christmas trees are grown for both the wholesale and retail market by hundreds of operators in Wisconsin, and the state ranks fifth in the nation in the sale of cut Christmas trees (Figure 6.15). There are discrepancies between statistics reported by the U.S. Department of Agriculture and the Wisconsin Christmas Tree Producers Association regarding production, which may indicate that many smaller growers, particularly those who focus upon "choose and cut" customers, are too small to be included in the official tallies. For example, an article published in 1995 by a representative of the Wisconsin Christmas Tree Growers Association reported that slightly over 3 million Christmas trees were harvested annually, with about 90 percent of those shipped out of state (Hovde 1995). The 1998 Census of Horticultural Specialties indicated 137 Wisconsin operators sold 1,263,000 Christmas trees, grown on 20,747 acres, with Scotch pine the most frequently cut tree. Scotch pine's popularity substantially fell over the next two decades.

Christmas tree cutting was reported by 184 Wisconsin operators in the 2014 Census of Horticultural Specialties (NASS 2015a), but the number of trees sold had fallen to 657,000, with 14,000 acres in production. Of the growers, 153 had retail operations, often of the "choose and cut" agritourism format, while 92 engaged in wholesale sales. Wholesale operators sold 81.4 percent of the Christmas trees cut. Fraser fir represented 44.6 percent of the trees cut, followed by balsam fir (36.2 percent), white pine (8.8 percent), and Scotch pine (5.8 percent). In contrast to these statistics, which only include those operators with at least $10,000 in sales, the 2012 Census of Agriculture (NASS 2014a) reported that 868 Wisconsin farms had 23,651 acres of "cut Christmas trees" in production in 2012, with 683 farms having cut trees. The Census of Agriculture's larger number, which includes all farms with at least $1,000 in sales, clearly illustrates that most Christmas trees growers are small operators. Indeed, 51 percent of Wisconsin's Christmas tree farms had under 20 acres. While both the number of farmers growing Christmas trees and the acres in production fell between 2007 and 2012, there was little change over the next five years. Wisconsin had 23,373 acres of cultivated Christmas trees spread among 859 farms in 2017 (NASS 2019c).

Geographically, Christmas tree production in Wisconsin focuses upon the central part of the state (Figure 6.16). Five counties, all with over 1,000 acres in production, grew 44.6 percent of the state's total acreage of Christmas trees in 2017. These leading counties include Jackson with 3,607 acres, Waushara with 2,789 acres, followed by Lincoln, Langlade, and Taylor (NASS 2019c). The greatest numbers of growers were in Langlade, Dane, and Marathon Counties, each with 40 or more producers, illustrating that both supplying trees to distant markets and catering to local residents sustain Christmas tree growers.

## Maple Syrup

Wisconsin produced 240,035 gallons of maple syrup in 2017, with 1,399 farms engaged in its production, an increase of 244 producers since 2012 (NASS 2014a, 2019c). Wisconsin's harvest ranked fourth in the nation, although it was less than 12 percent of that of Vermont, the nation's leading producer. The four Wisconsin counties with the greatest number of producers were Clark, with 187, followed by Marathon, Shawano, and Vernon (Figure 6.17). Marathon County led the state in its number of taps (154,226) and produced 17.5 percent of Wisconsin's syrup, while Shawano County with 108,544 taps ranked second in overall maple syrup production (NASS 2019c). Two major circumstances are clearly associated with Wisconsin's centers of maple syrup production. One is obviously the abundance of the sugar maple trees, but the other is the presence of Amish farmers. All four counties cited have Amish settlements, and they are particularly numerous in Clark and Vernon Counties.

### PRÉCIS

Forests played a huge role in bringing settlement into Wisconsin, even though the greatest populations were in the rapidly industrializing cities that processed the wood coming from the Northwoods. Clear-cutting of the northern two-thirds of the state greatly altered the landscape, reshaped the composition of the forest, and influenced today's use of the land. Forestry remains important today. It

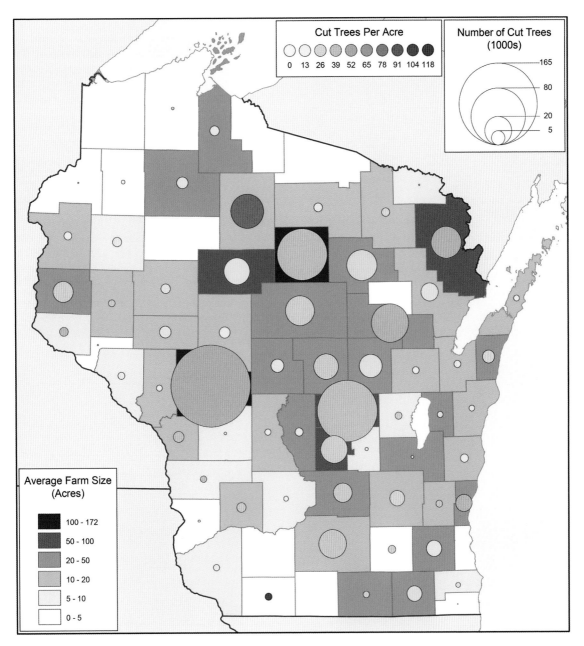

FIGURE 6.16. Christmas tree farms in Wisconsin in 2017. Map shows the number of Christmas trees cut per county, the average acreage of these tree farms, and average number of trees cut per acre. Data from NASS (2019c).

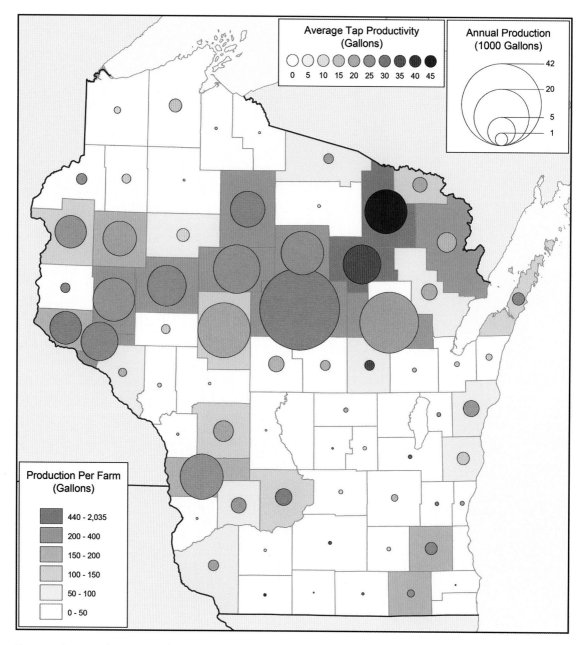

FIGURE 6.17. Maple sugar production in Wisconsin in 2017. Map shows the total production of maple syrup per county, the average tap productivity, and average output of maple syrup per farm. Data from NASS (2019c).

directly employs 60,480 Wisconsin workers and indirectly engages over 100,000 more (WDNR 2016). Forestry is the largest employer within 10 Wisconsin counties.

Forest cutting was not the only exploitative use of Wisconsin's abundant natural resources. Just as we saw in chapter 5, which explored the role of the fur traders and miners in shaping Wisconsin, farmers sought to clear the forests and plow the land as they put down roots in Wisconsin, as we will see in chapter 7. The production of a huge variety of raw materials, which are now harvested from second-growth forests or the tree farms in Wisconsin, continues to provide the foundation for much of the state's manufacturing industry, as discussed in chapter 11. Indeed, Wisconsin's manufacture of forest products results in $2.2 billion in exports annually (WDNR 2016).

*Chapter Seven*

# Settlement of Wisconsin 3

*Farming and Ethnic Settlers*

FARMERS WERE BY FAR the most numerous group drawn to Wisconsin by its natural resources. Their clearing of the vegetation and plowing of the soil were typically less exploitative than those miners and loggers who dug out the mineral wealth, cut the virgin forest, and then departed. Farmers, in establishing their farms, were making a long-term investment in the land, one that provided a justification for the creation of hamlets and villages that served their needs and dot the landscape of Wisconsin today. This chapter focuses upon those farmers who settled Wisconsin. They typically came after the initial mining and forestry development or after former Indian lands of southern Wisconsin were placed on sale.

Immigrants were the dominant agricultural settlers in much of the state. Their pattern of ethnic settlement was so conspicuous that the journalist Fred Holmes ([1944] 1990) wrote *Old World Wisconsin: Around Europe in the Badger State*, showing how the traveler could experience the ethnic diversity of Europe without ever leaving Wisconsin (Figure 7.1). While the agricultural settlement of Wisconsin was largely accomplished before the Great Depression of the 1930s—and in many areas retreat from the frontier had already begun—the ethnic patterns created in the nineteenth century remain etched upon today's cultural landscape.

## INITIATION OF AGRICULTURAL SETTLEMENT

Small farms were planted in the Mississippi River Valley near Prairie du Chien and the Lower Fox River Valley near De Pere in the eighteenth century by French Canadians to help feed those involved in the fur trade, yet these individuals were highly dependent upon hunting. Later, when lead drew thousands of miners to southwestern Wisconsin, the federal government withdrew that area's lands from those that could be sold to prospective farmers, instead entering into a type of lease arrangement allocating mineral deposits and forests—critical for the smelting of the ore—whereby miners paid a royalty on the lead they produced (Schafer 1932). While some of the miners planted gardens and some

FIGURE 7.1. The Wisconsin Historical Society's Old World Wisconsin outdoor living history museum features over sixty historic structures near Eagle, Waukesha County. This photograph shows the half-timbered Schulz stable with its thatched roof on an 1860s Pomeranian immigrant farm.

were fed from lands cultivated by squatters, lands were only sold by the federal government after treaties had been negotiated with their Native American populations.

White settlers flocking into Wisconsin's lead mining district, which was then still Indian territory, contributed to military conflict in 1827, causing infantry to be dispatched from St. Louis, Missouri; Fort Snelling, Minnesota; and Fort Howard, near Green Bay. A local militia formed under the leadership of Henry Dodge, who eventually became Wisconsin's first governor (Wyman 1998). While the Chippewa, Ottawas, Potawatomis, and Winnebago (or Ho-Chunk) ceded lands in the mining region in 1829, illegal mining continued on Indian grounds. In his frontier history, Mark Wyman (1998, 145–146) notes that "Dodge, with a large number of others, was mining well inside Winnebago territory (near the site of Dodgeville) . . . and rebuffed efforts to evict him and his 130 armed miners." Although the Black Hawk War began in Illinois, the Native Americans fled north and then west and were ultimately cornered and massacred near the junction of the Bad Axe River and the Mississippi River in 1832. That same year the Ho-Chunk were forced to relinquish additional land in Wisconsin.

### LAND SALES, LAND GRANTS, AND HOMESTEADING

Before land was offered for sale, it was surveyed using the congressional survey described in chapter 1 (see also Lippelt 2002). Lands that were already occupied, such as those around the French settlements that were laid out in the long-lot fashion, were considered private claims, as were lands resided upon by squatters. They were separately designated and excluded from public sale. The first public land surveys in Wisconsin were conducted in 1833, shortly following the cessation of Indian lands. In 1834 land offices were operating in Mineral Point and Green Bay, with one in Milwaukee opened by 1839. Surveying in Wisconsin continued through 1866, always following the signing of a treaty with an indigenous nation, which relinquished its claim to the territory. Often not willingly made, such ceding of Indian lands was in exchange for a promise of territory elsewhere along with some compensation and the granting of various treaty rights. Eventually 11 land offices were opened in Wisconsin, with most opening shortly after nearby lands were surveyed and made available for purchase and closing when most lands had been sold. The last federal land office in Wisconsin was in Wausau, which closed in 1925.

Lands surveyed into townships and sections were available for sale at the land offices, initially for a price of $1.25 per acre, with a minimum purchase of 80 acres required. Both individuals wanting to farm and establish settlements and land speculators sought the lands. As Joseph Schafer (1927, 62) explains in his history of settlement within southeast Wisconsin, "Usually, the speculator entered a newly surveyed region first. Armed with the land office plats and surveyor's notes, he traversed townships and sections, noting favorable situations here and there, and finally, at the land sale, he would bid in such tracts as promised to be salable at a good profit within a reasonable time. Then when the settlers came, they had the alternative of buying lands entered by the speculators or taking poorer tracts that the speculators had left." While it is not the purpose here to dwell upon the many intricacies involving land transfers from the federal government and strategies to thwart the abuses of land speculators (see Schafer 1927 and 1932 for a wealth of information regarding these sales), a variety of policies and programs were involved with the sales, and some shaped how the lands were settled. While many prospective farmers acquired their lands directly from the federal government, large numbers bought land from other entities—public and private—that had either been granted lands or purchased public land for the purpose of selling or reselling it.

#### Land Grants

Some land was transferred from the federal government to the state. The lands of section 16 of each survey township were considered as school lands, the title of which in Wisconsin was given to the state, with the sale of these lands used to support local schools. In total, slightly over 1 million acres were granted in Wisconsin for this purpose (Craig et al. 2015). Under the Swamp Land Act of 1850, the federal government granted 3,357,693 acres of such land to Wisconsin, with the hope that this would encourage their drainage or reclamation. The state also received smaller allocations of lands, such as 6,400 acres to help fund the building of the capitol building in Madison in 1846 and 301,931 acres under the Military Wagon Road Act of 1863 to finance the construction of a road from Fort Howard to Lake Superior (Craig et al. 2015).

Considerable acreage was transferred from the federal government to builders of canals and railroads to encourage and subsidize their construction. In 1856 and 1864 railroads in Wisconsin were provided land grants providing every other section of land extending 6 miles either side of the railroad right-of-way. Later the swaths were extended to 10 and then 15 miles. In total, federal land grants were provided to construct 973.5 miles of track in Wisconsin, with 2,874,000 acres of land given. In addition, 325,000 acres of land were allocated for the construction of canals in Wisconsin (Craig et al. 2015).

In 1862 Congress passed the Morrill Act, which established the land-grant college system of which the University of Wisconsin at Madison became a beneficiary. The State of Wisconsin received 240,000 acres to establish a college teaching agriculture and mechanical arts. Those lands were sold to the public for $1.25 per acre, the same price charged at the federal land office. An even larger acreage of Wisconsin land was granted to an out-of-state university. Cornell University in upstate New York was designated a land-grant university, but because there was no public land in New York to donate, the university received an allocation of 500,000 acres in the northern part of west-central Wisconsin. Although it eventually sold its lands, Cornell University still maintains the mineral rights. Smaller allocations of Wisconsin acreage supported land-grant colleges in several other eastern states.

With passage of the Homestead Act of 1862, unsold federal lands became open to homesteading in 1863. In Wisconsin an adult or head of household could homestead a quarter section of land, a parcel that measured one-half mile by one-half mile, containing 160 acres. Homesteads of 40 and 80 acres were also available. Rather than purchasing the land with cash, other than paying a filing fee of $18, the homesteader paid for the property with sweat equity, building and occupying a home on the property and cultivating the land for five years. In total, 29,246 homesteads were granted in Wisconsin, covering 3,110,990 acres, or 9 percent of the state's land area. Yet, because of when this program was enacted, little land was available to be homesteaded in southern Wisconsin.

## AGRICULTURAL SETTLEMENT OF WISCONSIN TO 1850

The availability of land in Wisconsin was a major factor in attracting new residents. The first major wave of agricultural settlers were Yankees, those individuals with roots in New England, particularly Vermont and Upstate New York. A combination of population growth in New England that spread westward across New York, depleting its available arable lands, and improved transportation into the Great Lakes region that came with the opening of the Erie Canal in 1825 spurred their movement. Settlers first turned their attention to lands in northern Ohio, closest to the western terminus of the canal. As the better lands were purchased, the wave of settlement proceeded into northern Indiana and southern Michigan and reached Wisconsin by 1832. Some settlers pushed overland, with even more settlers taking passage on sailing ships to reach Wisconsin. By the mid-1830s steamboats were operating upon the Great Lakes, and a person could travel from New York City to Wisconsin in eight days (Wyman 1998).

### American Agrarian Settlers

Most residents of that part of Michigan Territory west of Lake Michigan were in the lead mining area in 1830. That was shortly to change. By the time of the 1840 census, Wisconsin's population had increased tenfold, to 30,945, and population growth in southeastern Wisconsin was to rival and soon surpass that in the southwestern lead mining district. In that region, as a prominent geographer of a century ago explained, "the lands previously reserved from sale were placed upon the market . . . [and] rapid settlement in the middle forties changed the lead region into an agricultural section" (G-H. Smith 1929, 67–68). Although many Cornish miners eventually took up farming in the region, large numbers of settlers came from states west of the Appalachians. Indeed, Guy-Harold Smith (71) notes, "There were more people from Kentucky, Tennessee, and Missouri in the lead region in 1850 than all the remainder of the state."

Settlers in southeastern Wisconsin were primarily from New York or New England, or were the children of parents from those areas, who were born in northern Ohio or northern Indiana, illustrating a two-stage movement of many newcomers. Their selection of land proceeded rapidly. In some townships nearly all of the lands were taken the same year their entry became available at the land office. Nevertheless, preconceived biases regarding extensive prairies somewhat slowed their settlement. As Joseph Schafer described:

> Wherever practicable, timbered land was usually selected for a portion of the proposed new farm, say a fourth or a third, the balance being prairie or oak openings. If a tract of wet prairie—a slough or marsh—could be included, that would make the location ideal, for the combination of well drained plowland, hay land, and timber was what every one wanted. Even near the lake shore, in Racine and Kenosha counties, the larger prairies remained in the public domain some years after all the timbered lands with the prairies adjoining them had been taken up. Men from the well-wooded regions of New York, New England, Pennsylvania, and Ohio hesitated to build homes on the big, open, exposed prairies. (Schafer 1927, 121)

Yet even the less desirable prairies were typically taken by settlers within two to six years, given the rapidity of settlement in southeast Wisconsin.

### Early German Settlers

German immigrants played an early and dominate role in the settlement of Milwaukee and Ozaukee Counties, unlike the primacy of Yankees in Racine and Kenosha Counties. Foreign-born persons, of which three-quarters were Germans in 1850, outnumbered native-born Americans by a two-to-one margin in those towns that were to become Ozaukee County (Schafer 1927). Germans, who settled in distinct groups related to their *landsleute* or German state of origin, differed in their land selection behavior. They sought more densely forested lands and often valued accessibility to markets more heavily in their selection of lands than the soil's vegetative covering. In addition, "Germans who settled in the rural communities generally purchased small farms, from 40 to 80 acres, and this had the effect of creating a dense rural population in this area which was originally a maple forest. . . . No other part of Wisconsin has so dense a rural population as this German settled section" (G-H. Smith 1929, 77). Germans were the leading immigrant group within Wisconsin, followed by the Irish and English. Of the state's white population in 1850, 36.2 percent were foreign-born (USC 1853).

German influences differed spatially across Wisconsin, given their coming from multiple territories, which differed both politically and religiously. Because "Germans were a heterogenous group overall in terms of geographic, religious, and economic backgrounds, they often settled in homogenous clusters in rural areas" (Schlemper 2006, 189). The area north of Milwaukee in Ozaukee County, the southern part of the "German Coast," grew from settlements begun in 1839 near Mequon and Freistadt of "Old German North Germans . . . [fleeing] Prussian religious persecution which had already resulted in the suppression of worship according to the Old Lutheran ritual" (Schafer 1927, 90). Farther north in the German Coast, immigrants from Lippe-Detmold in northwestern Germany settled in Sheboygan and Manitowoc Counties. Other settlements in the 1840s brought Rhinelanders, Saxons, and Hanoverians, among others, to the region (Bawden 2006). In the heartland of the German Crescent of settlement—Jefferson, Dodge, Washington, Fond du Lac, and Calumet Counties— German immigrants were highly diverse. According to Timothy Bawden:

> The Old Lutherans who came from Pomerania between 1843 and 1845 were among the early German immigrants of this region. They settled in Kirchayn, Washington County; Lebanon, Dodge County; and Ixonia, Jefferson County. The Kirchayn group came from the Baltic regions of Pomerania, and the groups that settled the other two communities came from the farming district of Stettin and the Oderbruch. The towns of New Holstein, in Calumet County, and neighboring Kiel, in western Manitowoc County, were settled by immigrants from Schleswig-Holstein around 1848. (Bawden 2006, 85)

Another group of Germans settled within eastern Fond du Lac and southern Calumet Counties in the early 1840s. Unlike German Protestants who settled elsewhere in this heartland region, these Catholic settlers came from the Eifel region of the Rhineland north of Koblenz, settling a region that became known as Wisconsin's "Holy Land" (Schlemper 2004, 2006). Unlike most areas of rural settlement in Wisconsin, centralized rural villages developed around parish churches (Figure 7.2), giving both their names and religious character to the resulting cultural landscape, both in this region and in northwestern Dane County near Roxbury and Martinsville, which were also established in the 1840s (Zeitlin 2000). Representing Bavarian "dorf culture," these communities today are more in keeping with the European village-focused landscape than dispersed farmsteads that characterize many other regions of Wisconsin.

Irish settlement during the 1840s took place both around the periphery of Wisconsin's German "Holy Land" and in portions of Kenosha, Racine, Milwaukee, and Ozaukee Counties, spurred by famine from the Potato Blight that struck their island beginning in 1845. Most of Wisconsin's Irish ethnics are descended from those who immigrated over the following decade or who had already settled in the mining district of southwestern Wisconsin (D. G. Holmes 2004). In total, Wisconsin had about 21,000 Irish settlers in 1850.

## Spread of Agrarian Settlement

By 1850, settlement of Wisconsin had spread north from Illinois to Lake Winnebago in central Wisconsin, with a tentacle of settlement extending down the Fox River to Green Bay. The spread of settlement is clearly shown by the map (Figure 7.3) illustrating the dates of establishment of post offices as the frontier retreated. Along Lake Michigan settlement had spread north into southeastern Manitowoc County and west toward the Upper Fox River. The limit of settlement in southwestern Wisconsin

FIGURE 7.2. Fields surround rural settlement focused upon St. John the Baptist Catholic Church in Johnsburg, Fond du Lac County.

ran approximately along the trace of the Military Road atop Military Ridge, running eastward from Prairie du Chien south of the Wisconsin River. Wisconsin's 1850 population had reached 305,391, excluding the indigenous population that was not enumerated.

Farming was the primary occupation at the middle of the nineteenth century. Of the 78,139 males whose professions, occupations, and trades were enumerated in 1850, a total of 40,865 were farmers and an additional 11,201 were laborers (USC 1853). Nearly 3 million acres were reported as being land on farms, of which 1,045,499 acres were listed as improved, thus under cultivation. Three counties, Rock, Walworth, and Waukesha, reported over 100,000 acres of improved farmland (Figure 7.4). Wheat was the most important grain harvested, yielding 4,286,131 bushels. Oats were second highest, with 3,414,672 bushels, while Indian corn produced 1,988,979 bushels. In addition, 1,402,077 bushels of Irish potatoes were harvested. The state's farmers had 42,801 working oxen and 30,179 horses, as well as 64,339 "milch cows," as the term was then spelled (USC 1853). Wisconsin was a long way from becoming the nation's dairy state, having just 6.6 percent as many milk cows as New York, then the nation's leading dairy state. Wisconsin was known as a wheat-growing state, even though eight other states grew more wheat.

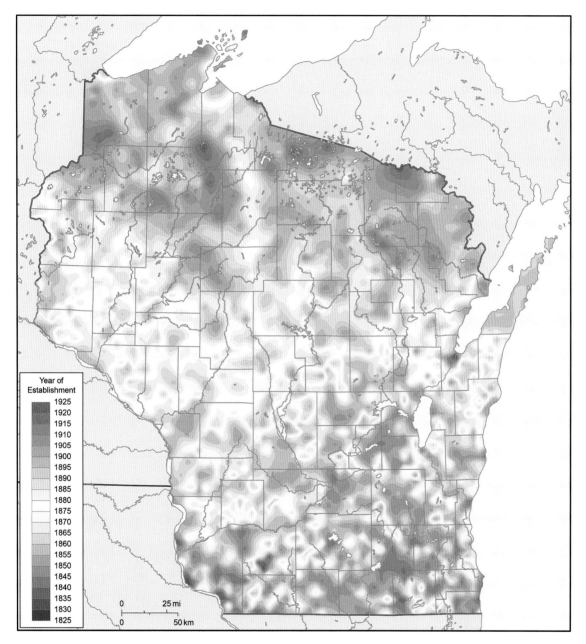

Year of
Establishment

| | |
|---|---|
| | 1925 |
| | 1920 |
| | 1915 |
| | 1910 |
| | 1905 |
| | 1900 |
| | 1895 |
| | 1890 |
| | 1885 |
| | 1880 |
| | 1875 |
| | 1870 |
| | 1865 |
| | 1860 |
| | 1855 |
| | 1850 |
| | 1845 |
| | 1840 |
| | 1835 |
| | 1830 |
| | 1825 |

FIGURE 7.3. Spread of agricultural settlement in Wisconsin between 1840 and 1920 shown by dates of establishment of post offices. Post office establishment dates from Moertl (1995).

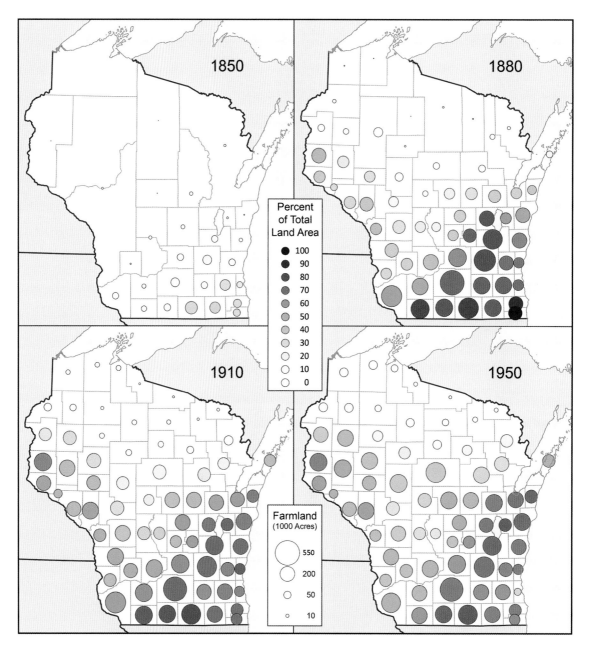

FIGURE 7.4. Maps displaying improved farmland in Wisconsin in 1850, 1880, 1910, and 1950, illustrating the expansion of farming up to the time of World War I, followed by a retreat in agriculture within northern Wisconsin during the Great Depression. Maps created using data from USC (1853, 1883a) and USBC (1914c, 1952b).

## SETTLEMENT BOOMS BETWEEN 1850 AND 1880

Two major waves of settlement of Wisconsin occurred during the three decades following 1850. The first major wave was fueled by European immigrants, and it was abruptly terminated by the Civil War. A second wave began with the cessation of hostilities and involved both a resumption of European immigration and the western movement of Americans (Ostergren 1997). By 1880 the settlement frontier had reached the edge of the vast virgin forest covering the northern third of the state. The state's population had reached 1,315,497, an increase of over a million persons since 1850, more than quadrupling in thirty years (USC 1883b). This growth had dramatic impacts upon the state's ethnic composition and its agricultural output.

### Settlement before the Civil War

During the 1850s the foreign-born population of Wisconsin grew by 166,456, with these individuals comprising 35.6 percent of the state's population in 1860. Immigrants from the various German states totaled 123,879 persons, or 45 percent of the foreign-born population and 16 percent of the total state population. The largest number of Germans, 52,983 individuals from Prussia, accounted for 43 percent of all Germans and outnumbered the 49,901 immigrants from Ireland, the second-largest source of Wisconsin's foreign-born population. Other nations that had provided at least 20,000 immigrants to Wisconsin included England and Norway (USC 1864b, 544). Considering that many Wisconsin natives were children of immigrants, households headed by a foreign-born parent were the norm in Wisconsin. Of the population that was born in the United States, 24 percent were from New York, just under the number who came from Germany.

The combination of immigration from overseas and westward movement of Americans resulted in the growth of Wisconsin's population from 305,391 to 775,881 over the decade (USC 1864b). By 1860 settlement covered all of the Lake Michigan shore of Wisconsin, except for the Door Peninsula; that area south of a line extending from Green Bay to Stevens Point to Trempealeau, except for poorly drained forest lands in Jackson and Wood Counties and northern Adams and Juneau Counties; and along the Mississippi Valley, including tongues of settlement extending up the Black, Chippewa, and St. Croix Rivers. The formation of new counties, well illustrated in the Wisconsin Cartographers' Guild's (WCG 1998) magnificent historical atlas of Wisconsin, largely corresponds to the frontier. Wisconsin had 58 counties in 1861. North of the settled areas, counties were much larger than those that were to be created by their partition over the next several decades.

With its growth of population and spreading settlement, Wisconsin's agricultural economy substantially expanded during the decade. The U.S. Census (1864b) counted 93,859 farmers in 1860, along with 31,472 farm laborers and 28,238 other laborers. (These were not listed separately a decade earlier.) Wisconsin's acreage of improved land in farms had grown to 3,746,167, with an additional 4,147,420 acres of farmland shown as unimproved. Fifteen counties had over 100,000 acres in cultivation, with the largest acreage of improved farmland being in Dane County, 279,124 acres. The number of working oxen had increased to 93,652, while horses totaled 116,180, as Wisconsin's farmers were shifting their reliance from oxen toward horses to pull their plows and wagons. Output of grain had dramatically increased by 1860, with the state's farmers harvesting 15,657,458 bushels of wheat, 11,059,260 bushels of oats, and 7,517,300 bushels of Indian corn. Wisconsin had become the nation's third leading wheat-growing state, only exceeded by Illinois and Indiana. Five Wisconsin counties—Columbia, Dane, Dodge, Fond du Lac, and Rock—each harvested over one million bushels of wheat (USC 1864a).

Wisconsin's involvement with dairying had also increased, with 203,001 milch cows enumerated. Yet, its national role in dairying remained small, and its output of butter exceeded its production of cheese by a factor of 12, while the national norm was closer to 4 to 1. Shortly, this was to begin changing.

### Settlement after the Civil War to 1880

During the Civil War immigration ceased, yet a major wave of European immigration resumed thereafter, before slowing during an economic depression beginning in 1873. Even though Wisconsin welcomed relatively few immigrants during half of the two-decade period, the state experienced major growth. Of the state's population in 1880, all but 0.21 percent were white, with persons who were foreign-born accounting for 30.8 percent of the total population (USC 1883b). By 1880 settlement had nearly expanded to today's limits of extensive cultivation, extending up the Door Peninsula and along the western shore of the Bay of Green Bay to the Michigan border. Settlement also extended north along the Wisconsin River into the Wausau region and along the Chippewa River into eastern Dunn County. Much of the previously unsettled lands between the southern sections of the St. Croix, Chippewa, and Black Rivers had become occupied. In 1880 the total land in farms in Wisconsin had reached 15,353,118 acres (USC 1883a), a figure that was over a million acres greater than Wisconsin's total farm acreage in 2017. In contrast, Wisconsin had 10,085,021 acres of cropland in 2017, of which 9,234,611 acres were harvested, compared with 9,162,528 acres of "improved" land in 1880 (NASS 2019c; USC 1883a).

### Immigrants Shape the Cultural Landscape of Wisconsin

Immigrants played a huge role in expanding Wisconsin's farmland. Of Wisconsin's 138,443 farmers enumerated in 1880, only 54,724, or 39.5 percent, were born in the United States. German-born farmers totaled 36,877, or 26.6 percent of Wisconsin's farmers, while those born in Sweden and Norway added another 8.9 percent, natives to Ireland provided 7.9 percent, and those immigrating from Great Britain accounted for 6.3 percent (USC 1883a). So dominant was the immigrant population that in his seminal work on the frontier, Frederick Jackson Turner (1921, 233) wrote, "Wisconsin's future is dependent upon the influence of the large proportion of her population of foreign parentage, for nearly three-fourths of her inhabitants are of that class."

### *Germans*

The massive wave of German migration to America, which came to its end in the 1880s, completed the formation of the "German Crescent" of settlement that remains visible on a map of dominant ancestry today. Given the location of lands available for farm establishment during the post–Civil War period, plus the targeted efforts of two railroads to sell their granted lands to these immigrants, the

*Chapter Eight*

# Agriculture in Wisconsin 1

## *Farms and Crops*

EARLY TWO CENTURIES OF FARMING have shaped Wisconsin's agricultural landscape, yet some of today's most widely grown crops would have been unknown to the state's farmers a century and a half ago. As we saw in chapter 7, most of today's farmland was already being cultivated in 1920, and most areas that presented physical challenges to farmers then continue to do so. Furthermore, the current distribution of farm settlements was determined when railroads transported most crops to market and most roads leading to the nearest railroad station were not yet paved, with many farm wagons being pulled by horses. Barns were erected to meet the contemporary needs of farmers, determined by the types of crops then in vogue and the technology then available, as well as the ethnic heritage of the farmer and financial resources that were available.

The threshing floor so critical to wheat growers in the nineteenth century was incorporated into many barns, space that was later used to store equipment or hay. Once erected, many well-built structures have been used for well over a century, but their uses may have altered. As farms grew in size and adopted new technologies, additional farm buildings were added to the farmstead, and redundant structures were allowed to collapse or were repurposed. Many dramatic changes occurred on Wisconsin's rural landscape from the 1820s through the middle of the twentieth century. This chapter focuses upon the contemporary agricultural industry of Wisconsin, largely displayed by the 2012 and 2017 Censuses of Agriculture (NASS 2014a, 2019c), and reviews major changes in farm structure, land tenure, and crop distributions that have occurred since 1950.

### WISCONSIN FARMLAND AND CROPLAND IN 2017

Farmland and cropland are not interchangeable terms. Farmland includes all lands that are part of a farm, whether used for cultivation of crops, pasturage of animals, or a managed woodlot, or simply not used for any particular purpose. Cropland is land that is cultivated, pasture that is in a rotation with various row crops, or pasture that is routinely cut and harvested for hay. Since 1950 Wisconsin's land in farms fell from 23,332,095 to 14,318,639 acres in 2017, while the state's harvested cropland declined from 10,112,027 to 9,234,611 acres over the same period. Between 2012 and 2017 farmland

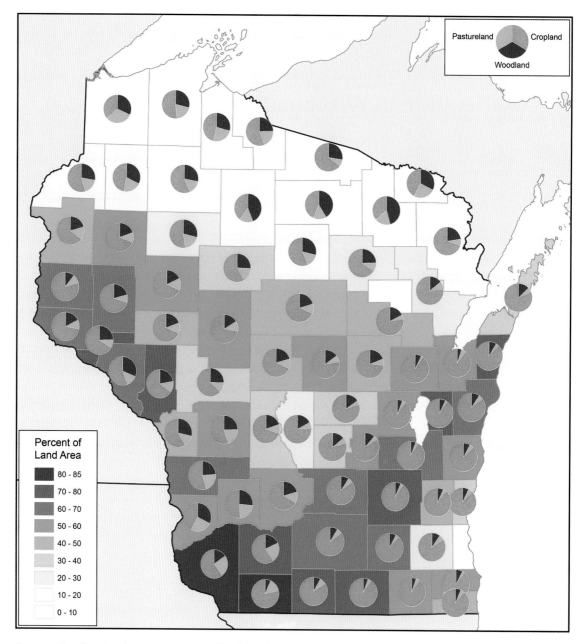

FIGURE 8.1. Farmland as a percentage of land area by county in Wisconsin in 2017, with the pie charts indicating the proportions of farmland that are used for cropland, pastureland, and forest. Data source: NASS (2019c).

declined by just over 250,000 acres, but harvested cropland increased 85,338 acres, as some pastures and conservation reserve lands were returned to cultivation (USBC 1952a; NASS 2014a; NASS 2019c). Thus, while total land in farms fell by 38.6 percent since 1950, actual cropland harvested declined by just 8.7 percent.

Much cropland loss is attributed to urban expansion in southeast, south-central, and east-central Wisconsin. Indeed, over the 62-year period ending in 2012, the acreage of harvested cropland fell from 40,817 to 2,887 in Milwaukee County. Its area is not disclosed in 2017. Large declines also occurred across much of northern Wisconsin, due to a combination of environmental and economic constraints.

The proportion of Wisconsin's land that is farmland varies regionally (Figure 8.1). The largest proportion is in Lafayette County, where 84.5 percent of the county's land area was in farms in 2017. In contrast, within all of the counties along Lake Superior and along the border with Michigan's Upper Peninsula—with the exception of Marinette County, plus Sawyer, Oneida, and Menominee Counties, that number is under 10 percent. In Menominee County, the lowest, the amount is not disclosed in 2017, but it was less than 0.25 percent in 2012. In Milwaukee County, farmland occupied under 3.0 percent of the county's land in 2012. Among the northernmost tier of counties, Vilas County had the smallest total farmland, 5,652 acres in 2017, accounting for 1.0 percent of its land area (NASS 2014a, 2019c). The extensive forest and short growing season of much of northern Wisconsin and the expanding urban development of southeastern Wisconsin are responsible for the limited farmland in those areas.

The share of all farmland that is used as cropland also varies across the state. In most of east-central, south-central, and southeastern Wisconsin, over 80 percent of total farm acreage is utilized for cropland. The highest figures in 2017 are in Outagamie and Walworth Counties, 87.0 and 87.2 percent, respectively. In contrast, the proportions are much lower across much of northern Wisconsin, as well as in much of west-central and parts of southwest Wisconsin. Statewide, 16.5 percent of Wisconsin's farmland is devoted to forage crops, including hay and haylage. However, this proportion is far higher in parts of northern Wisconsin, where the short growing season limits viability of many other crops. Thus, forage crops, including all varieties of hay, haylage, silage, and greenchop, occupy 95.2 percent of total harvested cropland in Florence County, 87.6 percent in Douglas County, and 64.4 percent in Forest County (Figure 8.2). Yet given their paucity of cropland, the percent of the total county land area devoted to hay crops is far smaller, ranging from 1.0 to 2.3 percent in these counties (NASS 2019c). Most of the of their other farmland is grazed, in woodlands, or simply unused lands that are reverting back to forest.

The smaller proportion of farmland used for cropland in much of west-central Wisconsin is mostly associated with steep slopes along valleys in the Driftless Area. For example, cropland comprised 47.0 percent of farmland within Crawford County in 2017. In adjacent Vernon County cropland was 58.3 percent of farmland, while in Buffalo County it was 56.1 percent (NASS 2019c). In all three counties cropland is largely confined to valley bottoms and to plateaus or ridge tops between the valleys, while woods cover the slopes (Figure 8.3). Locally, these woods are heavily tapped to produce maple syrup and support hardwood cutting. Because the steep slopes are highly vulnerable to soil erosion, contour

FIGURE 8.2. Large rolls of hay in field, with forests covering slopes in the distance, on a farm located east of Moquah, Bayfield County. Forage crops of all varieties accounted for 79.7 percent of the county's harvested cropland in 2017.

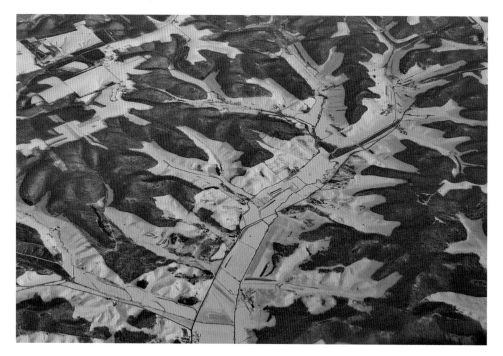

FIGURE 8.3. Aerial view of snow-covered fields in valley surrounded by forested slopes along Montana Ridge in the Town of Glencoe area of Buffalo County. Throughout much of the Driftless Area fields are located on the flatter areas of the valleys and on broad, relatively level ridge tops, shown in the upper left.

strip cropping was introduced to this area in 1933 by the agricultural experiment station in the Coon River Valley (Johansen 1971), the nation's first such large-scale demonstration of the practice. From there the technique spread throughout the region (Figures 8.4 and 8.5). Yet many slopes are too steep for such conservation practices, resulting in farmers leaving large portions of their farms tree covered.

## NUMBER AND CHARACTER OF WISCONSIN'S FARMS

The number of Wisconsin's farms has fallen considerably since 1950, when 168,561 were counted. The 2017 census tallied 64,793 farms, yet most are too small to provide anything other than a supplement to the farmer's income. Nearly a third reported sales of less than $2,500, with a total of 31,204 farms, or 48.2 percent, indicating that total annual sales of crops or animal products were less than $10,000 (NASS 2019c). Subtracting farm expenses—including the costs for seeds, fuel, fertilizer, equipment, property taxes, and other expenditures—leaves these individuals with minimal net income. Yet, nearly half of all farmers indicated that farming was their principal occupation.

The smaller farms are vital to Wisconsin's agricultural economy. As a Wisconsin Academy of Sciences, Arts and Letters (WASAL 2007, 137) report explains, "Using gross sale receipts as a measure, about three-fourths of Wisconsin farms return under $100,000 annually. Many of the smaller farms

FIGURE 8.4. Tractor applying chemicals on narrow sloping field plowed along the contour within Mormon Creek Valley near Kammel Coulee, south of St. Joseph in La Crosse County.

FIGURE 8.5. Contour strip cropping of corn and alfalfa in field of dairy farm southeast of Ridgeway, Iowa County.

specialize in unique products and service local/regional markets while the large farms often specialize in producing, with high efficiency, commodity products (bulk milk, grains, cheeses) traded in international markets." The 2017 agricultural census indicates that 38.7 percent of the principal operators of Wisconsin's farms, including those of all sizes, reported working at least 200 days annually off the farm in another occupation. In contrast with those farms with minimal income, 15,880 Wisconsin farms, or 24.5 percent of the total, reported farm sales of at least $100,000 (NASS 2019c). Sales of over $500,000 were reported by 4,700 Wisconsin farms in 2017, down from 4,830 five years earlier.

Considering just those farms with gross sales of $100,000 or more, the lack of a dynamic agricultural economy in much of northern Wisconsin is clearly shown. Only 8 farms in Vilas County sold $100,000 of farm products, as did 6 in Iron County, 13 in Forest County, and 3 in Florence County. None did in Menominee County. Even Milwaukee County, which is almost entirely urban, had 18 farms with sales of at least $100,000. In contrast, 19 counties each had at least 300 farms with sales of $100,000 or more. The largest number of these farms was in Grant County, which had 828. Other counties with 500 or more farms with such sales included Clark, Dane, Dodge, Lafayette, and Marathon (NASS 2019c).

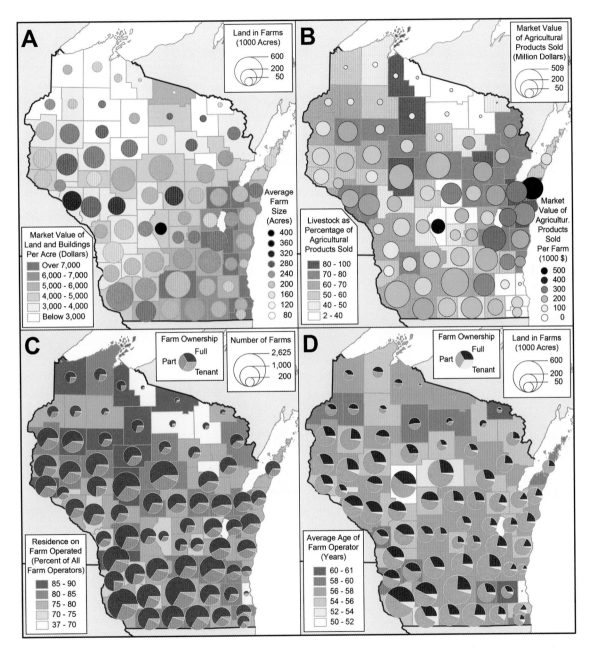

FIGURE 8.6. Characteristics of Wisconsin farms by county in 2017: Total land in farms and farm size by market value of holdings per acre; Market value of agricultural production and proportion of total sales coming from livestock and poultry products; Farm ownership (full, part, and tenant) by number of owners and residence on the farm; and Farm ownership (full, part, and tenant) by acreage in ownership category by average age of farm operator. Data source: NASS (2019c).

Just as farms vary considerably in sales, they also differ in acreage (Figure 8.6). The statewide average or mean farm size in 2017 was 221 acres, while the median was 90 acres, clearly illustrating how a few large farms can skew the statistics. There were 2,340 farms with 1,000 acres or more, while 5,923 farms covered less than 10 acres (NASS 2019c). The number of farms with at least 2,000 acres grew from 557 in 2007 to 721 in 2017 (NASS 2019b). Because of this growth in farm size, with the average size farm having grown from 194 to 221 acres over that 10-year period, many farmers—particularly those with the larger acreages—now lease or rent lands from other owners to supplement the acreage that they own outright. Thus, they are considered part owners. Over 70 percent of the land in farms within most counties in east-central and south-central Wisconsin is operated by part owners. Because the rented or leased land is often not contiguous to land owned by the farmer, and even when the farmer is able to purchase additional acreage it is likewise often not adjacent to the home farmstead, most of the larger farms in the southern three-quarters of Wisconsin are classified as fragmented farms.

With growth in farm size and usage of large tractors, including those that can plow and plant a dozen or more rows at a time (Figure 8.7), fewer farmers can now cultivate larger acreages. Farm consolidation reduces the number of farmhouses that are necessary, and the number of farms with

FIGURE 8.7. Farm tractor pulling unit that tills, plants seeds, and applies chemicals to the field. View from Fox River State Trail north of Greenleaf, Brown County.

livestock has steadily declined, rendering many barns unnecessary. Thus, we see far fewer barns in the countryside than were visible just several decades ago. It is not unusual to see an abandoned barn that has collapsed. On other farmsteads the barn remains, but it is now used to store equipment, while the farmhouse has fallen into ruin, as the farmer either lives in a farmhouse on another parcel of the fragmented farm or has become a sidewalk farmer, whose home is in a nearby village or city with a sidewalk out front.

Yet in those areas of Wisconsin with rapidly growing populations of Amish or Old Order Mennonite farmers, who together operate approximately 2,500 farms, we find the exact opposite. The Amish buy farms and often subdivide them into smaller farms of 80 or so acres, which can be tilled using horse-drawn plows. Thus, new farmhouses and barns are needed, and the rural farm population and population density around their communities grow. In some locales, they also influence the statistics describing farm characteristics. For example, other than those counties of northern Wisconsin in which over 70 percent of all farms are family or individual operated, as reported by the 2017 agricultural census (NASS 2019c), only in five other counties, which all have Amish settlements, is such a large share of farms family operated.

Thus, in exploring the patterns of farming in Wisconsin, we often see distinct differences among areas depending upon whether Amish or Old Order Mennonite farmers are locally prominent, other small farmers dominate (such as in northern Wisconsin), or large operations with substantial crop sales are prevalent. Both the numbers of small and large farm operations are increasing, while both the quantity of midsized farms and their total acreage have been declining. It is those midsized farms that have "been considered the heart of American (and Wisconsin) agriculture" (WASAL 2007, 138).

## CROP PRODUCTION SUPPORTS DAIRYING

The necessity of feeding well over a million dairy cows has strongly shaped crop production in Wisconsin (Figure 8.8). While crop combination patterns have changed since Weaver's research, which was described in chapter 7, much of what is grown in Wisconsin is either consumed on the farm where it is grown, or it is transported a short distance to the operator of a megadairy. Corn and hay of all varieties, including corn silage, haylage, and green chop, accounted for 71.5 percent of Wisconsin's total harvested cropland acreage in 2017 (NASS 2019c). If soybeans are included, which is reasonable considering that soy products and residues are often fed to dairy cows and that soybeans are grown in rotation with corn partly to maintain soil fertility, this amount jumps to 95.5 percent.

### Hay and Alfalfa

Hay of all forms was harvested from 2,372,478 acres in 2017, a greater acreage than any other crop save corn. Unlike other crops grown in Wisconsin, a farmer can often get four or five cuttings from a hayfield over the growing season, with the exact number depending upon the variety of hay being grown and the summer's weather conditions. In addition, hay may be cut and left in the field to dry before being baled into traditional sized bales that can be handled by hand, or compacted into large rectangular bales or hay rolls necessitating the use of a fork-lift. It may be simply gathered loose, by either traditional or modern methods (Figures 8.9 and 8.10). At times hay may be cut and harvested

FIGURE 8.8. Cornfield surrounding dairy farmstead northwest of Chippewa Falls. The silos are of a type promoted over a half century ago, while silage today is often stored in white plastic-covered rolls, visible on the ground to the left of the silos.

green and then stored as greenchop or haylage, similar to the way corn silage might be stored. Although silos were first introduced over a century ago to store silage, and a half century ago many farmers installed glass-lined Harveststore silos, today many farmers store their haylage and silage in large bunkers that are covered with plastic. Hay that is cut may be either tame or wild, and many farmers also grow various grains, such as rye, for forage or hay.

Most hay harvested in Wisconsin comes from the southern three-quarters of the state (Figure 8.11). Nevertheless, largely because the short growing season limits the growing of many other crops, the growing of hay accounts for a larger share of the total harvested cropland acreage in the northernmost two tiers of counties than in most of the rest of the state. For example, in 2017 hay accounted for 87.6 percent of harvested cropland in Douglas County, 95.1 percent in Florence County, and 64.4 percent in Forest County (NASS 2019c).

Alfalfa, or lucerne, as the crop is called in many parts of the world, is a high-yielding fodder crop with a large protein content. While much of Wisconsin's alfalfa is fed to dairy cows, it is also used to feed other livestock and poultry. Alfalfa was harvested from 1,042,016 acres in 2017, representing

FIGURE 8.9. Amish family operating a hay stacker with horse-drawn hay wagon in the Amish community of Augusta, Eau Claire County.

FIGURE 8.10. Freshly cut alfalfa, collected in the row in the foreground, is mechanically sprayed into a truck driven onto the field. Photograph taken in Town of Willard, Rusk County.

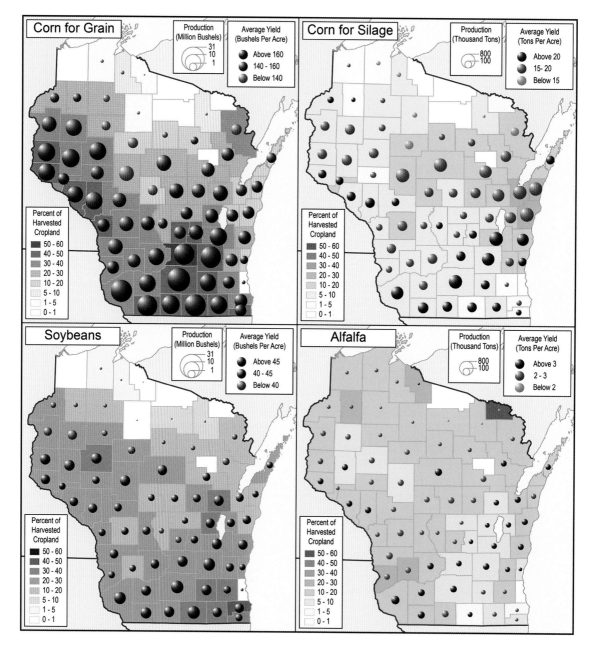

FIGURE 8.11. Harvested cropland by county devoted to Wisconsin's leading crops in 2017: corn for grain, corn for silage, soybeans, and alfalfa. Maps indicate percentage of harvested cropland allotted to these crops, production totals, and yields per acre. Data source: NASS (2019c).

11.3 percent of Wisconsin's harvested cropland. While statewide the acreage of alfalfa, cut and dried, equals 77.3 percent of the total hay acreage, it is considerably less in the northernmost counties. For example, alfalfa constitutes only 35.4 percent of the hay acreage in Ashland County and 43.9 percent in Iron County, with so little grown in two northern counties that nondisclosure rules prevented the 2017 census from reporting their alfalfa acreages. In contrast, alfalfa dominates the hay acreage in several counties of southwest and south-central Wisconsin. For example, alfalfa comprised 86.7 percent of the hay acreage in Grant County, 86.4 percent in Green County, 82.3 percent in Crawford County, and 81.9 percent in Rock County (NASS 2019c).

### Corn (Maize)

Corn (maize) is harvested both for grain and for silage. Corn being cut for silage is harvested when it is still green, and the entire corn stalks and cobs are chopped up for silage. Corn that is being grown for grain is allowed to dry in the field, and it is often cut in late autumn, and sometimes not until winter (Figure 8.12). Most corn harvested for silage is stored and fed locally to livestock, while corn harvested for grain has many uses. Some is used for animal feed, some is ground into cornmeal and used in a variety of human foods, including corn chips, breakfast foods, corn oil, and corn syrups. Over the past couple of decades increasing amounts of corn are being distilled into ethanol. Currently,

FIGURE 8.12. Combine harvesting corn for grain in field east of North Leeds in Columbia County.

ethanol production consumes 37 percent of Wisconsin's corn harvest. In addition, grain corn is a major export crop, with 10 percent being exported (WCGA 2018), before retaliatory tariffs disrupted sales, reducing the fall 2018 market price for corn ($3.66 per bushel in October 2018) to a level below the cost of production for most farmers.

Corn was harvested for grain from 3,074,502 acres in Wisconsin in 2017, being the state's most widely grown crop. Wisconsin ranked eleventh nationally in its corn harvest, down from eighth in 2012. Nevertheless, within eight counties of northern Wisconsin very little corn was grown, and what was grown comprised less than 5 percent of the harvested cropland. In contrast, within four counties, over 45 percent of the county's cropland grew corn for grain. Columbia County had the highest percentage with 52.3 percent of its harvested acreage growing corn for grain. When corn grown for silage or greenchop is added, the amount rose to 57.6 percent.

Substantial acreage is devoted to growing corn silage in a broad swath of Wisconsin extending from the southwest region, across south-central Wisconsin, through east-central Wisconsin. In addition, large acreages of corn silage are harvested in the northern part of central Wisconsin into west-central Wisconsin. Nationally, Wisconsin is the top-ranked state in its corn silage harvest (WASS 2017). The largest proportions of harvested acreage devoted to growing corn for silage in 2017 were in several counties of east-central Wisconsin: Brown, 24.5 percent; Kewaunee, 26.7 percent; and Manitowoc, 23.1 percent (NASS 2019c). As we will see in chapter 9, these counties are in the center of the state's largest concentration of megadairies. Wisconsin's farmers originally turned to producing corn silage because "an early frost could damage their corn before the grain had a chance to ripen, so they harvested most of the crop as silage" (Hart 1991, 163). Newer corn hybrids are less vulnerable, and the state's acreage of corn silage in 2017 was 12.7 percent below the amount grown in 1950. In contrast, the area devoted to corn for grain had grown by over 1.5 million acres, an increase of 103 percent (USBC 1952a; NASS 2019c).

## Soybeans

Soybeans were barely grown in Wisconsin in 1950. The state's growing season was simply too short. Two developments changed that situation. First, new hybrid varieties of soybeans were developed that needed fewer days between planting and harvest. Second, the growing season increased. Given global warming, between 1950 and 2006 the average growing season statewide increased by 12 days, with the "largest gains . . . in the northwest and central regions, where the growing season has been extended by two to three weeks" (WICCI 2011, 21). The consequences for soybean production in Wisconsin were tremendous. In 1949 Wisconsin harvested 273,702 bushels of soybeans for beans, while in 2017 the state harvested 101,917,737 bushels, nearly 375 times its earlier harvest (USBC 1952a, NASS 2019c). Soybeans were cultivated on 2,214,985 acres in 2017. Only hay and corn are grown on larger acreages in Wisconsin.

Soybeans are typically grown in rotation with corn in Wisconsin. Although the rotation enhances the yield of both crops, as the soybeans increase the fertility of the soil through the process of nitrogen fixation in their root zone, farmers also carefully monitor prices, or anticipated prices, in their decision of which crop to grow. Neither crop provides high net profits per acre. In some recent years, the price received per bushel of corn or soybeans resulted in a net loss, once the crop was sold. For farmers who also have a dairy herd, crop sales are less an issue than milk prices, yet increasingly Wisconsin's farmers are growing cash crops. Crop yields per acre vary annually depending upon weather conditions. During years of drought, such as in 1988 and 2012, yields can fall by 50 percent or more. They also differ depending upon type of soil, with some soils better suited for one crop than another.

Crop yields for 2017 and associated commodity prices for March 2018 illustrate some of the challenges farmers face when deciding what to grow. They often do not know the price they will receive until after their crop is harvested. Statewide, in 2017 the average yield per acre for corn harvested for grain was 174 bushels per acre, with yields ranging from a low of 82.6 bushels per acre in Ashland County to a high of 206 bushels per acre in Lafayette County. Within 12 counties, 5 of which were in south-central Wisconsin, production exceeded 180 bushels per acre (USDA-WFO 2018a). Given the low corn price, $3.25 per bushel—a price below the cost of production for almost all farmers, a farmer producing the statewide average harvest would have gross sales revenue of $565.50 per acre for growing corn. In contrast, the statewide average yield of soybeans was 47 bushels per acre, ranging from a low of 31.8 bushels per acre in Sawyer County to a high of 56.4 bushels per acre in Lafayette County (USDA-WFO 2018c). The price for soybeans was $9.35 per bushel. Thus, a farmer whose field yielded the statewide average soybean harvest would have gross sales revenue of $439.45 per acre. Yet by growing soybeans, the farmer assures a larger future corn harvest, and in some years soybeans bring higher sales revenue per acre than does corn. Given that 65 percent of Wisconsin's soybean crop is typically exported, its pricing is much more influenced by the export market and foreign producers than is the price of corn. As we saw in 2018 and 2019, it is highly vulnerable to retaliatory tariffs, such as those levied by China in response to American tariffs on Chinese goods. (The previous year China had purchased 60 percent of U.S. soybean exports, thus by October 2018 soybean prices had fallen to $8.70 per bushel.)

Soybeans are generally grown in the same regions of Wisconsin where corn is grown, although given its growing season needs, there is less growing of soybeans in the northernmost counties. In counties of northern Wisconsin where soybeans are grown, yields typically lag that of the more southern counties. Four northern counties grew no soybeans in 2012, while one had none in 2017. In contrast, 26 counties had at least 25 percent of their harvested cropland utilized for growing soybeans, including not only most of south-central and southeast Wisconsin but also 9 counties in west-central and northwest Wisconsin, along with three counties in east-central Wisconsin. The shares of harvested cropland used for soybeans in Racine, Waukesha, and Winnebago Counties were 44.1 percent, 38.4 percent, and 36.9 percent, respectively, all increases over their 2012 figures (NASS 2014a, 2019c).

## Other Cereal Crops

Other cereal or grain crops are grown in most counties, but their acreages are typically far lower than that devoted to corn and soybeans. Wheat, which was Wisconsin's most prominent crop a century and a half earlier, was grown on 200,613 acres in 2017, down from 261,519 acres in 2012. Most of this crop is winter wheat, planted in the fall and harvested in early summer, giving farmers time to reseed their fields to obtain a cutting of alfalfa or some other forage crop later that fall. The greatest acreages of wheat are in south-central and east-central Wisconsin, with over 10,000 acres harvested in 2017 within Dane, Dodge, Fond du Lac, and Manitowoc Counties (Figure 8.13). Yet in none of these counties did this constitute even a tenth of their total harvested cropland (NASS 2019c). Highest average yield per

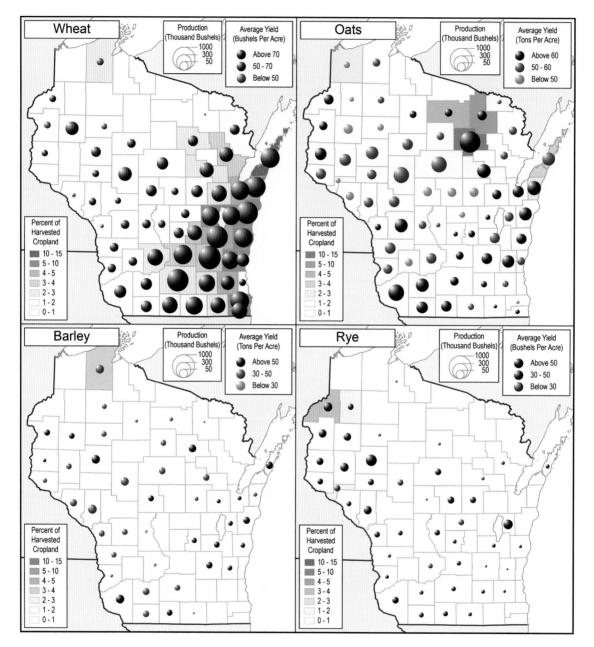

FIGURE 8.13. Harvested cropland by Wisconsin county devoted to production of small grains in 2017: wheat, oats, barley, and rye. Maps indicate percentage of harvested cropland allotted to these crops, production totals, and yields per acre. Data source: NASS (2019c).

acre in 2017 was in Dane County, 83.2 bushels, nearly twice the value for the lowest-yielding county reported (USDA-WFO 2018d).

Oats are grown on only half the acreage devoted to wheat, but they are much more evenly distributed across the state, excepting the northern tier of counties and southeast Wisconsin. Although 2,760,794 acres of oats were grown for grain in 1949, in 2017 oats were harvested from only 88,290 acres, down from 130,374 acres in 2012 (USBC 1952a; NASS 2014a, 2019c). Only two counties reported growing over 5,000 acres: Clark with 5,802 acres and Langlade with 9,232 acres. Crop yields in 2017 averaged 59 bushels per acre statewide, yet four counties reported harvests exceeding 70 bushels an acre (USDA-WFO 2018b).

Even smaller quantities of barley are grown, with 16,734 acres harvested in 2017. Only Clark County, with 1,165 acres, reported growing over 1,000 acres (NASS 2019c). Sunflower seeds were grown on 1,820 acres statewide, while no sugar beets were grown, down from 12,379 acres a century earlier.

## FACTORS DETERMINING WHAT CROPS ARE GROWN

Many factors determine what crops are grown in Wisconsin, as with any location. As we saw in our discussion of the cereal crops, all of which are field crops, the growing season strongly influences where certain crops are grown, and it is largely responsible for the northernmost two tiers of counties being largely non-agricultural. Many crops require certain types of soils for successful cultivation. For example, potatoes (which will be discussed shortly) grow best on sandy soils and are unsuitable for poorly drained or clay soils. Different crops have varying demands for labor. All of the cereal crops require little effort on the part of the farmer between the time of their planting and harvesting. While herbicides and pesticides may need to be applied to control weeds and insects, this may be done by a professional crop duster. Even harvesting is often done by custom harvesting crews, given that many farmers see no reason to make the major investment in a combine, which costs $400,000 to $600,000 if purchased new, that would be used just a few days a year at most. In contrast, custom crews move their equipment from south to north across the country, given the different dates of arrival of the harvest season, keeping their combines in use during much of the summer and fall.

### Labor and Processors

The availability of labor is a major issue for many growers of vegetables and fruits, as well as certain other specialty crops. Are there sufficient local residents willing and able to harvest the crop? Are migrant workers or immigrant laborers available? Are there facilities available locally to inexpensively house such workers? The shift of many orchard operators and berry growers to "u-pick" operations is simply one sign of a labor shortage, but many farmers find agritourism profitable. Other farmers have shifted to crops whose planting and harvesting is largely mechanized, such as field crops of grain.

The availability of processors for various crops is also critical. For example, Wisconsin grew no sugar beets in 2012 or 2017, yet it harvested 8,444 acres in 1949. During the 50-year period until 1920, Wisconsin had nine beet sugar factories, yet the last such factory closed in 1955 (Apps 2015). Both Michigan and Minnesota currently grow sugar beets, and both have sugar beet–processing facilities. Wisconsin has no such facility. Given the necessity of processing sugar beets shortly after their harvest and the distance of Wisconsin from sugar mills in south-central and western Minnesota and in the Saginaw area of Michigan, Wisconsin farmers cannot economically grow the crop. In contrast, farmers in central Wisconsin are well positioned with respect to canning and freezing plants that process their vegetables.

### Markets and Proximity to Urban Areas

Markets for various agricultural products are also critical. Some crops are highly perishable, and being grown close to market reduces transportation time and expense. Large urban places provide far more customers than rural regions where many residents may have space for their own gardens. Thus, many greenhouse and horticultural crops are grown near their customers, such as takes place in southeast Wisconsin. Ethnic heritage of the local population can also influence demand for particular crops, shown by the growing of cabbage in Kenosha and Racine Counties to meet the German ethnic demand for coleslaw and sauerkraut.

Urban proximity has both advantages and serious disadvantages. Nearness to customers assures that locally produced crops can be marketed fresh with minimum transportation time and expense. On the other hand, closeness to urban populations brings serious issues regarding nuisance and odor complaints, trespass, and crop pilferage. Movement of tractors and farm wagons on suburban streets slows traffic and leaves unwanted clods of dirt. Nearby residents object to chemical spraying of crops and manure spreading. The presence of large numbers of stinky farm animals can be even more objectionable, as is discussed in the next chapter.

Proximity to urban growth centers also creates a demand for land to expand tract housing and build industrial and commercial facilities, raising land prices. Such demand causes property taxes to rise, increases that are only somewhat mitigated by agricultural assessments. In his study of farming near suburbanizing areas by Appleton and Green Bay, Gerald Ottone (2006, 59) noted "evidence that proximity to residential districts posed problems for farm landowners; among the factors was 'impermanence syndrome' [where the] farmer, his property nearly surrounded in the 1990s by subdivisions, felt it unwise to make heavy investments on this property." Given these issues, agricultural retreat by many types of farmers occurs near large urban areas.

For many producers, it is a combination of several factors plus constraints of competition that determine their agricultural production. Can competitors elsewhere produce their crop and ship it to area supermarkets more cheaply than local producers? Can producers in certain areas of Wisconsin grow and ship their crops to other areas of the United States—or the world—more cheaply than growers in those locations? We begin by looking at vegetable production, for which access to processing facilities and market is critical, given the crops' perishability.

## SPECIALTY CROP PRODUCTION IN WISCONSIN

Specialty crops are legally defined under the Farm Bill of 2014 (Public Law 113–79) as "fruits and vegetables, tree nuts, dried fruits, horticulture, and nursery crops (including floriculture)." As a group these crops bring greater value per acre than field crops, yet they typically require considerably more labor in their growing and harvesting. We first look at potatoes, the most extensively grown vegetable crop in Wisconsin, which in 2017 averaged yielding $5,263.50 in sales per acre.

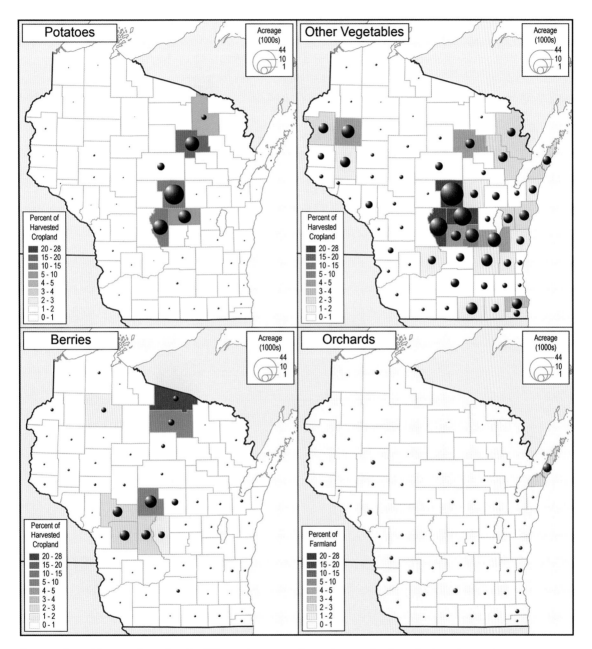

FIGURE 8.14. Harvested cropland by Wisconsin county devoted to production of selected crops in 2017: potatoes, other vegetables excluding potatoes, berries (mostly cranberries), and orchards (mostly apple and tart cherry). Maps indicate percentage of harvested cropland or farmland related to these crops and total acreage. Data source: NASS (2019c).

## Potatoes

Wisconsin is the nation's third-largest producer of potatoes, which are well suited to growing on the sandy soils that generally correspond to the outwash plains once covered by Glacial Lake Wisconsin (Figure 8.14). Given that the tubers grow underground and must be dug from the soil when harvested, sandy soils that neither misshape the potato nor hinder its harvest are preferred. In 2017, of the 70,110 acres of potatoes harvested in Wisconsin, 27,659 acres were in Portage County, which ranked first in Wisconsin and tenth in the nation among counties growing potatoes. Other major potato-producing counties included Adams, with 11,692 acres; Langlade, with 10,630 acres; and Waushara, with 7,212 acres (NASS 2019c). Major processing and warehouse facilities are located at Plover in Portage County.

The desirability of growing potatoes upon sandy soils necessitates that their growers install irrigation systems (Figure 8.15), given that such soils have limited water-holding capacities, making the fields vulnerable to even short dry periods. Statewide, only 4.9 percent of Wisconsin's harvested cropland is irrigated, yet the sandy soils of central Wisconsin's potato and vegetable producers are often irrigated. Portage County not only has the greatest acreage of potato fields in Wisconsin, but it

FIGURE 8.15. Sprinkler irrigation system misting potato field on sandy soils along the Tomorrow River State Trail east of Plover, Portage County.

also has the largest acreage of irrigated fields in the state, accounting for 21.2 percent of the total. Irrigated fields comprise 46.4 percent of its harvested cropland (NASS 2019c). Similarly, there are large areas of irrigated fields in Adams, Langlade, and Waushara Counties, also known for their potato production.

### Sweet Corn, Peas, and Snap Beans

Central Wisconsin is also the focus of much of the state's production of vegetables for canning and freezing. Because of the difference in harvest dates, with peas and snap beans (or green beans) being harvested considerably earlier in the summer than sweet corn, the same processing facilities can be utilized to preserve all of these crops. Thirty-one facilities can or freeze vegetables in Wisconsin, with Seneca Foods Corporation alone operating nine (see Figure 11.10). The geographic patterns of growing the three crops are quite similar.

Portage County grows more sweet corn than any other Wisconsin county, and it ranked fourth among the nation's counties in 2017. Its 18,257 acres represented 32.2 percent of the state's total acreage of sweet corn grown. Adams County, with 9,486 acres, was Wisconsin's second-largest grower of sweet corn, followed by Waushara, Green Lake, and Fond du Lac Counties (NASS 2019c).

Green peas were grown on 22,029 acres statewide in 2017, substantially less than the 37,162 acres grown in 2012. The leading county for growing green peas was Portage, followed closely by Dodge, Fond du Lac, Adams, and Waushara Counties. Five years earlier Door County was the second-largest producer, with 4,624 acres, which fell to 5 acres in 2017, as the bankruptcy of a nearby canning company reduced the market for this perishable product.

The statewide acreage from which green beans or snap beans were harvested in 2017 was 63,842, down about 7,500 acres since 2012. Wisconsin's leading grower was Portage County, which grew 16,794 acres of snap beans, followed by Adams County, which grew half as much, Waushara, and Barron Counties (NASS 2014a, 2019c). Besides utilizing the same processing facilities, these crops are often grown in rotation with sweet corn for the same reason corn and soybeans are rotated. The snap beans and peas are legumes known for their nitrogen fixation in the soil, enhancing the output of sweet corn that is subsequently planted on the field.

### Cabbage, Carrots, and Cucumbers

The total acreage devoted to cabbage, cucumbers, and carrots in Wisconsin in 2017 was nearly 15,000 acres, with cucumbers and pickles grown on nearly 5,903 acres. Racine and Outagamie Counties together accounted for 79.1 percent of Wisconsin's harvested acreage of cabbages in 2012, while disclosure rules hid their total for 2017 (Figure 8.16). Cucumbers, most of which are used for processing into pickles often on the grower's farm, are highly concentrated in Waushara County, where 14 farms harvested 2,980 acres, 50.5 percent of the state's total (NASS 2014a, 2019c).

### Tobacco and Hops

Several other specialty crops warrant discussion. A century ago Wisconsin was a prominent grower of tobacco, with 9,787 farms growing 41,445 acres in 1919. By 2017, the number of growers had fallen to

FIGURE 8.16. Harvesting cabbage in field southeast of Plainfield in Waushara County.

108, a decrease of 73 over the previous five years. Over that time the tobacco acreage fell from 810 to just 478 (NASS 2014a, 2019c). Hops are another crop that spectacularly declined in Wisconsin, yet it had just as explosively expanded production between 1860 and 1865, when New York growers experienced huge crop losses because of the hop louse (Apps 2015). In 1867 Wisconsin produced over 11 million pounds of hops, of which Sauk County was the largest producer, which reportedly accounted for one-fifth of the world's production. Yet with oversupply, the price dropped by 90 percent during 1867. At the time of the 1950 census, hops was not listed as a crop for Wisconsin. The 2012 census shows 22 farms that harvested 46 acres of hops, with 19 of those acres in Grant County. By 2017 33 farms harvested 107 acres statewide (Figure 8.17). While this acreage is miniscule compared with the 3,191 acres that were dedicated to growing mint for oil, the rapid growth in craft brewing in Wisconsin strongly supports an increasing interest in this crop.

### Ginseng

In 1949, 10 Wisconsin farmers grew a total of four acres of ginseng. This crop underwent a tremendous increase in production during the late 1990s, with hundreds of farmers developing shaded fields covered with wooden latticework or mesh supported by wooden poles, providing an environment

FIGURE 8.17. Rows of hops strung from wires between poles, with a sign advertising the product. Hop production is experiencing a resurgence in Wisconsin, as shown by this field north of Mather in Jackson County.

suitable for the crop's growth (Figure 8.18). Growing ginseng is time demanding, and "the seeds must remain dormant for eight to twenty months before they will germinate" (Hart 2001, 12). Prices were high, reaching well over $80 per pound of dry root, which encouraged even more farmers to plant the crop. Unfortunately, by the time the plant produced a viable crop, typically three to four years after planting, with fully grown plants with mature roots requiring five years (Chu 2001), the run-up in production flooded the market, causing a precipitous decline in both price and production.

In 1998 Wisconsin had 1,230 ginseng growers, of which 1,018 were in Marathon County. State-wide, there were 4,967 acres of ginseng, and 81 percent of the crop was exported to Hong Kong. Yet, Wisconsin production had already peaked in 1996, when 1,338 growers sold 2,229,901 pounds of dry root (WASS 1999). Sales fell from over $100 million annually to just over $20 million in 2005.

Even during its heyday most growers had relatively small fields, yet fields that generated far more revenue than most other crops. A geographer who studied the ginseng industry explains:

The land that is normally devoted to ginseng farming is very small compared to dairying or field crops. Since there is a high return on a small acreage, most ginseng farmers grow ginseng as a supplementary

FIGURE 8.18. Ginseng growing in a shade-covered field north of Marathon City in Marathon County.

income to other crops, milk cows, or livestock. The average size of ginseng gardens in Marathon County is only one to two acres; there are some exceptions, with some farmers devoting as many as 5 acres to ginseng farming. Another reason for small-scale plots is that once a field has been used to grow ginseng, it cannot be used again. (Chu 2001, 4)

Simply put, many ginseng aficionados believe that a field previously used to grow the crop produces a less potent product. Wisconsin-grown ginseng has the reputation of being the highest quality that is grown in North America, higher than that of its biggest competitor, British Columbia, keeping some Wisconsin growers active.

By 2012 the number of Wisconsin growers had fallen to 68, who harvested 267 acres scattered over eight counties (NASS 2014a). Five years later Wisconsin's harvested area had more than tripled to 962 acres, with 101 producers. Seventy-six farms grew ginseng in Marathon County in 2017, harvesting 780 acres, representing 74.3 percent of the nation's ginseng acreage (NASS 2019c). Marathon County maintains its claim as the "Ginseng Capital of the World." In 2017 the First International Ginseng Festival was held in Wausau, bringing nearly 500 attendees from abroad, including many Chinese.

While press reports at that time indicated renewed optimism among ginseng growers, with prices running between $65 and $85 per pound for a product widely recognized by buyers for its quality and used for medicinal purposes, particularly within East Asia, the Hong Kong and Chinese market for ginseng collapsed in 2018 after punitive tariffs were imposed in response to American tariffs on Chinese products (Elegant 2019).

## Hemp

After the 2017 census of agriculture was conducted, Wisconsin reauthorized the growing of hemp, which had been banned because it matures into marijuana. When hemp was grown in Wisconsin nearly three-quarters of a century ago, it was harvested for its fiber, often used to make rope, before its narcotic-rich flowers developed. Hemp is now sought for its cannabidiol or CBD oil, for which health and therapeutic uses are claimed (N. Johnson 2019). Given expectations that profits per acre will substantially exceed that from growing corn, 247 Wisconsin farmers gained permission to plant a total of 1,875 acres of hemp in 2018, and 1,405 growers were licensed in 2019, with estimates that over 10,000 acres would be planted (B. Adams 2019). It remains to be seen whether hemp production takes a trajectory similar to what has been seen with ginseng, but its growing is simply another indication that many Wisconsin farmers are continually changing what they grow.

## Fruit Orchards in Wisconsin

The Door Peninsula has long been known for its orchards of apple and cherry trees (Figure 8.19). Located between the Bay of Green Bay and Lake Michigan, whose waters delay spring blossoming and thus reduce the likelihood of a late-season frost destroying a season's crop, Door County had 11,861 acres of orchards in 1950, 35.2 percent of the state's total. Adjacent Kewaunee and Brown Counties had over 1,000 additional acres of orchards. Thus, the bulk of Door Peninsula's orchard acreage was in Door County, particularly the northern half of the county. In 1950 Door County had 171,642 apple trees and 940,403 cherry trees, almost all of which were tart cherries. It had 597 cherry growers. By 2012 Door County had 53 growers with 2,429 acres of the state's total 2,577 acres of tart cherry orchards. It produced only 1.8 million pounds in 2012, the consequence of a late-season frost and summer drought, with output between 2013 and 2016 ranging from 9.3 million to 13.6 million pounds, far less than the 20.6 million pounds harvested in 1949 (NASS 2014a; USBC 1952a; WASS 2017). Wisconsin's production continues to fall. The 2017 census found that 1,945 acres of the state's 1,982 acres of tart cherry orchards were on the orchards of 24 Door County growers (NASS 2019c).

Once reliant upon 10,000 migrant workers during harvest season, with even German prisoners of war utilized during World War II, Door County's cherry growers eventually turned to machines or u-pick operations. John Fraser Hart, America's preeminent agricultural geographer of the last half of the twentieth century, who maintained a summer cabin in Door County, eloquently described its cherry growing challenges. "The cherry business in Door County has been whipsawed by regular cycles of overproduction. Eager growers have planted more trees when prices have been good, but then, when the new trees have begun to bear, the fruit has glutted the market and depressed prices below the cost of producing it" (Hart 2008, 64–65). While a federal marketing order now controls the

FIGURE 8.19. Cherry trees covered with blossoms in orchard northwest of Baileys Harbor, Door County.

amount of cherries that can be marketed in a good year, with stored (often frozen) berries sold in years of poor harvests, changing styles of food consumption have also hurt demand. Americans now consume fewer fruit pies, yet more calories from other processed sweets.

Door County in 2012 had only 468 acres of apple orchards. As Hart (2008, 59) notes, "At one time, Door County was also an important apple-producing area, at least by Wisconsin standards, but most of the old apple orchards are now derelict." Statewide, apple orchards covered 5,520 acres in 2012, down nearly 900 acres from five years earlier (NASS 2014a), yet Door County still had a larger acreage of orchards than any other Wisconsin county. By 2017 Door County's apple acreage had fallen to 400, while that in Crawford County rose to 371 acres (NASS 2019c). The steep slopes overlooking the Kickapoo River Valley near Gays Mills in Crawford County have long been noted for their apples, and this area may soon overtake Door County's orchards. Trempealeau County now has the third-largest acreage of apple orchards.

## Cranberries

Cranberries have been commercially harvested in Wisconsin since the 1860s, when cranberry bogs in Waushara County north of Berlin were ditched and diked. Early producers also harvested wild cranberries, which grew extensively in marshes of central Wisconsin (Apps 2015). By the beginning of the twentieth century Wisconsin grew one-eighth of the nation's cranberry harvest on about 2,500 acres. Even then it was noted that there were "thousands of acres of marsh lands suited to cranberry-growing in Wisconsin" (Whitbeck 1913, 57).

Wisconsin had 21,514 acres of cranberry bogs in 2017, an increase of over 10 percent during the previous decade. As the world's leader in cranberry production, Wisconsin accounted for 48.9 percent of the nation's harvested cranberry acreage in 2017, but it yielded 5,372,000 100-pound barrels of cranberries, 64.2 percent of the nation's harvest (NASS 2019b; WASS 2018). Most cranberry growers in Wisconsin plant their perennial vines in bogs and marshes within the Central Sand Plains region. Just four counties, Jackson, Juneau, Monroe, and Wood, account for over three-quarters of Wisconsin's cranberry acreage. Wood County alone has 28.8 percent of the state's total.

Cranberry growing requires periodic flooding of the bogs, which are surrounded by earthen dikes (Figure 8.20). Planted upon acidic sandy soils, the crop must be irrigated during the growing season, and the fields are flooded during harvest season, allowing the lightweight berries to float to the surface, where they can be readily harvested by raking. The fields are also flooded during winter, "so that the vines freeze in a protective layer of ice . . . [that] shields the plants from cold winter temperatures

FIGURE 8.20. Aerial view of cranberry bogs showing the dikes separating the fields just south of Lake Dubay in the Town of Eau Pleine, northeast of Junction City in Portage County.

and winds that can quickly kill unprotected vines" (Berlin 2002, 27). The continued acquisition of fresh sand to spread on the cranberry beds and water for irrigation and flooding requires substantial acreages of land to support the area in actual production. It has also raised concerns about excessive losses of natural wetlands. As Cynthia Berlin (27–28) explains, "While cranberry growers in Wisconsin maintain about 120,000 acres of land, only about 15,000 of those are used directly for growing cranberries. The remaining acreage is reserved as "support land," which consist of interconnected natural and man-made wetlands and woodlands that provide growers with a resource to meet the high water demands of the cranberry crop. . . . [M]any areas of support lands are cris-crossed with extensive networks of ditches and dikes." Wisconsin law provides cranberry growers with extensive rights to ditch and drain wetlands to enhance their cranberry production (Berlin 2002). Total cranberry acreage has increased by nearly a third over the past fifteen years.

Cranberry producers in Wisconsin are highly involved with the Ocean-Spray Cooperative, which, given its name, indicates its origin serving growers primarily in Massachusetts, the nation's second-largest cranberry grower. Yet, its Wisconsin members, whose cranberry acreage is about 70 percent of the state's total, now dominate its fruit marketing. Over the past decade the share of Wisconsin's cranberry harvest that is exported has risen to 33 percent (WSCGA 2013), yet exports have recently been threatened by retaliatory tariffs. Major cranberry-processing facilities are located in Tomah, Babcock, and Warrens.

Within marshlands near where cranberries are grown in Jackson, Monroe, and Wood Counties, sphagnum moss is harvested. Widely used in the horticulture industry for plant seeding, plant baskets, mulching, and decorative soil covering, among other uses, the largest sphagnum moss harvester in North America has its plant and warehouses near Millston in Jackson County. After the naturally growing perennial is cut, three to seven years of regrowth are required before the next cutting can occur.

## Grapes and Wine Production

Wineries producing vintages from Wisconsin-grown grapes have increased dramatically in number over the past several decades, yet Wisconsin's early history of wine making is impressive. Agoston Haraszthy, a Hungarian immigrant, began growing wine grapes on slopes facing the Wisconsin River just east of Prairie du Sac in the 1840s. Discouraged by the response of European grapes to Wisconsin's cold winters and enticed by the Gold Rush, he traveled west and is credited with establishing California's wine industry (Apps 2015). A large stone winery building and farmhouse were constructed in the 1850s on the site of Haraszthy's operation by Peter Kehl, who together with his descendants operated the winery until 1899. After a hiatus of nearly three-quarters of a century, the Wollersheim Winery was established on the site of the historic winery, vines were planted, and wine making resumed, including using the original wine cave excavated by Haraszthy in the 1840s. The Wollersheim winery, the state's largest, has 30 acres of vineyards (Figure 8.21), obtains additional grapes from other growers, and also makes brandies from white grapes and apples.

Today the Wisconsin Grape Growers Association lists over 100 wineries in the state. While many Wisconsin wineries have long made fruit wines, use of locally grown grapes has recently expanded. The area of grapes grown statewide grew from 479 acres in 2007 to 917 in 2017, with much, but not

FIGURE 8.21. Grape vines within vineyard of Wollersheim Winery in Dane County's Town of Roxbury, immediately across the Wisconsin River from Prairie du Sac.

all, of the grapes grown destined for fermentation. Given the necessity of growing cold-resistant American and French American hybrid grapes because of the state's climate, the most common red varieties in Wisconsin are Marquette, Frontenac, and Marechal Foch, while the most common white cultivars are Frontenac Gris, Brianna, and La Crescent (Karakis, Cameron, and Kean 2016).

Wines are often described mentioning their terroir, or the character of the environment where they were grown. A variety of factors, including the geology of the rocks upon which the soil has formed, the depth of the soil, the slope and drainage of the land, and the microclimate of the vineyard, including the orientation of the slope to the sun—such as exposure to morning versus afternoon sun—all shape the local terroir. While wine character is widely acknowledged as being shaped by its terroir, terroir also influences a variety of other agricultural products, including cheese.

Wisconsin is loosely considered as having five wine regions, yet three areas of Wisconsin lie within designated American Viticultural Areas (AVAs), which share similar physical settings. These include the Upper Mississippi River Valley AVA, "the largest designated appellation in the world . . . [covering] southeastern Minnesota, southwest Wisconsin, northwest Illinois, and northeast Iowa" (Karakis, Cameron, and Kean. 2016, 267); the Lake Wisconsin AVA, which covers 50 square miles along the

Wisconsin River, including the Wollersheim Winery; and the Wisconsin Ledge AVA, which extends from the tip of the Door Peninsula to central Dodge County. It covers about 3,800 square miles lying between the Horicon Marsh, Lake Winnebago, and the Fox River on the west and Lake Michigan, including the Niagara Escarpment with its outcrops of dolomitic rock.

Given the increased national taste for wine and projected adverse consequences of global warming upon wineries located in hotter climates (G. Jones, Reid, and Vilks 2012), wine production is likely to expand within Wisconsin. Although extreme winter temperatures and climate variability will remain a challenge for Wisconsin grape growers, climate researchers note that the region of the United States suitable for the growing of premium wine grapes is likely to dramatically decline during the current century.

Climate change consequences are not limited to wine grapes. Growers of a wide variety of fruits and vegetables will need to adjust to a warming climate, which will be marked by unseasonable cold spells, as well as longer dry periods broken by more intense rainstorms. Not only will those changes directly influence plant growth, but they will also permit the spread of insects and plant diseases that were previously limited by Wisconsin's cold winters.

## ORGANIC AGRICULTURE IN WISCONSIN

Wisconsin ranks second nationally in its number of organic farmers, with only California having more organic producers. The 2017 Census of Agriculture counted 1,537 certified organic farms in Wisconsin. An additional 145 farms are listed as USDA National Organic Program organic producers exempt from certification. In total these farms generated $248.6 million in sales in 2017, slightly over twice their sales in 2012. While California has slightly over twice as many organic farms as Wisconsin, it has approximately five times the acreage of organic agricultural land and eleven times the sales (NASS 2017, 2019b).

Half of Wisconsin's organic agricultural sales revenue comes from cows' milk, yet this focus necessitates substantial acreage dedicated to growing organic fodder for those animals, as well as for milk goats and egg layers. Organic hay, of which alfalfa comprised nearly two-thirds of the total, was grown on 45,025 acres in 2016. Haylage was obtained from 22,458 acres of organic fields. Organic corn for grain was harvested from 27,855 acres, while corn silage was cut from 8,000 acres. Other organic cereal crops included 4,171 acres of barley, 6,410 acres of oats, and 3,810 acres of wheat. In addition, 7,102 acres of organic soybeans were grown (NASS 2017).

Organic fruits and vegetables are produced on far less acreage in Wisconsin, yet the gross sales of these products rival the total for all organic field crops. While 6,433 acres were devoted to growing organic vegetables in 2016, they yielded $24.5 million in sales. Organic berries, of which cranberries dominated the total acreage, were grown on 185 acres, yielding $3.3 million in sales (NASS 2017). Although the 2016 survey did not separately report the various organic vegetables, the 2014 organic survey did. Organic vegetables that comprised the greatest areas harvested included snap beans with 1,002 acres, green peas with 633 acres, and sweet corn with 901 acres (NASS 2015b).

Acreages of organic fields are not reported separately by county, unlike the number of organic farms and their total organic product sales. Sixty-eight of Wisconsin's counties had at least one organic farm in 2017, with 14 counties having over $5 million in sales of organic products, up from just 5 counties in 2012. Vernon County had the greatest sales, $25,524,000 followed by Grant County with $25,211,000. Sales in Lafayette, Marathon, Monroe, and Trempealeau Counties exceeded $10 million (NASS 2019c).

There is a strong correlation between those areas with the largest organic crop production and Amish settlements, even though many of the larger organic growers are not Amish. The largest cooperative of organic growers in Wisconsin, Organic Valley, which was founded in 1988 in the Driftless Area of western Wisconsin, became "the largest organic farmer cooperative in North America" (Eisen 2015). Although it has grown from that region to have members across the country, Amish and Mennonite farmers comprise 44 percent of its members. Amish farmers also played a role in its initial establishment, just as they have in the opening of several produce auction houses, such as ones near Fennimore in Grant County, between Kingston and Dalton in Green Lake County, south of Cashton in Vernon County, and near Loyal in Clark County, which facilitate the marketing of their fresh fruits and vegetables (Cross 2018).

Farmers' markets provide a venue for many organic farmers, plus other produce growers and livestock producers, to directly sell their products to customers. The U.S. Department of Agriculture's (USDA 2019) National Farmers Market Directory listed 312 such markets in Wisconsin in 2019, spread over 65 of the state's counties. Such sales directly to consumers accounted for 10 percent of total sales by organic farms in Wisconsin in 2014, while 85 percent was sold through wholesale markets. The remaining 5 percent was sold directly to retail shops and institutions (NASS 2015b). Not surprising, it was several counties of northern Wisconsin—Florence, Forest, Menominee, and Oconto—plus Juneau and Marquette Counties that lacked such farmers' markets. Direct marketing of organic produce to consumers, such as through farmers' markets, roadside farm stands, and community-supported agriculture operations, provided lower net returns to growers than did wholesale marketing, with farmers "dissatisfied with their profitability," while farmers selling in other markets "are significantly more likely to be dissatisfied with their quality of life" (Silva et al. 2014, 428). Both farmers' markets and community-supported agriculture are growing in the state.

In Wisconsin one finds involvement of not only the Amish but also the Hmong at many farmers' markets. Although some of the produce sold by the Hmong comes from community gardens and may not be reported in the census statistics, within some of Wisconsin's midsized cities Hmong vendors operate a quarter to a third of all of the produce tables, selling both vegetables long familiar to Wisconsin residents and items that are associated with Asian cuisine. The 2017 agricultural census showed 361 Wisconsin farms with Asian operators, clearly missing many Hmong gardeners who sell their vegetables at the local farmers' markets. That same census indicated that 591 Hispanic-operated farms had 128,549 acres, five times the 25,144 acres farmed by Asian producers (NASS 2019c). While these farmers comprise under one percent of the state's total, Hispanic farm workers are instrumental in harvesting many of the state's crops, as well as milking its cows, as described in the next chapter.

# Chapter Nine

# Agriculture in Wisconsin 2

## *Dairying and Livestock Production*

WISCONSIN HAS LONG BEEN KNOWN as America's Dairyland. Its foundations as dominant player in the nation's dairy industry date from the late nineteenth century. Pioneering research was conducted by agricultural scientists at the University of Wisconsin and promoters disseminated information and encouragement to the state's farmers (Lampard 1963). Agricultural historians have noted the importance of Wisconsin-based publications, such as *Hoard's Dairyman*, in promoting the industry and educating farmers regarding care, feeding, and housing of successful dairy herds. Agricultural research and education, together with the development of agricultural experiment stations and extension efforts consistent with promoting the Wisconsin Idea, brought the University of Wisconsin's expertise to the state's farmers, conveying new advancements. For example, the work of Professor Stephen Babcock led to the development of a way to readily determine butterfat content of milk in 1890, critical to facilitating sales of milk and cream to the state's cheese factories and creameries (Apps 2015). A century ago the majority of Wisconsin's farmers had at least several milk cows. The number of Wisconsin farms with milk cows peaked in 1935, when 180,695 farms with milk cows were counted, representing 90.4 percent of all of the state's farms. The size of the state's dairy herd reached its maximum a decade later, when 2,360,000 cows were counted (Cross 2001).

The 1950 Census of Agriculture reported that Wisconsin had 142,977 farms with milk cows, representing 84.8 percent of all farms (USBC 1952a). Wisconsin's 2,075,571 milk cows accounted for 9.8 percent of the nation's total. Not only did Wisconsin have far more milk cows than any other state, but its cows were more productive than the national average, giving 12.7 percent of the nation's milk. Although the average farm had 15 cows, they provided milk to 1,279 cheese factories in 1950. Wisconsin was the nation's largest manufacturer of cheese. The stereotypical image of numerous red barns scattered across Wisconsin's rural landscape harks back to this time (Figure 9.1), when the small herds could easily be accommodated in barns that dated to the nineteenth century and pasturing of herds outdoors was the norm every summer for many dairy farmers.

FIGURE 9.1. Dairy cows in barnyard beside traditional red barn east of Luxemburg in Kewaunee County. Note the addition of the milking parlor to the barn in this relatively small operation. Seeing cows outside the barn is fairly uncommon today.

## CHANGE RESHAPES WISCONSIN'S DAIRY INDUSTRY

Before we explore the contemporary pattern of dairy production in Wisconsin, it is essential to review a variety of factors that fundamentally reshaped both the nation's and Wisconsin's dairy industry over the past seven decades (Cross 2006). Wisconsin is no longer the nation's leading milk producer, its number of farms with milk cows has fallen by over 135,000, it has 800,000 fewer milk cows, and it has lost 1,150 of its cheese factories. What has happened and why did it happen?

Even with all of these declines, Wisconsin remains a major milk producer, second only to California. It remains the nation's largest cheese producer. It leads the nation in number of organic farms with milk cows. It produces much more milk today than it did in 1950, even with only 60 percent as many cows. The increase in milk production, at the same time the number of cows plummeted, is a testament to advances in technology, breeding, feeding, and animal care that have enabled productivity per cow to steadily increase. In 2017, the typical Wisconsin milk cow produced 23,725 pounds of milk annually (WASS 2018), nearly three times the average yield in 1950.

## Challenges of Dairy Farming

Dairy farming has long been known as an occupation that necessitates the farmer milking the cows at least twice a day (and typically three times daily for the past several decades), beginning chores well before sunrise and finishing long after sundown, seven days a week, with never any vacation. Indeed, John Fraser Hart (1991) entitled his highly informative chapter about dairy farming as "Twice a Day for the Rest of Your Life." This continuous hard work without relief, combined with uncertain earnings given highly fluctuating milk prices and input costs, together with having to increase the size of one's operation to obtain a viable income, discouraged many farm children from wanting to follow in their parents' footsteps. At the same time, family size decreased, with fewer children available to be interested or to even assist with chores on the farm.

Both agricultural census statistics and farmer surveys portrayed unsustainable conditions on many dairy farms. Average age of dairy farmers kept rising. In 1950 the average age of all Wisconsin farmers was 47.7 years; by 1974 it was 50.2 years, and by 1997 that age had reached 52.2 years. In that year 6,772 dairy farm operators were 55 years of age or older, representing 31.9 percent of all dairy farmers. One-eighth of all dairy farmers were at least 65 years old, and of these 2,663 farmers, 1,866 had fewer than 50 cows (NASS 1999). These farms were simply too small to provide a living wage to the modern farm family without a major financial investment to upgrade facilities and increase the herd size. Five years later the average age had risen to 53.0 years, yet the proportion of dairy farmers who were 55 or older had fallen to 28.2 percent (NASS 2004). Although over 10 percent of the operators were above 65 years of age, the increase in number of Amish dairymen, as discussed later in this chapter, had begun to skew the statistics downward.

In his prescient comments early this century, Hart wrote: "The undersized dairy farms of Wisconsin and Minnesota cannot provide an acceptable standard of living. A farm couple might be able to hang on, if the farm is bought and paid for, by living frugally on their capital, but their children will not come home to work on a 50-cow dairy farm. Even a 100-cow dairy farm is a last-generation farm, and it will go out of business when ma and pa stop milking" (Hart 2003, 97). We have seen this happen. Small farms are more likely to cease dairying than larger farms. In addition, many farms had not made investments in a variety of modern technologies and processes that enhance milk output. For example, in the 1990s under 60 percent of dairy farmers maintained individual milk production records for each cow in their herds—up from 20 percent three decades earlier. Only a quarter used total mixed ration machinery, and one-eighth used milking parlors, all practices known to enhance milk production (Jackson-Smith and Barham 2000). Strong correlations between herd size and the way cows are housed and their milk output were becoming apparent, and they remain so today.

## Differences in Dairy Operations

By 2010, 60 percent of all Wisconsin dairy farms had stanchion barns, in which the cows are restrained within a narrow stall for purposes of milking. Forty-five percent had freestall barns (Figure 9.2), in which cows have freedom of movement within a specific area and are taken to the milking parlor, typically three times a day. In addition, 17 percent used a variety of other types of "housing including tie stall barns and loose housing with bedded pack" (NASS-W 2010a, 2). The total exceeds 100 percent because some dairy farms had more than one type of housing. Small dairy herds, those with under 50 cows, are almost always housed in stanchion barns, which characterized almost all barns in the mid-twentieth century. Seventy percent of dairy herds with 100 to 199 cows are housed in freestall barns, and this barn type is used by over 97 percent of those with 200 or more cows (NASS-W 2010b). Freestall barns have many advantages. "Besides being more comfortable for the cows than the barns of old, freestall barns make the work of the farmer easier and more efficient. Manure drops through slots in the floor and flows into holding ponds . . . , eliminating much scraping, shoveling . . . [and] feed is efficiently distributed by machines" (Janus 2011, 86). In addition, within freestall barns cows in different "stages of lactation can be separated and treated according to their needs" (87). These benefits are enhanced as freestall barns are much larger than the older stanchion barns.

Many practices distinguish the larger and smaller dairy operations. Among those with fewer than 30 cows, 4 percent used sexed semen to artificially inseminate their cows. In contrast, among dairy operations with over 500 cows, 61 percent did. Among those with herds of under 30 cows, 77 percent grazed their cows, while no operators with 500 cows did so. Even among farms with 200 to 499 cows, only 9 percent grazed their cows (NASS-W 2010a). All of these differences contribute to smaller herds

FIGURE 9.2. Modern megadairy freestall barn with soybean field, located south of Forest Junction, Calumet County.

being considerably less productive than larger herds. In 2010 the average milk cow in herds of under 30 cows gave 15,900 pounds of milk annually, compared to an average yield of 28,600 pounds per cow in herds of at least 500 cows (NASS-W 2010a).

Given these differences in productivity, dairying in Wisconsin is much different today than what it was in 1950. Now there are far fewer dairy herds, yet today's herds are far larger and much more productive. Cows are much more likely to be kept within the barn rather than being seen grazing in a pasture, notwithstanding efforts to promote management-intensive rotational grazing for environmental reasons (Brock and Barham 2009). Many smaller to midsized dairy farms, particularly those producing organic milk, have adopted this grazing practice, in which cows graze a particular area for a day or two before being moved to another section of the field with fresh grass. Larger operators with industrial-sized dairy farms lack the pastures to do so, even if they were so inclined.

## Continuing Losses of Producers

The number of milk producers in Wisconsin continues to decline. When surveyed in 2010, 26 percent of the state's dairy producers anticipated discontinuing milking within the next five years (NASS-W 2010a). Over the five-year period from the August 2010 date of the survey, the number of dairy herds dropped from 12,668 to 9,848, a net decline of 22.3 percent. This includes both those farmers exiting dairying and newcomers entering dairying. The rate of decline accelerated at the end of the decade, as dairy farmers were beset by overproduction and low milk prices exacerbated by the loss of many export markets in retaliation for punitive tariffs imposed by the U.S. president. The state empaneled its second Dairy Task Force in 2018 (the first was in 1985) to explore the industry's woes and to recommend "actions to maintain a viable and profitable dairy industry in Wisconsin" (WDATCP 2019d, 10) given milk price volatility, regional inequalities of milk production and demand, and changing farm structure. During its year-long study Wisconsin lost 800 milk cow herds.

The results of the 2017 agricultural census, which showed Wisconsin had 9,037 farms with milk cows (NASS 2019c), were badly outdated when finally released in late April 2019. In April 2018 Wisconsin had 8,649 dairy cow herds (WASS 2018). A year later Wisconsin had 7,898 dairy cow herds, of which 7,070 produced Grade A milk (WASS 2019). During 2019, Wisconsin lost 818 dairy cow herds, an average of over 68 herds monthly.

### SPATIAL PATTERN OF DAIRY FARM LOSSES

The loss of dairy farms is not spread evenly across Wisconsin (Figure 9.3), and the greatest proportional losses are not necessarily located in those areas with the greatest numerical losses. Examination of dairy producer licenses (WDATCP 1989, 2009, 2019a, 2019b) show that of Wisconsin's 1,249 civil towns, 1,186 had at least one dairy herd in 1989 and 1,075 had at least one in 2009; yet by March 2019 that number had fallen to 1,004. By November 2019 it was 990.

## Towns without Dairy Farms

Towns that lost their dairy herds are mostly located across northern Wisconsin, clustered in a smaller area of central Wisconsin, and scattered near large cities in southeast Wisconsin, where many of the previous towns had themselves been annexed into expanding cities or had incorporated as a village or

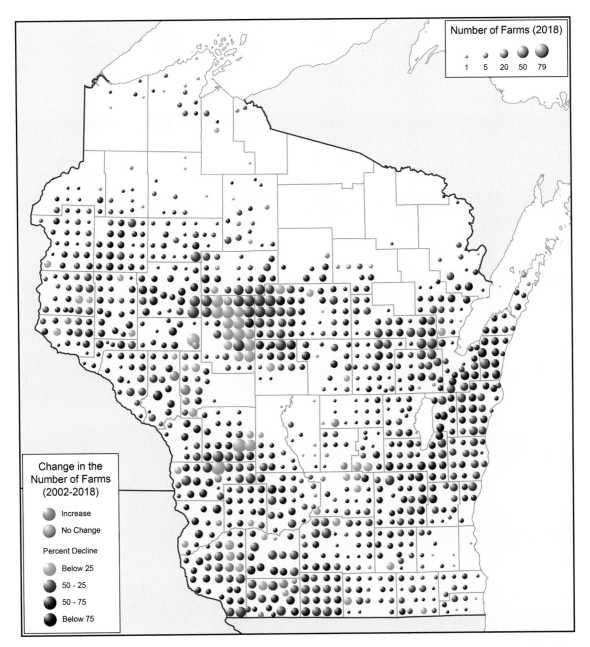

Number of Farms (2018)

1  5  20  50  79

Change in the
Number of Farms
(2002-2018)

Increase

No Change

Percent Decline

Below 25

50 - 25

50 - 75

Below 75

FIGURE 9.3. Change in number of dairy herds per town in Wisconsin between 2002 and 2018. Calculated by the authors from unpublished dairy producer license data obtained from Wisconsin Department of Agriculture, Trade, and Consumer Protection in 2002 and 2018.

city. In 1989 three large, yet disconnected, clusters of towns lacking any dairy farms were apparent across northern Wisconsin: one largely within southeast Douglas County and southwest Bayfield County; a larger cluster extending from southern Iron County into Forest County; and a smaller group including Menominee County and parts of Marinette and Oconto Counties. By 2009 a broad swath of towns, extending across northern Wisconsin from Minnesota to Michigan's Upper Peninsula and then stretching farther south from the Michigan border, lacked any dairying. Since then, that swath has become more prominent.

In 1989, 6 contiguous towns in central Wisconsin, in northern Juneau, Wood, and Jackson Counties, lacked any dairy herds. In addition, 5 other central Wisconsin towns, including 3 east of the Wisconsin River, had no dairy herds. By 2009 this zone lacking any dairy farms had expanded into a contiguous cluster of 31 towns that straddled the Wisconsin River. Motivating factors for the losses differed between northern and southeastern Wisconsin.

## Northern Wisconsin

Dairy farmers in northern Wisconsin were most likely to succumb (Figure 9.4), given the short growing season, distance to the nearest cheese factory or other purchaser of their milk, smaller choice

FIGURE 9.4. Former dairy barn with large bales of hay in field in Town of Aurora, Florence County. Florence County's last commercial dairy farm ceased shipping milk in March 2019.

of milk buyers and haulers, and, for some milk producers, no choice at all. In addition, this region has fewer businesses that support dairy farmers, whether selling implements or feed or providing veterinary services. The vulnerability of the region's dairy farmers was highlighted by the closing of Ashland County's only cheese factory in 1990, necessitating long-distance shipments to alternative milk processers, given the lack of any cheese factories or creameries in adjacent counties. Not surprisingly, Ashland County's towns of Agenda and Chippewa saw their numbers of dairy farms drop from 14 to 1 and from 25 to 4, respectively, between 1989 and 1999 (Cross 2001). By 2019, one dairy farm remained in the Town of Agenda, none within the Town of Chippewa. Another example of the exodus of dairy farming from northern Wisconsin is the Town of Dairyland, in the southwest corner of Douglas County. A large town, including four public land survey townships and occupying 144 square miles, it lost its last dairy herd by 2001, leaving it with a name that falsely hints at its agricultural engagement. By June 2019 no dairy herds remained in Florence, Forest, Menominee, and Oneida Counties.

## Southeast Wisconsin

The abandonment of dairy farms within the Milwaukee metropolitan area is well documented. Loyal Durand noted steep decline, beginning in the 1940s, that transformed Milwaukee County, which had once been "near the top of the leading dairy counties of the nation" (Durand 1962, 198), and subsequently spread into surrounding Waukesha and Ozaukee Counties. Today just one dairy herd remains in the county, within the city of Milwaukee, housed at the Milwaukee County Zoo. Waukesha County had 17 dairy herds in 2019. Kenosha County had 18 (WASS 2019).

A study undertaken in the mid-1990s investigated the decline of dairy farming within the seven counties constituting southeast Wisconsin, exploring "entry and exit behavior of dairy farmers at the civil town level" (Cross 1995, 76). At that time those farmers exiting greatly outnumbered those entering dairying within most of the region, except those towns "around the northern and western perimeter of the southeast agricultural reporting district. Many towns had no entering dairy farmers, and former dairy farms [were] becoming the sites of shopping centers and residential subdivisions" (80). Yet, not all exiting dairy farmers were leaving agriculture, nor were all of those leaving actually quitting dairying. Some of the lands of the former dairy farms remained in agriculture, growing crops. Other dairy farmers moved. Between 1989 and 1994 52 dairy operators relocated their herds. While some moves were to another town or county within southeast Wisconsin, 29 operators moved their herds to other parts of Wisconsin. Eight of the moves were to locations at least 100 miles distant (Cross 1995). In the 25 years since the study, southeast Wisconsin lost 72 percent of its dairy farms, slightly greater than the overall statewide loss.

Losses of dairy herds occurred throughout the rest of the state. The largest number of herds were lost within those areas that had the greatest number of herds, with a few exceptions, while the greatest proportional losses typically occurred in towns that had relatively small numbers of herds, where the loss of one or two herds would constitute a large percentage—or even all—of that locale's dairy herds. Yet, while dairy farms are disappearing from parts of Wisconsin's rural landscape, some areas are conspicuous because of their persistent numbers.

## Old Order Exceptions

Analysis of patterns of dairy losses between 1989 and 1999 discerned 22 towns that had at least 30 herds yet had lost fewer than 20 percent of their herds during the decade (and 6 had lost under 10 percent), while Wisconsin overall lost nearly 40 percent. At the time, it was noted that most of these towns, primarily located in Clark County, as well as in Taylor, Monroe, Vernon, and Sauk Counties, were areas with growing Old Order Amish populations (Cross 2001). The connection between the Amish and dairying in Grant County, where stability was also noticed, was not apparent then, but it soon became so when new settlements of Pennsylvania Amish were identified, given distinctive surnames among Amish from Pennsylvania (Cross 2003, 2004). By 2018 Wisconsin had 1,160 Amish dairy farms. Given that the Amish are religiously motivated to embrace an agrarian lifestyle—or at least one that is rural based—we cannot equate their goals and lifestyles with other small dairy farmers.

Contemporary dairy farming in Wisconsin involves several distinct types of operators. To illustrate Wisconsin's dairy farming, the spatial distribution of three groups of dairy farms are reviewed: (1) small to mid-sized dairy farms whose operators are not Plain People; (2) large dairy farms; and (3) dairy farms operated by the Amish and the Old Order Mennonites. While all of these dairy farms require markets for their milk, fundamental differences among their operators influence where their milk is sold and how their cows are fed and housed.

## CONTEMPORARY DISTRIBUTION OF DAIRY HERDS AND COWS

Dairy cows are present in approximately four-fifths of Wisconsin's towns, yet certain areas are distinctly favored. Furthermore, the spatial arrangements of dairy farms and milk cows are not uniformly similar (Figure 9.5), as the areas where the smallest dairy herds are concentrated differ from where we find most large operations. The distribution of dairy cows across Wisconsin can be likened to a donut—one that does not extend all the way to Lake Superior or Michigan's Upper Peninsula, yet whose hollow center nicely marks the absence of dairying in portions of central Wisconsin. While *Agricultural Atlases*, produced to accompany the reports of the last several agricultural censuses, display the distribution of dairy cows and dairy farms, they only map dairy farms with at least 200 cows (NASS 2014c, 2019a). This is unfortunate, for it completely ignores the state's Amish and Old Order Mennonite dairy farmers, who together operate nearly one-fifth of the state's dairy herds. It also ignores the majority of all of the other Wisconsin dairy farms.

Wisconsin had 8,757 dairy herds with fewer than 100 cows in 2012. An additional 1,584 had between 100 and 199 cows (NASS 2014a). The 2017 Census of Agriculture reported that 6,125 Wisconsin's dairy herds had fewer than 100 cows, and 1,529 had between 100 and 199. These operations comprised 84.7 percent of all the state's dairy farms (NASS 2019a), down from 89.6 percent in 2012. These herds included 616,129 cows in 2012, fully 48.5 percent of the state's entire dairy herd. Put another way, at that time Wisconsin had more dairy cows in its herds of under 200 cows than all other states, except California, had within all of their herds. Although the number within herds of fewer than 200 cows had fallen to 470,396 by 2017, it was 36.7 percent of Wisconsin's total dairy herd. It exceeded the total number of cows in herds of all sizes within all but five states (NASS 2014a, 2019c).

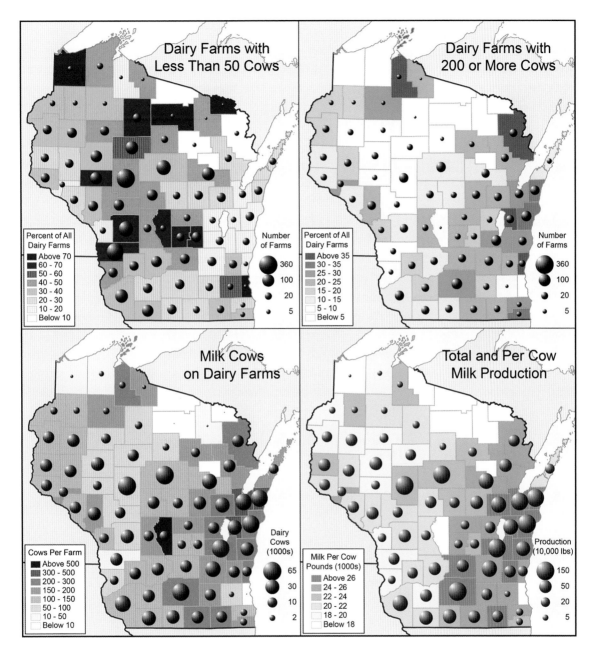

FIGURE 9.5. Distribution of dairy farms and dairy cows in Wisconsin in 2017. Top two maps indicate number of small (<50 cows) and large (200+ cows) dairy farms per county and the percentage of dairy farms in the county of these sizes. Bottom maps indicate the number of dairy cows per county, the mean number of cows per herd per county, and the total and average milk yield per county. Milk production data was not disclosed for several counties. Data source: NASS (2019c).

## SMALL TO MIDSIZED DAIRY FARMS (EXCEPTING THE OLD ORDERS)

Small dairy operations represent a mixed group. Some farmers for a variety of reasons have a token number of milk cows. The 2017 agricultural census (NASS 2019c) showed that Wisconsin had 874 farms with fewer than 10 milk cows (Table 9.1). They represent 9.7 percent of the state's farms with milk cows, averaging 2.8 cows each. Regardless of whether they are Amish or not, these operators cannot rely upon their milk cows to provide much of their income.

Wisconsin had 2,343 farms with 10 through 49 milk cows in 2017, representing 25.9 percent of the state's dairy herds. Altogether, these small dairy operations had 70,751 milk cows, 5.5 percent of the state's total and a little over half the number reported in 2012. Among herds of this size, some represent dairy farms whose owners are semiretired or will shortly exit dairying. Others represent new dairy farmers who are starting fresh, rather than taking over a sizeable operation from their parents. It is likely that nearly half of the dairy farms of this size represent Amish or Old Order Mennonite farmers, who dominate dairying in parts of Wisconsin.

Dairy farms with 50 through 199 cows accounted for 49.1 percent of all farms with milk cows in 2017. These operations were home to 397,177 milk cows, 31.0 percent of Wisconsin's total. The largest number of dairy farms of this size was in Clark County, with 422 herds, undoubtedly many of which are operated by that county's large Amish and Old Order Mennonite populations. Marathon County had 296 herds, not surprising given that it is also the state's largest county, followed by Grant County with 222 herds and Chippewa County with 148 herds of 50 to 199 cows. Nine other counties had at least 100 dairy farms of this size: Green, Lafayette, Dodge, Fond du Lac, Shawano, Dane, Vernon, Barron, and Iowa Counties (NASS 2019c). There are Amish dairy farms in all but two of these counties.

TABLE 9.1 Dairy farm characteristics 2007, 2012, and 2017

| Size of dairy herds | Number of Dairy Farms in Wisconsin by Herd Size | | |
| --- | --- | --- | --- |
| | 2007 | 2012 | 2017 |
| 1 to 9 cows | 598 | 586 | 874 |
| 10 to 19 cows | 736 | 712 | 495 |
| 20 to 49 cows | 4,502 | 3,278 | 1,848 |
| 50 to 99 cows | 5,567 | 4,181 | 2,908 |
| 100 to 199 cows | 1,685 | 1,584 | 1,529 |
| 200 to 499 cows | 798 | 815 | 930 |
| 500 to 999 cows | 194 | 256 | 281 |
| 1,000 to 2,499 cows | 71 | 106 | 133 |
| 2,500 or more cows | 7 | 25 | 39 |
| Total | 14,158 | 11,543 | 9,037 |

SOURCES: NASS 2009, 2014a, 2019c

However, with the possible exception of Clark County, the majority of dairy farms of this size in these counties are neither Amish nor Old Order Mennonite.

Among Wisconsin's small to midsized dairy farms, the vast majority grow most, if not all, of the feed for their herds. As noted in the last chapter, fields of alfalfa and corn, much of which is cut green for silage, are found on most dairy farms, with these farmers only purchasing feed during years of drought or other adverse weather conditions (Cross 1994). Soybeans, or products from processing the beans, are frequently included in the feed rations for dairy cows. Dairy farmers who wish to maximize milk output must carefully determine the appropriate mix of roughage and energy within their cows' diet. Roughage comes from grass, hay, alfalfa, and silage, while concentrated energy is provided by grain mixed into the diet. While the smaller dairy farms rely upon family labor, those farms with 100 or more cows typically have hired hands to help with the milking and other chores.

## MEGADAIRY DEVELOPMENT IN WISCONSIN

Megadairy development in Wisconsin considerably lagged its growth in California, yet it transformed Wisconsin's dairy industry in the past couple of decades (Cross 2011b). In describing these large operations, we need an appropriate definition of a megadairy (Figure 9.6). Some simply define it as being a "very large dairy farm," while others view megadairies as having at least 1,000 animals. Yet, the Census of Agriculture only reports the number of 1,000-cow dairy farms at the state level. The largest category reported at the county level has 500 or more cows. Megadairies may also be linked with concentrated animal feeding operations (CAFOs), which are defined as having at least 1,000 animal units. Yet animal units are not equivalent to that number of animals. If one is raising chickens, roughly 100,000 chickens are equivalent to 1,000 animal units—with that number of chickens being greater if they are broilers and smaller if they are layers. In contrast, 700 dairy cows comprise 1,000 animal units.

### Definitions of Megadairies

Because greater environmental permitting and regulations apply to dairy farms with 1,000 animal units, given the large quantity of manure and waste that is produced, it makes sense to consider any dairy CAFO as being a megadairy. In 2013 the Wisconsin Department of Natural Resources indicated there were 212 CAFO permits for dairy operations, a number that had grown to 289 in early 2019 (WDNR-RM 2019). Several of these operations are divided into more than one herd, so over half to two-thirds of dairy farms with at least 500 cows should be considered a megadairy, exceeding 1,000 animal units. Given that the largest size category reported at the county level by the Census of Agriculture includes at least 500 cows, this is what is used as equivalent to a megadairy in this book. In 2012 the Census of Agriculture reported that Wisconsin had 387 dairy herds of at least 500 cows, a number that had grown to 453 in 2017 (NASS 2014a, 2019c). Of these dairy farms, 106 had herds of 1,000 to 2,500 cows, with 25 having more than 2,500 cows.

Although not as common in Wisconsin, it has been proposed that the word "kilofarm" be applied to farms based upon their acreage (Hart and Lindberg 2014). In 2017 Wisconsin had 2,302 farms with at least 1,000 acres. Even if the operator of a megadairy has a smaller acreage, the feed for its

FIGURE 9.6. Aerial view of modern megadairy operation north-northeast of Sherwood in Calumet County. Two semitanker milk trucks are parked alongside the large freestall barn. Two bunkers of silage are visible, as are separate buildings to accommodate some of the cows.

herd will undoubtedly require the harvest from a kilofarm acreage. Indeed, many of the largest mega-dairy farms have longstanding arrangements with nearby farmers to grow the desired alfalfa, clover, and corn. Given the appetite of their cows, many megadairies also seek food-processing waste to be fed to their cows. The lead author has seen trucks from a megadairy farm lined up outside of a vegetable cannery and being loaded with the waste pods from green peas that were being processed. Husks from sweet corn canning and freezing plants can be fed to the cows. Distillers grain and sometimes other coproducts that remain from distilling corn into ethanol are widely used to feed dairy cattle and may account for up to 30 percent of the feed ration for dairy cows (RFA 2014). Distillers grain has three times the protein and fat content of unprocessed corn.

Besides inputs from the fields of nearby farmers, megadairies also need agreements with those farmers to take their manure, to be spread on the fields, fertilizing the soil and disposing of the waste. Unfortunately, as several megadairies and those nearby farmers whose fields received the manure have discovered, the waste is not eliminated (Barrett and Bergquist 2020). Extensive ground water contamination has ensued in parts of Kewaunee County and nearby counties (Muldoon et al. 2018).

Yet, while megadairies produce prodigious amounts of manure, economies of scale have enabled some of them to invest in biodigesters to convert the waste into methane, used to produce electricity and to heat farm buildings. Construction was pending approval in late 2019 of a $66 million digester, the state's largest to date, in Brown County's Town of Wrightstown to serve seven dairies (Kaeding 2019).

Another way of looking at large farms considers the market value of their land and buildings. In 2017, 849 farms in Wisconsin had estimated values upward of $10 million, with another 1,742 farms being valued between $5 million and $10 million (NASS 2019c). Megadairies require huge financial investments in modern freestall barns, the largest of which may be over 1,000 feet long, and milk houses. Freestall barns almost universally characterize dairy farms of 500 cows. The cost to construct a freestall barn was estimated between $2,000 and $3,000 per cow in 2015, to which electrification and plumbing expenses must be added. Thus, a freestall barn to accommodate 1,000 cows could cost $3 million. Additional expenses would be incurred to prepare the site, to build the milk house and install the milking system, and to construct facilities for stockpiling feed and providing for manure storage and handling (Kammel 2015). Given these expenses, it is easy to see that most Wisconsin farms with a market value above $10 million are focused upon producing milk. And they produce it on a grand scale. Whereas small to midsized dairy farms are visited once a day by a milk truck that may stop at a dozen farms before taking its load to the processing plant, alongside the milk house on a megadairy one sees parked one or more semitruck tankers, into which milk is directly pumped without first being stored in tanks at the milk house. A herd of 800 milk cows will fill a large semi-truck tanker every day. Wisconsin's largest megadairies can fill a dozen such trucks every day.

## Megadairy Employees

Regardless of how precisely we define a megadairy, their numbers of cows are far too large for any family—even one with a dozen children—to provide all of the necessary care. The farm will require hired labor. Few farms with under 100 cows hire any nonfamily labor. Among those with at least 200 cows, 90 percent hire such labor, as do all of those with over 500 cows. Increasingly that labor is Hispanic, with the larger the dairy farm, the greater the likelihood that Hispanics are hired. For example, among dairy herds with 301 to 600 cows, typically 3 U.S.-born workers and 4 immigrant employees are hired to work full-time, while on dairy farms with over 1,200 cows, an average of 6 native-born employees and 18 immigrants are working full-time (Harrison, Lloyd, and O'Kane 2009b). The largest megadairies employ far more workers. For example, Holsum Dairies in Calumet County has about 75 full-time employees to care for 6,800 cows (Veldman 2018). Large operations can afford individuals with specialized skills, such as one or more veterinarians, as full-time employees.

A 2008 survey found that over 40 percent of all dairy employees in Wisconsin were Hispanic. Of these, 88.5 percent were Mexican (Harrison, Lloyd, and O'Kane 2009b, 2). Although many vegetable farms have used Hispanic migrant workers seasonally since the 1930s, immigrant labor on dairy farms largely dates from 2000. Thus, one increasingly sees stores and churches in many rural Wisconsin communities with signage in Spanish. As survey results indicated, Wisconsin's immigrant dairy workers "often live on or very near the farms where they work, and these farms are often geographically isolated from towns and cities" (Harrison, Lloyd, and O'Kane 2009a, 1). Fear by those immigrants who lack legal status, and by others who encounter ethnic bias even though they are U.S. citizens, reduces the participation by many Hispanics in the activities of nearby rural communities. Furthermore, the threat of deportation further restricts the engagement of these workers off the farms and exacerbates the labor shortage facing dairy farm owners (Hall 2017).

### Megadairies Concentrated in East-Central Wisconsin

Megadairies are increasingly concentrated within east-central Wisconsin and to a somewhat lesser extent in south-central Wisconsin, although each of the state's nine agricultural reporting regions has at least several. The 2017 agricultural census indicated that east-central Wisconsin, a nine-county region, included 30.7 percent of the state's 500-plus cow herds, but 38.0 percent of the state's cows in herds of such large size. Four counties in east-central Wisconsin had gained at least ten herds of 500 or more cows between 2002 and 2017, including Brown, Fond du Lac, Kewaunee, and Manitowoc Counties, with Manitowoc gaining the most, 17 (NASS 2019c). Only Dane County equaled the expansion in Manitowoc County, and only one other county—Grant—outside the east-central region had gained 10 500-plus cow herds.

Because of environmental permitting requirements, dairy CAFOs must disclose both current and planned herd sizes. Thus, the locations of the largest megadairies can be mapped with greater precision than would be possible using just census statistics (Figure 9.7). In 2009 the Town of Wrightstown in Brown County, south of Green Bay, had five CAFO permitted dairy farms, the most megadairy operations of any town in Wisconsin. Ten dairy farms in east-central Wisconsin had at least 2,000 dairy cows at that time, and several additional megadairies had received Wisconsin Pollutant Discharge Elimination System permits to expand their operations to that level (Cross 2011b).

By early 2019 Wisconsin Department of Natural Resources (WDNR-RM 2019) and Wisconsin Department of Agriculture, Trade and Consumer Protection (WDATCP 2019a) records indicated that 100 dairy CAFO permits had been issued to holders of dairy producer licenses within the nine counties of east-central Wisconsin (Figure 9.8). These permittees had 114 dairy herds, and 39 of these herds had over 2,000 milk cows. Manitowoc County, with 22, had the most dairy CAFO permittees with a licensed dairy herd, but only 4 of its herds exceeded 2,000 cows. Kewaunee County had 10 herds of at least 2,000 cows, the most in the state. Its largest megadairy, Kinnard Farms, Inc., had 10,029 animal units, the equivalent of 7,020 milk cows, split between two herds. Brown County had 16 CAFO dairy permittees with milk producer licenses, with 6 having herds of over 2,000 cows. That county's largest megadairy, Wiese Brothers Farms, located near Greenleaf, had 10,463 animal units, the equivalent of 7,324 milk cows and heifers.

Nine CAFO dairy permittees in Calumet County totaled nearly 25,000 cows. Holsum Dairies LLC has the most animal units (13,865) of any CAFO dairy in Wisconsin, including both dairy cows and dairy goats, split between its herds in the towns of Chilton and Rantoul. Its Drumlin Dairy has nearly 8,000 goats, while the nearby Chilton Dairy has the "capacity to milk between 6,500 and 9,000 goats" (Cushman 2019a). The state's largest megadairy herd remains the Rosendale Dairy, located northwest of the village of that name in Fond du Lac County. It had 8,400 cows, housed in two contiguous barns that extend a distance of 1,200 feet. In total, Fond du Lac County had 15 dairy CAFO permittees. Immediately to the east, Sheboygan County had 10.

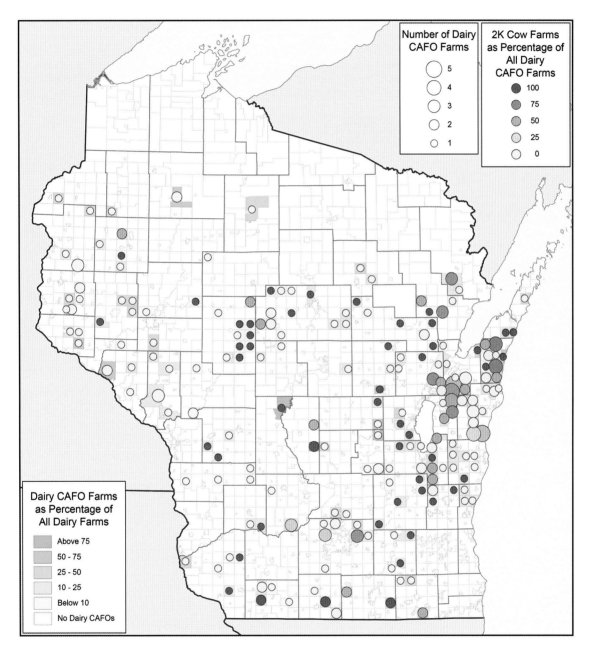

FIGURE 9.7. CAFO or megadairy herds per town in Wisconsin in 2019, with map showing the percentage of all dairy herds in the town that are CAFOs and the proportion of the dairy CAFOs that had at least 2,000 cows. Calculated by the authors from dairy producer license and CAFO data (WDATCP 2019a; WDNR-RM 2019).

FIGURE 9.8. Mound of silage covered with plastic by modern freestall barn. Note the array of ventilation fans at the gable end of the barn, located east of New Franken, Brown County.

East-central Wisconsin contained 34.6 percent of the state's 289 CAFOs with licensed dairy herds in March 2019. This region had 40.6 percent of Wisconsin's dairy herds with over 2,000 cows, yet several very large dairy farms are located in adjacent counties. In Shawano County, Matsche Farm Inc. had 10,966 animal units, equivalent to 7,676 cows, in early 2020, while to the north in Marinette County, B&D Dairy Farm near Pound had 6,960 cows in 2019. Oconto County had a CAFO with 4,069 cows in early 2020 (WDNR-RM 2020). Shawano and Oconto Counties had seven and six dairy CAFOs, respectively, in 2019.

## Megadairies Expand into Central Wisconsin

Large dairy CAFOs are also found in both south-central and central Wisconsin. For example, both Clark and Dane Counties had 10 CAFOs with dairy cows in 2019. Marathon County had 12. Its largest, Van Der Geest Dairy Cattle Inc., had 5,570 animal units, equivalent to 3,899 cows, in January 2017 and was projected to reach 6,248 cows by 2020. A far larger CAFO has opened in Adams County, where Milk Source, owner of Fond du Lac County's Rosendale Dairy, operates its similarly sized and configured New Chester Dairy. Besides this 8,400-cow operation and one raising heifers, together comprising

13,276 animal units in 2019, a second large dairy CAFO in Adams County to have nearly 4,500 cows was permitted and then repermitted in February 2017, although its application to operate high-capacity water wells met strong local and environmental resistance. Across the Wisconsin River in Juneau County, Central Sands Dairy near Nekoosa had the equivalent of 4,315 dairy cows when inventoried in 2019 (WDNR-RM 2019).

## Megadairies Increasingly Dominate Milk Production

Given the number of CAFO dairy herds and their sizes, over half of the milk cows in 14 counties are in CAFOs. Within east-central Wisconsin, Kewaunee County CAFOs had 43,565 milk cows in 2019, or 82 percent of the total milk cows in the county. In contrast, just two years earlier the 2017 census showed that 73.3 percent of Kewaunee County's cows were in herds of 500-plus cows. In Brown County 78 percent of its cows are in CAFOs, while in Calumet County it is about 80 percent. Establishment of large dairy CAFOs elsewhere has also resulted in their dominance. For example, Marinette County had only four CAFOs in 2019, but two of them had over 2,000 milk cows. Its largest one has 9,943 animal units, the equivalent of 6,960 cows, so its CAFOs are home to over 80 percent of the county's dairy cows.

Even more dramatic has been the impact upon Adams County. The 2012 Census of Agriculture counted only 617 milk cows, and their numbers were not disclosed five years later. Yet by 2019 the county had three dairy CAFOs, one of which operated two herds, even though there were only 13 dairy herds in the county. Adams County's largest CAFO had 9,293 cows and heifers (WDNR-RM 2019). Thus, dairy CAFOs are not just found in areas with many other dairy farms, but sometimes account for a large, if not total, share of a town's dairy herds. Clearly, CAFOs with their multiple large freestall barns have begun to reshape Wisconsin's dairy landscape.

## AMISH AND OLD ORDER MENNONITE DAIRY FARMS

In the mid-twentieth century, the presence of the Amish in Wisconsin was negligible, with only 2 small settlements existing in 1960. By 2019, Wisconsin had 56 settlements, which together had a population estimated at 22,020 (Young Center 2019). Only three other states have more Amish residents than Wisconsin. The expansion of these Plain People, an Anabaptist population whose religious practices and lifestyle are intricately linked to agricultural pursuits or, if not farming, to a rural-based activity such as woodworking that is focused around a farmstead where horses are tethered and kitchen gardens are planted, has fundamentally altered the dairy farming landscape that one sees in many areas of Wisconsin (Cross 2016).

## Reasons for Amish Expansion

Rapid population growth within long-established Amish settlements in other states led to escalating property values, forcing their residents to take non-agricultural jobs, to farm ever-smaller tracts of land, or to sell their farm and buy a larger farm elsewhere where lands were cheaper. Given the dramatic exodus of many dairy farmers in Wisconsin, lands were readily available at prices the Amish considered reasonable (Cross 2004, 2007). Because the Amish eschew the use of tractors in their fields, instead using horses to pull their plows, harrows, wagons, and buggies, the 160-acre farm is typically too large for an Amish farmer. Indeed, the size of Amish dairy farms in southwestern Wisconsin averages 81.4 acres, roughly one-quarter the size of farms of other operators (Brock, Barham, and Foltz 2006, 3). Thus, the sale of a farm elsewhere often allowed the Amish newcomer not only to purchase a Wisconsin farm but to subdivide it, providing lands for several Amish farmsteads. Although many Amish farms resemble the general farm of the past, raising crops and several varieties of livestock, dairying was—and remains—an important activity for the typical Amish farmer in Wisconsin (Figure 9.9). Unlike the larger Amish settlements in Pennsylvania, Ohio, and Indiana, where the majority of workers now work in non-agricultural pursuits, over half of Wisconsin's Amish families remain in farming (Cross 2018).

The growing prominence of the Amish in Wisconsin's dairy farming occurred simultaneously with the dramatic decline in the total number of dairy farms in the state. In 1989, Wisconsin had 35,600 dairy farms, of which the Amish ran no more than 475. By 2002, Wisconsin's total number of dairy farms had fallen to 17,594, with 865 having Amish operators (Cross 2004, WASS 2002). Ten years later, there were at least 1,026 Amish dairy farms in Wisconsin, while the total number of dairy

FIGURE 9.9. Dairy cows in pasture with farm buildings, including a windmill, in background. Relatively few dairy farmers graze their cows today, other than the Amish. Photograph in Town of Jefferson, Monroe County, part of the Cashton Amish settlement.

herds had fallen to 11,637 (Cross 2014; WASS 2012). Over the decade, "even though a smaller proportion of Amish families had dairy herds . . . the proportion of Wisconsin's dairy farms that had Amish operators rose from five to nine percent between 2002 and 2012, a result of increased numbers of Amish farmers and diminished numbers of other operators" (Cross 2014, 52). By 2018, Wisconsin had 1,160 Amish dairy herds, including those of cows, goats, and sheep. Wisconsin's total number of dairy herds, of all three types, was 9,004. Thus, 12.9 percent of the state's total dairy herds resided on Amish farms. Of those dairy farms producing cows' milk, 11.1 percent were Amish operated, while the Amish had 201 dairy goat herds, 56.5 percent of the state's total (Cross 2021a). Including all operators, Wisconsin had 12.4 percent of the nation's milk goats, more than any other state (WASS 2018).

Wisconsin also has about 460 dairy farms run by Old Order Mennonites, another Anabaptist Plain People (Cross 2021b). Those of the Groffdale Conference, Wisconsin's largest, are referred to as the "Horse and Buggy Mennonites," who travel in horse-drawn buggies like the Amish, yet use tractors with steel wheels in their fields. They, along with the Weaverland Conference and Meadow Springs Conference Old Order Mennonites who use automobiles, primarily reside in northern Clark County, overlapping into western Marathon and Chippewa Counties, and in Grant County.

## Amish Dairy Farm Landscapes

The Amish are locally conspicuous on the dairy farm landscape. In 2018 Amish dairy herds were found within one-sixth of those Wisconsin towns that had at least one dairy farm. Within 12 towns, the Amish operated over 75 percent of all of the dairy herds, while their herds accounted for over 50 percent of the town's herds in another 30 towns in 2018 (Figure 9.10). Within 51 Wisconsin towns the Amish ran at least half of the dairy farms. They had between 25.0 and 49.9 percent of the dairy herds within an additional 59 towns. The largest clusters of towns where the Amish dominate dairying are from southwest Wisconsin through west-central Wisconsin into the southern part of north-central Wisconsin. This includes those Amish dairymen in Grant and Lafayette Counties who are largely producers of Grade A milk, as well as those around the Vernon and Monroe County settlements of Cashton, Hillsboro, and Wilton who produce Grade B can milk.

## Amish Milk Issues

Amish dairy farmers often encounter obstacles in marketing their milk. While all Amish use horse-drawn equipment on their farms, their various settlements are governed by differing codes of conduct, called the *Ordnung*. For the dairy farmer, they govern the type of mechanization that is appropriate for use in the barn to milk the cows, to cool the milk, and to store and transport the milk (Cross 2004, 2014, 2018). Among less conservative Amish, one will see electric generators to run refrigeration units that cool and agitate the milk and the use of milking machines. Solar-powered electric fences may be seen. These individuals often meet the requirements necessary to produce Grade A milk that is suitable for fluid consumption. Other Amish settlements eschew the use of generators or even gas-powered refrigeration units and rely upon well water or blocks of ice cut during the winter to cool their milk that is stored in ten gallon cans. These settlements produce Grade B can milk, which is only suitable for manufacturing purposes. Grade B can milk is transported in its cans to the cheese

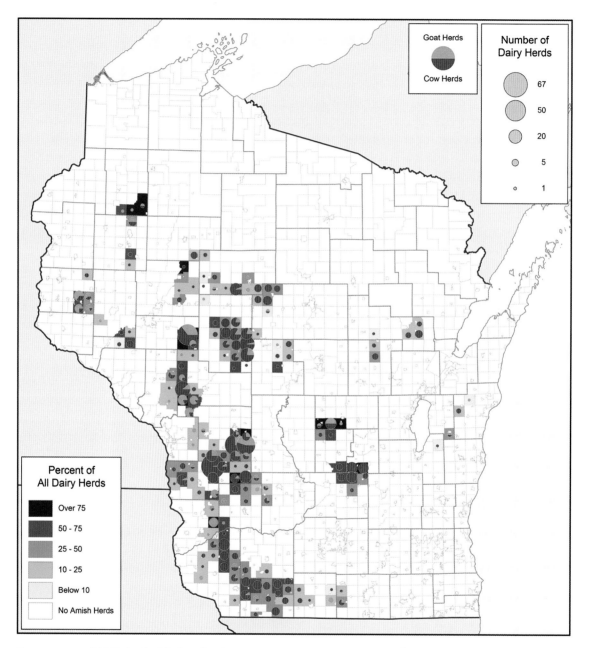

FIGURE 9.10. Old Order Amish dairy herds per town in Wisconsin in 2019, with map showing the share of these herds that produced cow's milk and goat's milk and the percentage of all dairy herds in the towns that were on Amish farms. Calculated by the authors from dairy producer license data (WDATCP 2019a).

factory, and the Amish accounted for nearly all of the state's 587 Grade B can cow's milk operators in April 2018 (WASS 2018). One-third of Grade B dairy farms use bulk tanks, rather than cans, yet are also subject to usage restrictions.

Because various Amish settlements produce different grades of milk, those that rely upon can milk face the greatest obstacles in marketing their milk. Most large dairy processors will not accept can milk. Producers of Grade A milk receive a greater price for their milk, and because their communities typically use milking machines, their herds are larger than those in settlements producing can milk. To provide a market for their can milk, several Amish settlements have either built or acquired their own cheese factory or have established a relationship with a local milk buyer. The two largest Amish settlements in Wisconsin have their own cheese factories and only produce can milk. Over 230 Amish dairy farms in the Cashton settlement, which sprawls across the Monroe and Vernon County line from south of Cashton to Ontario, and the nearby Hillsboro and Wilton settlements plus several smaller ones, send their milk to a factory that the Amish built, which is now operated by Old Country Cheese Factory. The Amish in the Kingston settlement, which is situated in southwestern Green Lake, southeast Marquette, and northern Columbia Counties, purchased an old cheese factory, expanded and modernized it, and operate the Kingston Cheese Co-op in Salemville themselves, focusing upon the production of bleu cheese and gorgonzola. There are sufficient Amish can milk producers in southern Clark County that a local milk processor, the Lynn Dairy in the small hamlet of Lynn, built a new facility to specifically handle can milk. While all dairy farmers have experienced challenges over the past several years, Amish producers of can milk have been the most stressed (Cross 2021a, 2021b). The number of can milk producers statewide fell by 127 between April 1, 2018, and April 1, 2020.

## OVERPRODUCTION OF MILK AND DAIRY LOSS

Federal Milk Marketing Orders establish minimum prices for raw Grade A milk in 11 milk marketing areas. Most of Wisconsin is within the Upper Midwest Marketing Order, except for Crawford and Grant Counties that are in the Central Federal Order. Historically, milk prices within each marketing order were based upon their aggregate distance from Eau Claire, such that the greater the distance, the higher the minimum price. Today's prices are calculated blending what processors within the region pay every two weeks relative to the utilization of that milk. Minimum prices are calculated monthly, and all producers within the revenue-sharing pool receive the same price per hundredweight, adjusted for butterfat and protein content, regardless of the end use of their milk—that is, fluid consumption, soft products, hard cheese, or butter and powdered milk (Cushman 2019b; Nepveux 2019). When demand is high, dairy farmers received a pooled price that exceeds the minimum. Unfortunately, many dairy farmers cannot cover their costs when they receive the minimum prices.

Wisconsin's dairy farmers enjoyed record high prices for their milk in the mid-2010s, which averaged $26.60 per hundredweight in September 2014 (NASS-W 2020). (A hundredweight is 8.6 gallons of milk.) Because of enhanced demand for cheese and other dairy products, partly attributed to a burgeoning export market, nearly $1.25 billion was invested in buildings, processing, and handling equipment in Wisconsin dairy plants making natural cheese and other dairy products between 2004 and 2008. Fifteen percent of the milk used in Wisconsin's dairy plants came from out-of-state. As demand for milk was expected to grow, the Wisconsin Milk Marketing Board envisioned another $2 billion being invested in dairy plant infrastructure over the following five years.

In contrast with 2014, the following five years were difficult for many dairy producers, notwithstanding increases in production. Milk prices plummeted. In 2018 monthly milk prices ranged between $15.40 to $17.80 per hundredweight (NASS-W 2020). This is before the cost of hauling the milk is deducted. Depending upon location within Wisconsin, the average hauling charges in 2018 varied from under 10 cents per hundredweight in two counties to a high of 90 cents in Iron County (WASS 2019). Costs varied depending upon distance to a cheese factory or other processing facility, the number of competing buyers and milk haulers, and how the milk was hauled. Even if the dairy farmers paid nothing for the transport of their milk, the average producer sold their milk at a loss, given its cost of production. In 2016, the average dairy farmer in Wisconsin lost 96 cents for every hundredweight of milk sold (Wokatsch 2018). Even if trade disputes that have resulted in tariffs upon cheese exports are resolved, the fact remains that America's dairy farms are producing more milk than the market can absorb. In 2019 the Wisconsin Dairy Task Force 2.0 recommended a variety of actions to resolve this issue, by raising demand and focusing upon Wisconsin's advantages over other producers (WDATCP 2019d).

FIGURE 9.11. Row of calf shelters, or calf hutches, near dairy barn south of Ripon, Fond du Lac County.

## BEEF CATTLE AND VEAL PRODUCTION IN WISCONSIN

Beef and veal production in Wisconsin are intimately connected to the dairy industry. Many dairy farms, particularly the larger operations, utilize sexed semen during the artificial insemination process to concentrate upon the birth of heifers, essential for maintenance of their milking herds. The majority of small to midsized operations do not. Thus, on these farms similar numbers of male and female calves are born. The females are worth approximately six or seven times the value of male calves, which can be acquired cheaply. Male calves can be raised for veal, which brings a higher price per pound than generic beef—although a smaller yield given their size, or raised for beef, even though they are not the more highly sought Angus beef variety.

Raising of veal and beef demands less labor than feeding and milking a dairy herd. Wisconsin has steadily lost dairy farms, yet many individuals giving up their milking chores continue raising bovines, but focus upon steers. In 2017, 24,293 Wisconsin farms reported having "other cattle," a broad category that includes "heifers that had not calved, steers, calves, and bulls" (NASS 2014b, B-16–17). While heifers are destined to become milk producers before being slaughtered when their milk output declines, male Guernseys, Holsteins, and Jerseys are raised for their meat. All of these dairy breeds together constitute dairy beef. By the barns of many former dairy farms, and on some smaller operating dairy farms, one often sees rows of small kennels to separately house the calves being raised for veal (Figure 9.11). The 2017 agricultural census reported that 10,710 Wisconsin farms sold 754,923 calves weighing less than 500 pounds during the year. In contrast, 23,505 farms sold 1,158,206 cattle, including calves weighing over 500 pounds (NASS 2019c).

Beef cows were counted on 13,954 Wisconsin farms in 2017 (Figure 9.12), nearly 5,000 more than the number of farms with dairy cows. "Cattle on feed sold" were reported by 3,198 Wisconsin farms in 2017, with 70 operations having sold 500 or more cows. The U.S. Department of Agriculture defines "cattle on feed . . . as cattle and calves that were fed a ration of grain or other concentrates that will be shipped directly from the feedlot to the slaughter market and are expected to produce a carcass that will grade select or better" (NASS 2014b, B-4–5). Feedlot operations with at least 500 head sold annually were scattered across 27 counties, with the largest numbers, 8 and 9, respectively, being within Grant and Portage Counties.

Wisconsin Department of Natural Resources records in July 2019 showed nine CAFO beef operation permittees. Seven were located in the southern third of the state. At the time of permit application, the largest beef operation had 2,300 head of beef cattle, while another operator, focusing upon calves, had over 4,000 calves (WDNR-RM 2019). Nearly 62 percent of the 1,913,129 head of cattle sold during 2017 came from the 1,759 Wisconsin farms that sold 200 or more head. Operators of this size are found in 65 counties. The only counties without at least one farm selling over 200 head are in northern Wisconsin, with the exception of urban Milwaukee County (NASS 2019c).

## SHEEP, SWINE, AND OTHER LIVESTOCK IN WISCONSIN

Wisconsin plays a relatively small role in hog and sheep production, ranking nineteenth and twentieth nationally, respectively, in their output. Both the number of farms reporting hogs and pigs and their inventory of animals fell between 2012 and 2017. Statewide, 2,198 farms had hogs or pigs in 2017,

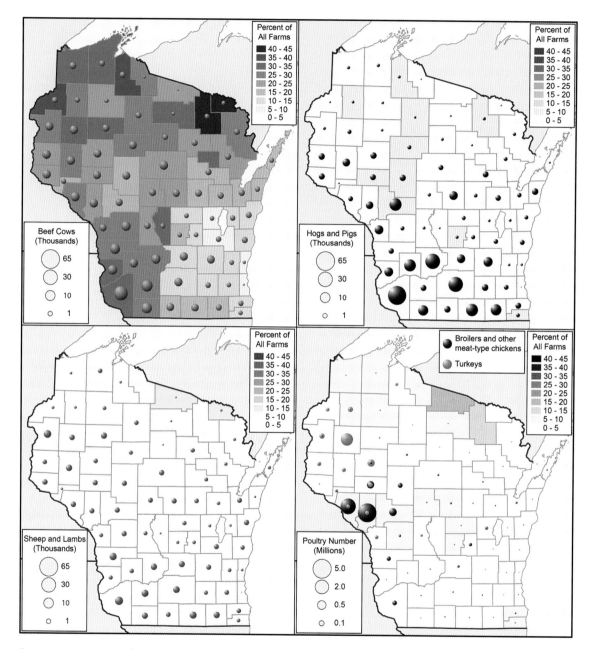

FIGURE 9.12. Livestock production by Wisconsin county in 2017: beef cattle, hogs and pigs, sheep and lambs, and poultry. Maps indicate percentage of farms within each county raising the animals and their inventory. Data source: NASS (2019c).

with 804,586 animals sold in 2017, generating revenues of $92 million (NASS 2019c). Of these sales, 720,229 hogs and pigs were sold by farms that sold 1,000 or more porkers. Fourteen swine CAFO permittees operated in Wisconsin in 2019. The largest such operation, Wolf L & G Farms near Lancaster in Grant County, had 4,250 animal units, the equivalent of 10,625 swine weighing more than 55 pounds, in 2017. Four additional swine CAFOs were in Grant County, while adjacent Crawford, Lafayette, and Richland Counties each had one (WDNR-RM 2019). In total, 10 of Wisconsin's swine CAFOs were in southwest Wisconsin, while 3 of the other 4 were in west-central Wisconsin. Wisconsin's hog operations are considerably smaller, fewer, and less odoriferous than those in Iowa and Minnesota, given the state's more stringent environmental regulations.

More Wisconsin farms had sheep and lambs than hogs in 2017, with the number of farms with sheep having grown, and their inventory having slightly increased, over the previous five years. Of the 2,845 farms with sheep and lambs, only 26 had at least 300 animals, and none exceeded 1,000. Only two counties, Dunn and Grant, had more than two herds of over 300 sheep. Although at least a few sheep are found in 70 of Wisconsin's counties, the largest numbers were in Grant and Polk Counties, which had 5,563 and 4,357 respectively, likely associated with those county's Amish populations. Statewide, sales of lambs and sheep brought $8,573,000 in 2017, while 325,345 pounds of wool fetched $413,000 (NASS 2019c).

Although Wisconsin leads the nation in its number of milk goats, its raising of goats for their meat involves fewer than half as many animals. Furthermore, sales in 2017 were less than five years earlier, as 582 farms sold 7,581 meat goats in 2017, generating revenues of just below $1 million. In contrast, Wisconsin's number of milk goats had nearly doubled, totaling 83,570 in 2017, spread over 1,029 farms (NASS 2019c). However, only 356 of Wisconsin's farms with milk goats had dairy producer licenses in April 2018. Of these, 201 were on Amish farms.

## MINK AND OTHER FURS IN WISCONSIN

Wisconsin has long led the nation in its production of mink pelts, even though wearing of furs has waned over the past several decades. While the former fur factories and shops retailing mink coats and stoles in Berlin are now closed, Wisconsin produced 1,091,000 milk pelts in 2017, 33.0 percent of the nation's total, with sales of $39.6 million (WASS 2018). Although output fell 14 percent during 2018—the third straight year of declines, Wisconsin still led the nation in pelt production, with 31.3 percent of the total (NASS-W 2019b). The state's largest mink farm is located in Sheboygan County.

## POULTRY: CHICKENS, EGGS, AND TURKEYS IN WISCONSIN

Wisconsin's position in poultry and egg production is close to its population rank. It ranks twentieth in the nation for raising chickens and seventeenth in producing eggs. In contrast, neighboring Iowa leads the nation in eggs. Wisconsin produced 1.7 percent of the nation's eggs in 2017 (WASS 2018).

### Eggs

Production of poultry and eggs is now concentrated into giant operations that are vertically integrated and conducted at the industrial scale. Long gone are the days when many farms had a few chickens that produced eggs. While many Amish farms display signs advertising their eggs, most eggs today are laid in huge chicken houses that accommodate tens to hundreds of thousands of layers. The 2017 agricultural census reported five Wisconsin operators that had over 100,000 layers in 2017, plus an additional four with over 50,000 layers. Of those egg producers with over 100,000 layers, four were in Jefferson County and one was in Washington County (NASS 2019c). However, Wisconsin CAFO records show that the state's largest egg producer is located just south of Jefferson County.

The S and R Egg Farm's facilities east of Whitewater in Walworth County had 49,361 animal units in 2019, the equivalent of 4,013,000 laying hens. It is over four times larger than the next two largest facilities, located in Jefferson and Dane Counties (WDNR-RM 2019). It included over half of the state's 7,081,000 layers in June 2019 (NASS-W 2019a). Centrality of location near both consumers and crop growers who can efficiently transport feed grains and soybean products to their facilities is essential to egg producers, who produce a product that is both fragile and perishable, while at the same time inexpensive.

### Broiler Chickens

Thirty-eight Wisconsin farms sold at least 500,000 broilers in 2017. Statewide, 53,438,462 broilers— the term used to describe all varieties of chickens raised for their meat, whether fried, grilled, roasted,

FIGURE 9.13. Chicken houses of a Gold'n Plump producer north of Buffalo City in Buffalo County.

boiled, or broiled—were sold in 2017 (NASS 2019c). The industrial-scale chicken houses are operated by independent farmers, who feed and raise chicks provided by large chicken processors, such as Gold'n Plump—now part of Pilgrim Foods (Figure 9.13), whose chickens are raised on 400 family farms in Wisconsin and Minnesota. Typically, the chicken producer supplies the farmer with the chicks and desired feed. From delivery of the chicks to dispatch of the grown birds to the slaughterhouse takes seven to nine weeks, enabling chicken farmers to raise multiple flocks each year.

Twenty-six of Wisconsin's farms that sold more than 500,000 chickens in 2017 are located in Trempealeau County. Another eight are in adjacent Buffalo County. These two counties accounted for 90.2 percent of all broilers sold by Wisconsin farmers in 2017 (NASS 2019c), illustrating the dominant role that industrial-sized vertically arranged chicken producers have upon the poultry production. The Gold'n Plump slaughterhouse operated in Arcadia is centrally located to the chicken farms.

The story is similar when considering turkeys, whose leading producer-processor is Jennie-O Turkey Store. The counties leading production are slightly farther north. Jennie-O Turkey Store raises turkeys at 31 locations, mostly in Barron County and northern Dunn County. Of the 5,579,620 turkeys raised in Wisconsin in 2017, 53.1 percent came from Barron County (NASS 2019c). Fifteen percent came from adjacent Chippewa County, while numbers from Dunn County were not disclosed. Jennie-O Turkey Store's Wisconsin processing plant is located in the city of Barron.

Ducks and pheasants are raised in disproportionately large numbers by Wisconsin farmers, although in far smaller numbers than chickens and turkeys. Wisconsin was the nation's leading grower of pheasants in 2017, selling 1,791,196 birds, 23.0 percent of the nation's total (WASS 2019). Most pheasants were raised in south-central and east-central Wisconsin, although 107 farmers statewide sold pheasants in 2017. Wisconsin's production of ducks fell precipitously between 2007 and 2012 and only rose slightly by 2017. Even with an over 90 percent drop in production, Wisconsin still ranked fifth in the nation, even though it raised under 1 percent of the nation's ducks (NASS 2019c).

## FOOD-PROCESSING AND MANUFACTURING INDUSTRIES

Wisconsin's prodigious production of milk supports the manufacturing of cheese, produced on such a grand scale that only two foreign nations, Germany and France, make more cheese than Wisconsin. Dairying accounts for 12.7 percent of Wisconsin's economy, and it "supports one out of ten jobs in Wisconsin" (WDATCP 2019d). The state is also famous for its sausages and other products manufactured from the animals that it raises. These topics are discussed in further detail in chapter 11, which explores Wisconsin's manufacturing industries.

*Chapter Ten*

# Transportation in Wisconsin

SUCCESSFUL SETTLEMENT OF WISCONSIN and utilization of its abundant resources was predicated upon voyageurs, miners, lumbermen, and farmers having a viable way of transporting their goods to market. Dairy farmers produced milk that needed to be taken to cheese factories and creameries to be manufactured into cheese and butter, which were marketed to consumers far beyond the state's borders. All of these activities required some type of transportation. The settlement of the state by miners, farmers, and lumbermen was shaped by the available transportation.

Before discussing the transformation of Wisconsin's primary products into various manufactured goods and why various communities became centers of specific types of manufacturing, we review what transportation was available to both producers of the raw materials and shippers of the finished products. To economically manufacture and distribute any product, there must be reliable and affordable routes to bring the raw materials and workers to the factory. There must be customers, who either reside nearby or can be reached by shippers at minimal to moderate cost. The availability of appropriate transportation is critical to the development of a viable economy, whether it involves primary products, such as furs, lead, wheat, or maize, or secondary products, which include all goods that are manufactured from primary products. Suitable transport depends upon both the value and weight of the product as well as its fragility and perishability. For example, fresh milk is cumbersome and perishable. Thus, it was not amenable for long-distance transport before Wisconsin embraced the manufacturing of cheese. Cheese, on the other hand, is far less bulky and perishable, has a much higher value per pound, and can readily withstand shipment to distant places.

This chapter reviews the historical development of transportation routes and nodes in Wisconsin, starting with the routes of the voyagers, the digging of canals, and the establishment of the railroads, and finishing with highways, expressways, airline routes, and energy transmission. Much of today's transportation network is predicated upon decisions made in the nineteenth century. Thus, even when modes of transportation have dramatically changed, it is essential to understand their historical development. For example, our current systems of navigation and railroads are largely based upon a network built in the 1800s. The rail network reached its greatest density before 1920. The biggest changes since then are route abandonments, even as tonnage of cargo increased. Beginning over a century ago, infrastructure to transport energy, including electric power lines and gas pipelines, became necessary to supplement goods that were physically transported. Today's economy has become dependent upon the nearly instantaneous transfer of vast amounts of data, and at the present time, lack of access to high-speed internet service—whether by wi-fi or cable—seriously handicaps the economic development opportunities for certain areas. Even when considering more traditional transportation modes, not all areas of Wisconsin are well served.

## WISCONSIN'S WATERWAYS

Wisconsin's rivers were the routes of the voyagers. In the seventeenth and eighteenth centuries canoes, laden with European manufactured products when headed westbound and with furs when eastbound, needed to be unloaded and then reloaded at every major rapids or waterfall along the rivers. At divides between watersheds the goods and canoes were carried over a portage, such as the one between the Fox River and the Wisconsin River that became aptly named Portage. Locations away from navigable waters were difficult to access, requiring the use of pack animals to carry bundles of goods or to drag them on travois along crude trails.

Water transport was critical in the early shipment of lead from southwestern Wisconsin and in bringing many miners to the state, who either traveled up the Mississippi River or journeyed from the east to ports along Lake Michigan. With the exceptions of several of the lead-mining centers in southwestern Wisconsin, all of the important early urban centers that had developed by 1850 were located along navigable waters.

Commercial navigation was relatively short-lived on several rivers in western Wisconsin, given a combination of river channel characteristics and navigational hazards caused by floating logs. Steamboats operated from the Mississippi River to Eau Claire, and reached Chippewa Falls with difficulty, in the nineteenth century. The railroad bridge across the Chippewa River, now traversed as part of the Red Cedar State Trail south of Menominee, was originally constructed as a swing bridge, yet the river is marked by many shoals and sandbars and is today only utilized by small pleasure craft. Navigation along the Wisconsin River was similarly hindered by a variety of hazards, including sediment that forms numerous sandbars. The U.S. Army Corps of Engineers, after constructing 157 wing dams along the river and finding that dredging proved futile in maintaining a six-foot channel, abandoned its efforts to maintain the channel in 1887, closing the Lower Wisconsin River to commercial navigation. Construction of hydroelectric dams across the Wisconsin River, such as the one just upstream from Prairie du Sac, and the subsequent shutting of its lock cemented its closure to navigation.

### Canal Development in the Fox River Valley

One of the earliest efforts to construct a canal in Wisconsin ended in failure. The Milwaukee and Rock River Canal Company was chartered in 1838 to construct a canal linking Milwaukee with the Rock River via the Oconomowoc lakes. Congress approved a land grant of 166,000 acres to support the endeavor, and by 1842 the first mile of canal was dug. That is as far as it went, yet the dam built to divert water into the canal provided water power to early industries near the mouth of the Milwaukee River (Whitbeck 1921). The canal dug at Portage was only twice as long, but it was far more important.

Improvements were undertaken in the mid-nineteenth century to enable steamboats to operate in the Lower Fox River between Green Bay and Lake Winnebago, and then up the Upper Fox River to Portage, where a canal linked the Fox and Wisconsin Rivers. Given that the Lower Fox River tumbled over eight sets of rapids between Lake Winnebago and Green Bay, descending 170 feet in 35 miles, 17 locks were necessary to facilitate navigation (Figure 10.1). The greatest obstacles to navigation were the rapids at Grand Chute, Little Chute, and Kaukauna. A century ago a geographer wrote, "In the 9-mile stretch between the upper dam at Appleton and the foot of the Grand Kaukauna, the fall is 134 feet, or an average of 15 feet to the mile" (Whitbeck 1915, 19). The first lock at De Pere was completed in 1850. "By 1856 the Fox River from Lake Winnebago to Green Bay could be traversed with difficulty by boats drawing three feet of water" (31). That same year a small steamboat traveled from the Ohio River and up the Mississippi River and Wisconsin River during high water, went through the newly dug— but not yet finished—Portage Canal at Portage, and then traveled down the Fox River to Green Bay.

When its construction was completed in 1876, the Portage Canal was 75 feet wide and could accommodate boats drawing 6 feet of water. Nine locks were erected along the Upper Fox River, from

FIGURE 10.1. Appleton Lock Number 4 along Fox River Navigation Canal, viewed from East College Avenue. A former paper mill site being redeveloped by construction of hotel, office, and restaurant buildings is on the right.

the one at Portage, where the Wisconsin River was higher than the Fox River, to one just upstream from Eureka. Nevertheless, by that time it was abundantly clear that the Lower Wisconsin River was not fully suitable for navigation, given shifting sandbars, which were particularly problematic near the river's mouth. Furthermore, the state's rapidly expanding railroad network provided insurmountable competition for such cross-state navigation. However, steamboat service within the Lake Winnebago and Fox River system played a key role in that region's economic development. The locks along the Lower Fox River are 35 to 36 feet wide and 144 to 146 feet long. They were particularly amenable for stern-wheel steamboats.

Steamboats provided regular service between locations in the Lake Winnebago to Lake Poygan area, transporting cargo and passengers and towing rafts of logs during the last half of the 1800s. Steamboats and barges, often loaded with coal, plied the Lower Fox River. The demand for waterborne transportation was so great that steamboats were built to serve on Lake Winnebago even before the canal was completed linking it with Green Bay. Eventually 62 steamboats were built between Green Bay and communities along the Fox and Wolf Rivers, with 27 being built in Oshkosh (Whitbeck 1915). Some were constructed upriver in Omro, Eureka, and Berlin.

By the early twentieth century, use of the Fox River waterway and the canals was waning. It was noted in 1915, "No regular lines of steamboats now run on the Lower Fox. On the Upper Fox and the Wolf, 2 or 3 small steamers or power boats make more or less regular trips. One runs from Omro to Berlin twice daily and two run from Oshkosh up and down the Wolf. A line of barges, largely but not wholly engaged in carrying coal, runs from Green Bay to up-river points, mainly to Oshkosh" (Whitbeck 1915, 35). Excursion craft and pleasure boats became the waterway's primary users. Lockages at Portage so substantially dropped that the Wisconsin River Lock at Portage was sealed, and the Fort Winnebago Lock was demolished in 1951.

Navigation along the entire Lower Fox River continued until 1987. For its last several decades activity was limited to excursion vessels and, more commonly, private pleasure craft. Abandonment of the canal occurred given the lack of commercial traffic and because of concerns about invasive aquatic species traveling upriver into Lake Winnebago. The Rapide Croche Lock south of De Pere remains closed, but the remaining 16 locks were restored and reopened between 2006 and 2015. No longer managed by the U.S. Army Corps of Engineers, the Lower Fox Locks are now maintained by the Fox River Navigational System Authority, which restored the locks to maintain their historical integrity and to provide access for sports boaters and fishermen and women.

## River Navigation Improvements along the Mississippi River

Navigation on the Mississippi River progressed from canoes and rafts to steamboats by the 1820s. Although shallows, sandbars, and rapids—with the most treacherous ones being downriver from Wisconsin near Keokuk, Iowa, and Rock Island, Illinois—were formidable, the first steamboat traveled to Fort Snelling, Minnesota, in 1823. The following year Congress authorized the U.S. Army Corps of Engineers to survey navigation routes and to improve the channel for steamboats (Fremling 2005). Steamboat travel was busiest during the 1850s and 1860s but declined with the arrival of railroads, which both bridged the Mississippi River and built tracks paralleling it. Yet, river transportation provided an alternative to rail, and in 1878 Congress authorized the creation of a 4.5-foot navigation channel in the Upper Mississippi River. Over the following three decades, dredging, wing dams, and closing dams deepened the channel. Wing dams, which stretch from the water's edge toward the main channel, constrict the channel width and increase the water's ability to scour sediments, deepening the channel. Closing dams constrict the water during times of low flow, similarly narrowing and deepening the channel. In 1907 Congress authorized the deepening of the channel to 6 feet. Before it was completed the River and Harbors Act of 1927 authorized a 9-foot channel.

The current nine-foot channel was created by the Corps of Engineers during the 1930s. Between St. Louis, Missouri, and St. Paul, Minnesota, 29 dams raise the water level, and ongoing dredging maintains the requisite channel depth. Seven of these dams and associated locks are along the border between Wisconsin and Minnesota. They are located from just west of Hager City, Wisconsin, to Genoa, Wisconsin. Between these locations locks and dams are located at Alma, Fountain City, and Trempealeau (Figure 10.2). Others are northwest of Minnesota City, Minnesota, and at La Crescent, Minnesota, across the river from La Crosse. Three locks and dams are along Wisconsin's border with Iowa: at Lynxville, which is north of Prairie du Chien; at Guttenberg, Iowa, which is northwest of Cassville; and at the northern edge of Dubuque, Iowa, about two miles north of Wisconsin's border with Illinois.

FIGURE 10.2. Consist of barges, three barges wide, being pushed into Lock and Dam Number 6 along Mississippi River at Trempealeau.

The locks are 110 feet wide and 600 feet long, sufficient to accommodate 9 barges tethered together or 6 barges and their push boat. Given that the standard consist, or grouping, of barges today is 15, the barge consist is split each time a lock is encountered. The push boat shoves 9 barges, tethered 3 barges wide by 3 barges long, into the lock and then withdraws. After the barges are either raised or lowered and pulled from the lock, the push boat and the other 6 barges go through the lock. Following their lockage, they are reconnected using steel cables to form the 15-barge consist. The process takes about two hours each time the barges encounter a lock. Because of other river traffic and the necessity of waiting, the average delay per lock ranges from four to five hours. Although Congress passed the Water Resources Development Act in 2007, authorizing replacement of several locks with 1,200-foot locks to accommodate the 15-barge consists, it failed to provide funding.

Wisconsin's primary commercial ports along the Mississippi River are La Crosse and Prairie du Chien. The tonnage shipped from La Crosse is twice that of Prairie du Chien, which primarily deals with agricultural products and scrap metal. La Crosse receives and ships those items plus caustic soda, cement, coal, fertilizer, and highway construction materials, among others (WDOT 2018a, 105). Until recently five electrical power plants along the Mississippi River, located between Alma and Cassville, received shipments of coal by both water and rail, but three closed between 2014 and 2015.

FIGURE 10.3. Consists of barges with push boats traveling in opposite directions in the Mississippi River, upriver from Trempealeau. Note the double-tracked BNSF rail line along the river bank.

Dairyland Cooperative's coal-fired generating stations continue operating in Genoa and Alma. Today most barge shipments (Figure 10.3) on the Mississippi River in Wisconsin are through traffic, grain that originates from various elevators along the Minnesota or Mississippi Rivers in the Minneapolis–St. Paul area, plus shipments of coal and scrap metal. Such shipment by barge is slow, but far cheaper than shipment by railroad or highway. A single barge carries the equivalent of 15 hopper cars. A consist of 15 barges represents more cargo than is transported by two 100-car freight trains.

### Shipping in the Great Lakes

Sailing ships and steamboats played a major role in delivering settlers and cargo to developing cities along Lake Michigan. Over 50 steamboats, plus many more sailing ships, served ports along Lake Michigan during Wisconsin's territorial days. In a geography of southeastern Wisconsin written a century ago, the importance of water transport was clearly addressed. "The Great Lakes were the only convenient means of connecting the region with the more highly developed East. . . . Settlers from the East and from overseas poured into the new West, coming in large numbers by way of the Erie Canal and the Great Lakes. Many of these people landed in Milwaukee, Racine, Kenosha, and other

FIGURE 10.4. D. P. Wrigley Company's industrial building backing on Root River in Racine. This building was erected for the Emerson Linseed Oil Works in 1872, when schooners could access Lake Michigan from the river. Today the river provides dockage for pleasure craft.

lake ports and furnished a constant supply of labor which was a distinct advantage to manufacturers" (Whitbeck 1921, 31–32). Businesses located where they had access to water transport (Figure 10.4). Milwaukee, located where the Milwaukee River, draining from the north, met the Menomonee River just west of its mouth into Lake Michigan, clearly had the best port facilities. It rapidly became Wisconsin's most important city and lake port.

Navigational improvements were rapidly undertaken at the entrance to the inner harbor at Milwaukee. Channels were dredged across the sandbar at the harbor entrance and across the narrow spit that separated the Milwaukee River from its outer bay on Lake Michigan. It gained greater utility with construction of a breakwater. While Milwaukee never gained canal access to the interior, it gained railroad access. The Menomonee Valley, just south of downtown Milwaukee, developed into an area of many wharfs and rail yards.

The primary freight cargo handled by Milwaukee's harbor has shifted over time. During much of the 1860s and 1870s, Milwaukee was "the greatest primary wheat market in the country," annually shipping over 1,000 cargos of wheat and flour, largely by ship (Whitbeck 1921, 61). Receipts of coal grew dramatically during the 1890s, and by 1910 Milwaukee received over 600 cargoes of coal annually. In a contemporary description, a geographer noted, "The long water frontage afforded by the inner harbor at Milwaukee is ideal for the great coal docks. More wharfage is used for this purpose than any other. There are 28 coal docks in the city, and the great steel trestles with their power driven clam-shell buckets for unloading the coal are among the most conspicuous sights along the water front" (63). A century ago Milwaukee's docks also received large shipments of salt and hides for its tanning facilities. Of Milwaukee's port facilities at the time, those in the Menomonee Valley, created by dredging channels and filling marshland, were the most important, having extensive facilities for transshipment to railroads. The southern part of the harbor along the Kinnickinnic River was the second most important, given its coal tonnage, while the area along the Milwaukee River provided "docks of the steamship companies and tug lines, and for elevators and freight houses" (69–70).

Milwaukee's harbor remained critical to its economic activity well into the twentieth century. One-third to one-half of Milwaukee's freight moved by water, with coal dominating, "placing Milwaukee as the second-ranking port on the Great Lakes" (WPA 1954, 247). Railroad traffic was also funneled through the port.

Railroad car ferries operated across Lake Michigan, transporting entire freight trains to and from Milwaukee, Manitowoc, Kewaunee, and Menominee, Michigan, across the river from Marinette. These ferries, owned by the railroads and primarily transporting railroad cars, connected Wisconsin ports with Muskegon, Ludington, and Frankfort, Michigan, until the final movement of railroad cars took place from Kewaunee in 1990. Railroad ferries saved considerable transit time by avoiding congestion that has long plagued—and continues to frustrate—shippers who exchange freight cars with other railroads extending east or south from Chicago.

## Lake Superior Shipments of Ore and Grain

Until the completion of the St. Lawrence Seaway in 1959, shipping to ports along Lakes Michigan and Superior involved cargoes moving between ports within the Great Lakes or cargo that could be transshipped to other locations by canal or railroad. Besides the shipment of grain and coal, large quantities of iron ore were shipped on the Great Lakes, although relatively little was shipped to Wisconsin ports. However, the proximity of the twin port at Duluth and Superior to Minnesota's Mesabi Range, the nation's leading source of iron ore, and to the wheat fields of the northern Great Plains led to the development of extensive port facilities in both cities to ship these products. By 1940, "fully ten thousand vessels arrive and depart during the navigation season between spring and fall" (WPA 1954, 292).

The Great Northern Grain Elevator in Superior had a storage capacity of 12 million bushels and was considered "the largest working house grain elevator in the world" (WPA 1954, 296), given its facility for both storing and sorting grain into different grades, varieties, and ownership. It was one of many grain elevators and storage bins that lined the harbor (Figure 10.5). Superior's Great Northern Ore Docks were "the largest group of ore docks in one location in the world" (299), having loaded over 20 million tons of iron ore in a single year. Railroad ore cars dumped their ore into storage pockets, of which there were 1,352, from which the ore was dumped via chutes into ore carriers. This set of docks (Figure 10.6), parts of which remain in service, could simultaneously load 16 ships. The longest single ore dock, extending 1,600 feet into Chequamegon Bay, was at Ashland, which once had three ore docks. Ashland's last ore dock (see Figure 5.15), handling ore from the Gogebic Range, ceased operating in 1965, but its superstructure remained until 2014.

## Port of Milwaukee Languishes

With the completion of the St. Lawrence Seaway, ocean-going vessels drawing up to 26.5 feet could bring their cargoes directly to Great Lakes ports. Initially international shipping increased, yet many shortcomings of Milwaukee's port facilities and those in several other Wisconsin cities led to dramatically decreased activity. Many former slips are far too small to accommodate modern commercial shipping. Over 40 years ago two geographers noted the dismal prospects for Milwaukee's port. "Today the Port of Milwaukee faces serious problems. No major container facility exists, and with conversion of much general cargo to containers, the St. Lawrence Seaway's slow speeds have proven a liability. Most container ships find a two week journey too costly, and operate primarily to Atlantic and Pacific coastal ports. Many new container ships draw too much water or are too wide to enter the Seaway. Inland transport is by truck or container train. As a result of these technological changes, the Port of Milwaukee has seen its overseas trade plummet" (Thaller and Richards 1975, 120).

At the Port of Milwaukee today, at best only a ship or two will be docked (Figure 10.7). While a passenger automobile ferry departs Milwaukee for Muskegon, Michigan, twice a day during spring and fall, and three times a day during the summer, other shipping is far less regular. When the lead author visited the port a few years ago, only a ship delivering road salt was docked at Jones Island, the publicly owned outer harbor. Railroad sidings leading to the former elevators and warehouses along the inner harbor are overgrown with grass, with many not having been used for years. As noted in chapter 12, many of Milwaukee's former wharves and rail yards in the Menomonee Valley are being redeveloped.

Milwaukee's port is now a distant second to the Duluth-Superior port, handling less than one-tenth of its tonnage of cargo. The port of Duluth-Superior continues as a major shipper of grain, iron ore,

FIGURE 10.5. Large grain storage bins and elevator along harbor in Superior. Grain is transported to Gavilon Grain's Peavey elevator, which can hold eight million bushels, by railroad.

FIGURE 10.6. Great Northern Railroad's Allouez iron ore loading dock along Nemadji River at Superior.

FIGURE 10.7. Aerial view of Jones Island area of Milwaukee's harbor, with no ships at dock.

and coal from western mines, with outgoing traffic accounting for 86 percent of the cargo shipped through the port (WDOT 2018a, 105). The harbor also handles cement, fertilizer, forest products, scrap iron, steel, and wind turbine components, among other items. Its total amount was 35.1 million tons in 2018 (USACE 2019). Milwaukee's tonnage was 2.3 million, largely lacking the iron ore, grain, and forest products that Duluth-Superior handled, but receiving some coal. It also handled oversized machinery, not amenable to highway or rail transit. Green Bay was Wisconsin's third most important port in 2018 as measured by tonnage, with 2.2 million tons, including cement, limestone, petroleum products, and salt, among other items. Other Lake Michigan ports, such as Marinette, Manitowoc, and Sturgeon Bay, handled far less cargo, with the total for the three ports equaling 535,000 tons in 2014 (WDOT 2018a). Manitowoc, like Milwaukee, has retained cross-lake ferry service, but using the historic coal-fired S.S. *Badger*, which began service in 1953 primarily transporting railroad cars. It now only carries automobiles and trucks, along with passengers.

## RAILROAD TRANSPORTATION IN WISCONSIN

Railroad service began in Wisconsin in 1851, linking Milwaukee with Waukesha. That line reached Madison in 1854 and Prairie du Chien in 1857. Other lines were rapidly laid, connecting Chicago to

Milwaukee in 1855, running from Milwaukee to Horicon in 1855 and continuing to La Crosse by 1858, and running from Milton to Monroe in 1857. A railroad was built from Fond du Lac to Oakfield in 1852, reaching Milwaukee in 1859 (Whitbeck 1915, 1921). Wisconsin had 891 miles of track in service by the end of 1860 (Raney 1936). The next three decades saw tremendous expansion of Wisconsin's railroad network. The Wisconsin Central Railroad reached Ashland in 1877. Superior became the Lake Superior terminal for the Great Northern Railroad that stretched to the Pacific Coast. Ashland was reached by the Northern Pacific Railway, which also extended to the West Coast. By 1890 Wisconsin had 5,583 miles of track. The maximum extent of railroad lines in Wisconsin was reached in 1916, when Wisconsin had 7,693 miles (Raney 1936).

Wisconsin also had 679 miles of electrified interurban railcar service in southeastern Wisconsin and east-central Wisconsin in 1908, providing transit to passengers between the larger cities and outlying communities. Today's Ozaukee Interurban Trail runs 30 miles along the grade of the interurban line between Milwaukee and communities north, including Cedarburg, Grafton, Port Washington, and Belgium.

## Railroads Shape Settlement

Wisconsin's early railroad development linked existing communities, yet railroads spurred the settlement of much of Wisconsin. In the northern two-thirds of Wisconsin, railroads provided the path for settlement. Settlements that failed to attract suitable connections were bypassed figuratively and economically.

Railroads selected locations for their stations, platting settlements focused upon that station, typically upon lands granted to the line. Thus, across much of Wisconsin "railroads created towns in areas where none had existed, because without towns the railroad had no local business" (Hudson 1997, 213). Thus, most settlements established in Wisconsin from the 1850s until the end of the railroad expansion that coincided with World War I owe their origins to the railroads. Because wagon travel on dirt roads was time-consuming in the nineteenth century, rural railroad stations were often located five to seven miles apart, enabling farmers to travel to the station and back in a single day. Thus, one can spot strings of villages and cities across much of central and northern Wisconsin aligned along railroad lines, even if that line is now abandoned.

Similar to what Hudson (1985) noted across the northern Plains, many examples of T-shaped settlements exist along railroads in Wisconsin. For example, Wood County's Milladore and Auburndale are both aligned along the railroad (now the Canadian National, previously the Wisconsin Central and the Soo Line) running toward Marshfield. The primary business street in both communities runs parallel to the railroad, with their businesses along the northern side. Running perpendicular to the railroad are the residential streets, such that instead of streets running N-S or E-W, they are oriented NNE-SSW or ESE-WNW, as the grid is aligned with the railroad.

Railroads also shaped Wisconsin's cities, and locations at the junction of major lines gained particular prominence, such as Fond du Lac, Madison, and, particularly, Milwaukee. The prominence of Milwaukee has been attributed to its railroad facilities. "Milwaukee men made the railroad, and the railroad was a most important factor in the making of Milwaukee. . . . Had the earliest important railroad system of Wisconsin focused upon . . . any other lake port than Milwaukee, it is probable that Milwaukee would not have been much, if any, larger than the other lake ports of Wisconsin" (Whitbeck 1921, 80). While it is true that Milwaukee had a suitable harbor, those of nearby lake ports were also relatively small and "probably could have been dredged and improved to meet the growing needs" (81). It has been claimed that the political and economic clout of Byron Kilbourn, one of the founders of both Milwaukee and two railroads from that city, was instrumental for Milwaukee becoming an early focus of Wisconsin's railroad network.

## Type of Railroad Service

Few incorporated communities lacked service by at least one railroad a century ago, although the network was denser in southern Wisconsin. The Driftless Area, with its many steep-sided valleys, had a greater share of its incorporated places without rail service, but all of them had fewer than 1,000 residents. Most of Wisconsin's railroad lines had standard-gauge track, with 4 feet 8 ½ inches between rails, but in some of the more rugged areas of southwestern Wisconsin 3-foot narrow-gauge railroads

FIGURE 10.8. Narrow-gauge steam locomotive displayed by water tank in Fennimore, Grant County. Fennimore was served by this narrow-gauge railroad from 1878 until 1926 and was a location where freight and passengers were exchanged between narrow- and standard-gauge cars. While the steam locomotive is emblazoned on the city's water tower and signs, the city no longer has any railroad service.

were built (Figure 10.8). Most were subsequently upgraded to standard gauge, but 16 miles of 3-foot narrow-gauge tracks operated in northern Grant County until 1926.

All of Wisconsin's then 71 counties had rail service a century ago. For many cities, frequent service was available. For example, Oshkosh saw 32 passenger trains and 48 freight trains a day during the first decade of the 1900s (WPA 1954). The past century has seen a huge shrinkage of Wisconsin's railroad network. Whereas once most of the traffic, both of passengers and freight, was relatively local, given that road access was poor, highway improvements and the acquisition of automobiles by the majority of residents provided alternative transportation. Yet Wisconsin's trackage had only fallen to about 6,200 miles by the mid-1960s. At that time it was noted that Wisconsin had "only a moderate amount of 'bridge' traffic, i.e. freight passing through the state but whose origin and destination are outside the state" (Finley 1965, 189). The massive wave of railroad abandonment was to soon begin.

## Wisconsin's Contemporary Railroad Network

Today's railroad network is vastly different from that of even a half century ago (Figure 10.9). The mileage of track is less than half as extensive as it was in 1916, with abandonments leaving twelve

FIGURE 10.9. Canadian Pacific Railway freight train on track paralleling La Crosse River State Trail at West Salem, La Crosse County. The Canadian Pacific runs on tracks formerly operated by the Milwaukee Road, while the bike trail uses the abandoned route of the Chicago and Northwestern Railroad.

Wisconsin counties without any rail service (Figure 10.10). Railroad cars are now much larger, with most railroads having maximum carloads of 286,000 pounds, and some can handle 315,000 pounds, while the boxcar of a century ago could handle 40 to 50 tons (100,000 pounds). Boxcars are used far less often today. Very little rail haulage, less than 2 percent, is from one location within the state to another today. Overhead haulage now exceeds within-state tonnage by a factor of 30, and it also exceeds traffic that is either inbound or outbound to Wisconsin manufacturers or distributors.

The major railroads of today concentrate upon moving bulk commodities or unit trains from one destination to another, without frequent breakage of car sets while in transit. While at one time railroad locomotives gathered boxcars a few at a time, and sometimes a single carload from many sidings, to assemble the cars that comprised the train, small sidings are no longer served. The Class 1 railroads, the larger operators who serve large portions of the nation, generally will not serve any siding that cannot provide at least 100 carloads annually. A spur line would need that many carloads per mile of track (WDOT 2018a).

Four Class 1 railroads currently serve Wisconsin: the Canadian National, the Union Pacific, the Canadian Pacific, and the Burlington Northern Santa Fe (BNSF) Railway. Their presence is the outcome of major shifts in ownership and mergers that occurred since the late 1970s. The Canadian National purchased the Wisconsin Central Railroad, which itself had acquired much of the former Soo Line, the Green Bay and Western, and the Fox Valley and Western, formerly part of the Chicago and Northwestern in east-central Wisconsin. The Canadian Pacific acquired the mainline of the former Chicago, Milwaukee, St. Paul and Pacific—the Milwaukee Road—when it went into bankruptcy. The Union Pacific entered Wisconsin with its purchase of most of the former Chicago and Northwestern in 1995. The BNSF had long had operations in far western Wisconsin, where one of its predecessor lines, the Chicago, Burlington, and Quincy Railroad, routed its main line between Chicago and Minneapolis along the eastern side of the Mississippi River.

These four railroads are focused upon overhead or bridge traffic in Wisconsin, moving commodities to and from out-of-state shippers. Besides the BNSF, the Canadian Pacific connects the Minneapolis–St. Paul area with Chicago, with its line crossing into Wisconsin at La Crosse. The Canadian Pacific also provides the trackage that Amtrak's passenger service utilizes across Wisconsin (Figure 10.11). The Union Pacific also links Chicago to the Minneapolis–St. Paul area, but its line is farther north, going through Eau Claire en route and crossing the St. Croix River near Hudson. The Canadian National, which operates 43.8 percent of Wisconsin's track miles, spans the state from Superior to the Illinois border southwest of Kenosha. It also has a main line running through Chippewa Falls to Minneapolis–St. Paul, and it extends northeast into Michigan's Upper Peninsula (WDOT 2020). From Superior the Canadian National extends across northern Minnesota toward Winnipeg, Manitoba, while from Chicago it connects east into Ontario and south to New Orleans. Some trans-Canadian rail traffic today moves across Wisconsin, given that routing transcontinental trains south of the Great Lakes saves a day in transit time compared to the traditional Canadian National and Canadian Pacific routes north of the Great Lakes.

As of 2020, Wisconsin has about 3,300 miles of operating rail lines, down 300 miles from 2014 when there were "3,600 miles of rail lines [with] four hundred miles of that trackage is double tracked,

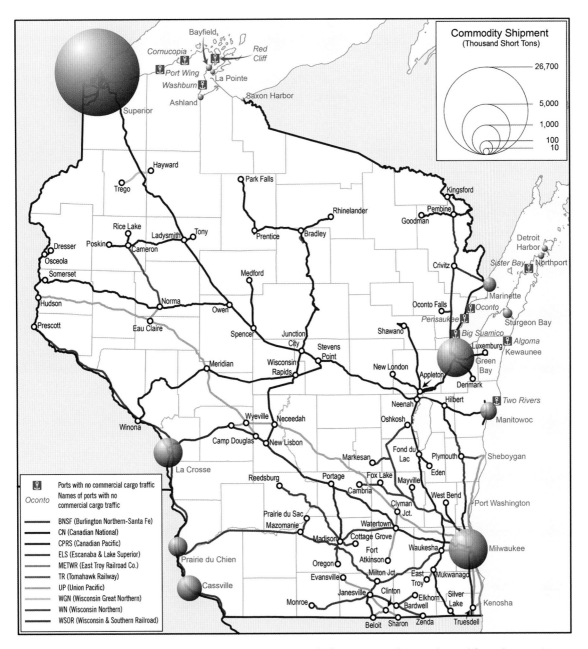

FIGURE 10.10. Ports and railroads in Wisconsin in 2020, with the tonnage of cargo shipped from the state's river and lake ports in 2018. Railroad data from WDOT (2020) and commodity shipment data for the ports from USACE (2019).

FIGURE 10.11. Amtrak passenger train at Milwaukee's Intermodal Station, which serves both rail and bus passengers. Milwaukee has better rail passenger service than any other metropolitan city in Wisconsin.

with BNSF leading the way with most of its system double tracked" (WDOT 2014, 3–29). During the past decade abandonments and service discontinuances were reported. "CN petitioned to abandon the segment from Argonne to Crandon in 2014 and announced discontinuation of service between Rhinelander and Goodman in 2017. Following flooding in 2016 that caused substantial bridge damage, CN embargoed the line from Morse to Ashland in mid-2016, and officially discontinued service on that segment in 2017" (WDOT 2018b, 20). In portions of southern Wisconsin railroad lines are state owned but contractually operated by the Wisconsin and Southern Railroad Company. Established upon the bankruptcy of prior operators, this railroad provides local shippers, including some too small to interest Class 1 operators, with access to them through interconnections.

Freight carried today greatly varies depending upon the type of railroad. The Wisconsin Department of Transportation (2018a, 89) reports, "Primary commodities by weight, moved by rail included coal, crude and petroleum oil, natural gas, chemicals and allied products, nonmetallic minerals, and farm products." Yet many trains today, particularly on the Class 1 lines, are unit trains, lacking such diversity. Thus, one sees a train of over 100 coal cars, traveling from a mine in Montana or Wyoming to a single power plant, or tank cars carrying Balken crude from North Dakota to a refinery. Seasonally, unit trains of covered railroad hopper cars transport grain. Throughout the year one sees unit trains loaded with cargo containers, now often double-stacked, bringing containers from West Coast ports, ranging from California to British Columbia, bearing such labels as COSCO, standing for China Overseas Shipping Company, among many others. Almost all of this traffic is across the state, given that the biggest intermodal facilities, where containers are transferred to or from trucks to trains, are in the Chicago and Minneapolis–St. Paul areas. Wisconsin has only two intermodal facilities that handle containerized freight. One is operated by Canadian National Railway in Chippewa Falls, while the other caters to Ashley Furniture in Arcadia. In contrast to the paucity of containerized intermodal facilities, Wisconsin has 95 freight rail transload facilities, where cargo is unloaded and transferred to or from trucks for delivery or shipment, respectively (WDOT 2014, 5–11).

## HIGHWAY TRANSPORTATION

The tonnage of freight moved by highway in Wisconsin considerably exceeds that which is transported by railroad, yet much of this traffic is intermodal freight. Because of the need to move containerized cargo to intermodal facilities, much of Wisconsin's highway traffic is actually destined to a railroad. The Wisconsin Department of Transportation explains:

> Wisconsin also suffers major impacts from freight shippers moving goods over its highway system to access large railroad intermodal facilities in Chicago. West bound intermodal freight traffic from Minnesota and the Dakotas often travels east by truck on I-94 and I-90 through Wisconsin before it is transferred in Chicago to west-bound trains. Truck volume on these interstate routes is high—around 10,000 vehicles per day—and is expected to grow faster than passenger vehicle traffic over the next 20 years. (WDOT 2014, 5–22)

Trucks currently account for just under 60 percent of all freight tonnage moved in Wisconsin, and one-seventh of that traffic crosses the state. A greater tonnage of truck cargo leaves Wisconsin than comes into the state, a consequence of the state's major manufacturing sector.

Most freight traffic travels upon Wisconsin's 11,800 miles of state highways, U.S. highways, and Interstate highways. Local traffic is accommodated by 103,000 miles of county highways, town roads, and community streets (WDOT 2018a). The Interstate highway system, which provides multilane limited access freeway traffic, largely dates from the past half century and has been continuously updated and expanded since first built. In contrast, many of the town and county roads evolved from far older rights-of-way along section line boundaries, which transitioned over time from dirt trails to graveled and then paved roadways. We look at these local roads first and then return to the expressway system that handles so much of the state's truck and automobile traffic.

### Road Development

Even before statehood, Wisconsin residents and officials recognized the need to develop road connections linking distant parts of the state. The Military Road, built in 1835 to connect Fort Crawford in Prairie du Chien with Fort Howard near Green Bay, went via Dodgeville, Madison, Fort Winnebago near Portage, and Fond du Lac. Segments of rural roads still bear the Military Road name east and northeast of Lake Winnebago, and the Military Ridge State Trail running from Madison to Dodgeville is so named for the ridge that gained its name from its proximity to the Military Road. Although this road when built was primitive and rough, other roads were improved by the laying of logs or planks over sandy or muddy surfaces, such as the Old Plank Road that ran from Fond du Lac to Sheboygan. It is now partially marked by the Old Plank Road State Trail. By the 1850s seven plank toll roads had been built from Milwaukee, with others constructed elsewhere in southeastern and south-central Wisconsin.

As noted in chapter 1, section line roads typically ran north-south and east-west, necessitating many right angle turns at intersections if one was traveling in any direction other than a cardinal compass direction. Over time, many of the sharp corners were rounded, facilitating higher-speed travel and safer intersections, but the origins of the county and town roads remain conspicuous. In the Driftless Area local roads are far less tied to the grid, instead typically following valleys or ridgelines.

The network of the local roads is considerably denser in southern Wisconsin than in much of northern Wisconsin, particularly in those areas that are largely non-agricultural. County and town roads were originally designed to provide access to the nearest settlement or to link a settlement with the nearest railroad station. The development of the state's dairy industry in which over 180,000 farms had milk to transport to over 2,000 cheese factories plus creameries in the 1930s provided strong incentives for the creation and maintenance of an all-weather road network linking dairy farms to milk processors. Today, far fewer farms have milk to ship, and the state's taxation of transportation fuels has been largely frozen for over three decades, resulting in many rural roads in significant need of resurfacing and bridge upgrades (Cohen 2020). Nevertheless, Wisconsin's rural road network is not only denser, but more likely to be paved than what is found in neighboring states.

With the coming of the automobile, local roads were improved and linked together to facilitate through traffic, although until the middle of the twentieth century railroad service was far superior to intercity highway travel. In 1912, before U.S. highways were designated, the forerunner of the Yellowstone Trail Association mapped the nation's first transcontinental automobile route, which ran through Wisconsin on a route linking Kenosha, Milwaukee, Fond du Lac, Oshkosh, Stevens Point, Marshfield, and Eau Claire. In 1926 U.S. highways were designated, although their building and maintenance were state or local responsibilities. Later, their construction expense became largely funded by the federal government, although maintenance remains a local responsibility. The northernmost east-west U.S. Highway 2 links Michigan's Upper Peninsula with Minnesota, while U.S. Highway 8 connects those two states a little farther south. U.S. Highway 10 extends from the ferry terminal in Manitowoc to the mouth of the St. Croix River at Prescott. Others, including U.S. Highways 12, 14, and 151, extend diagonally across the state, while U.S. Highways 41, 45, 51, 53, 61, and 63 stretch from south to north across Wisconsin. Over time the exact routes of these highways have varied.

### Today's Expressways and Critical Highways

The Interstate Highway System extends 875 miles across Wisconsin. As shown in Figure 10.12, heavily traveled interstate highways run from Milwaukee south to Chicago, southwest to Beloit, west to Madison continuing into Minnesota, and north along two routes to Green Bay. One route connects

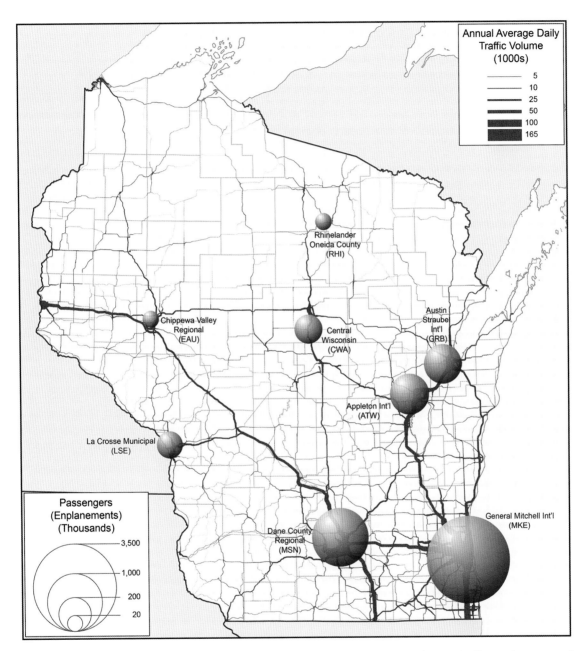

**Annual Average Daily Traffic Volume (1000s)**

———— 5
———— 10
———— 25
———— 50
———— 100
———— 165

Rhinelander
Oneida County
(RHI)

Chippewa Valley
Regional
(EAU)

Central
Wisconsin
(CWA)

Austin
Straube
Int'l
(GRB)

Appleton Int'l
(ATW)

La Crosse Municipal
(LSE)

General Mitchell Int'l
(MKE)

Dane County
Regional
(MSN)

**Passengers
(Enplanements)
(Thousands)**

3,500
1,000
200
20

FIGURE 10.12. Wisconsin's major highways and commercial passenger airports. The map indicates the routes of interstate highways, U.S. highways, and state highways in Wisconsin, and their average daily traffic volume. Passenger enplanements in 2018 are shown for airports with scheduled passenger service. Data sources: FAA (2019) and Caliper (2020).

communities along Lake Michigan; the second travels just west of Lake Winnebago and serves the Fox River Valley cities. Another highway extends through Madison north to Wausau. Interstate highways and the U.S. highways, along with some of the state highways, are considered by the Wisconsin Department of Transportation (2018a, 74) as being critical highways, designated Corridors 2030, that "are considered vital to mobility and economic development in the state." These critical highways include Wisconsin's interstate highways, the multilane U.S. highways, and State Highway 29, which together comprise the backbone routes, plus "approximately 2,340 miles of predominantly two-lane highways connecting all other significant economic centers to the Backbone system" (WDOT 2018a, 74).

Critical highways run 3,930 miles across Wisconsin, as most of the state's highways and roads lack the quality of Corridors 2030 highways. While additional mileage is included on Wisconsin's designated long truck operators' map (WDOT 2018a), indicating those routes that semitrucks or tractor-trailers can utilize, not all permit the 53-foot trailer. In addition, most county and town roads, plus some of the state highways, have weight restrictions for trucks, particularly at bridges. In 2018 when nearly 2,000 bridges were reinspected, 184 were found needing lower weight restrictions, creating considerable hardship to many farmers. This is particularly problematic in the Driftless Area, where farm trucks had become too heavy, when fully loaded, to transport manure to fields or milk to the cheese factory. This has necessitated many lengthy detours or having partly loaded trucks making multiple trips (Hubbuch 2018).

Automobile traffic is unevenly served across Wisconsin, shown by the distribution of divided highways. Yet multilane expressways do not always translate into easy transit, given congestion that can clog highways around cities such as Milwaukee and Madison. Divided highways do not serve parts of Wisconsin. One of the largest cities without such roadways is Ashland, which is also the largest city in Wisconsin without nearby access to a commercial airport. Locally, topography and lack of bridges creates difficulties for those wishing to travel by automobile or truck (Figure 10.13). Wisconsin's near exclusive reliance upon highways increases traffic and delays. It also hastens the decline of the highway surfacing, a problem that is exacerbated by limitations upon collecting sufficient taxes to adequately maintain the highways.

## AIR TRANSPORTATION IN WISCONSIN

Eight Wisconsin airports had scheduled commercial passenger service in early 2020. Six also had regular air cargo service. The status of airline service has changed dramatically during the past half century, and several Wisconsin manufacturing companies played a major role in establishing and shaping Wisconsin's air transportation network.

### Airlines Established

When Finley (1965) published his *Geography of Wisconsin* in the mid-1960s, 20 airports in Wisconsin were served by North Central Airlines. That airline, established by executives of Four Wheel Drive Auto Company in Clintonville, eventually grew to serve over 60 airports before it became Republic Airlines, following a merger. As it later merged into first Northwest and then Delta Airlines, 12 of the Wisconsin cities it served into the 1970s no longer have commercial passenger service.

FIGURE 10.13. *Pride of Cassville* tugboat and barge ferry along the Mississippi River. The nearest bridge across the Mississippi River is 31 miles to the north or 35 miles to the south of Cassville, Grant County, and the ferry does not operate in the winter or when the river level is high.

A few decades after the founding of North Central Airlines, Kimberly-Clark Corporation, then headquartered in Neenah, also established its own airline, Midwest Express. The company had been providing air transportation for its employees between the nearby Appleton airport and several of its distant manufacturing locations. In 1984 it began carrying paying passengers on a scheduled service. It rapidly expanded and established Milwaukee as its hub. It eventually merged into Frontier Airlines. Appleton business interests in 1965 used another route to improve service from their city by starting a feeder airline, Air Wisconsin. Initially offering commuter flights only to Chicago, it grew into one of the largest, if not the largest, regional airlines in the nation. For several decades it has flown connecting commuter flights for specific airlines, currently doing so for United Airlines. Its headquarters remains in Appleton, and it has expanded its maintenance facilities at Appleton International Airport.

### Wisconsin's International Airports

Three Wisconsin airports have international status, indicating the presence of a U.S. Customs office and the ability to receive flights arriving from foreign countries. While Appleton's international status is primarily to facilitate the arrival of foreign aircraft at its Gulfstream Aerospace Services center, which employs nearly 1,000 workers, Milwaukee is focused upon international passenger flights. In early 2020, before the pandemic disrupted service, Milwaukee's General Mitchell International Airport had nonstop services to cities in Canada, Mexico, Jamaica, and the Dominican Republic, plus 30 cities within the United States. By far the busiest Wisconsin airport, excluding the huge activity during one week a year when Oshkosh's Wittman Field hosts the Experimental Aircraft Association's annual fly-in, Mitchell International Airport handles 200 flights a day. It had 3,548,817 enplanements in 2018 (WDOT 2019). The other international airport is Green Bay's Austin Straubel Airport, which in early 2020 had no scheduled international flights. It offered direct flights to four hubs: Chicago, Minneapolis, Detroit, and Atlanta, plus seasonal service to two others. From Appleton, whose passenger numbers had grown 28 percent since 2017, direct flights connected to those hubs as well as Denver, plus six vacation destinations. Both Appleton and Green Bay each had less than one-tenth of Milwaukee's passenger traffic (FAA 2019).

Milwaukee is classified as a midsized hub, yet because of its proximity to Chicago, it is sometimes considered that city's third airport, drawing customers from the northern portion of that metropolitan

FIGURE 10.14. Aerial view of Central Wisconsin Airport located across Interstate Highway 39 from Mosinee in Marathon County. Situated 15 miles south of Wausau and 21 miles north of Stevens Point, this airport was served by three airlines connecting to three hubs in early 2020. It was the state's fifth busiest passenger airport.

area. Because of the nearness of Chicago's O'Hare Airport, Chicago serves as a hub for flights from most of Wisconsin's commercial airports, not Milwaukee.

## Wisconsin's Other Airports

Wisconsin's other commercial airports include Central Wisconsin Airport in Mosinee, midway between Wausau and Stevens Point (Figure 10.14); Chippewa Valley Regional Airport that serves Eau Claire; Dane County Regional Airport in Madison; La Crosse Regional Airport; and Rhinelander–Oneida County Airport. Madison's airport had direct flights to 24 destinations, 8 of which are seasonal, and service from five carriers in early 2020. They together enplaned 1,043,185 passengers in 2018 (FAA 2019). In contrast, the other airports have rather limited service. For example, in early 2020, La Crosse had flights to only 3 destinations, while Rhinelander–Oneida County Airport had 2 flights a day to a single hub. In addition, several areas of Wisconsin are served by commercial airports located within 10 miles of the state line, including those in Duluth, Minnesota; Ironwood, Michigan; Iron Mountain, Michigan; and Dubuque, Iowa.

Wisconsin lacks intrastate air flights. Indeed, there are no scheduled commercial passenger flights from any other Wisconsin airport to Milwaukee, yet all but one (Rhinelander) of Wisconsin's airports have scheduled service to Chicago O'Hare Airport. Likewise, the state's capital has no flights to any other Wisconsin airport.

The Wisconsin Department of Transportation (WDOT 2015, 2–14) notes that "88 percent of Wisconsin's current population lies within a 60-minute drive time of a commercial service airport." Furthermore, residents in the Fox Valley have two such airports only 33 miles apart, giving them a choice within that amount of drive time. For passengers utilizing general access airports for charter or air-taxi service, 95 percent of the population is within 75 minutes from such general access airports. While most of Wisconsin's population is so covered, serious shortcomings remain. First, there are portions of northern and central Wisconsin that lack access to any commercial air transport within a reasonable travel distance. Second, many areas of Wisconsin, including those closer to the airports with scheduled service, lack public transportation to their airport. Research conducted in Wisconsin has noted "rural areas that are closer to airports are preferred residential areas" (Chi 2012, 2723), and "airport accessibility . . . is positively associated with suburban population growth" (2726). Thus, concerns about air transport accessibility undoubtedly influence the economic activities in such regions, whether it is based upon manufacturing or tourism.

## PUBLIC TRANSIT IN WISCONSIN

Wisconsin residents rely upon automobiles for almost all of their transportation needs, which has many undesired consequences. Many elderly individuals and disabled persons are not able to drive. The lack of public transportation leaves them place-bound and reliant upon friends or neighbors for transportation to shop or visit their doctors. Many impoverished persons, who cannot afford the cost of a vehicle and the requisite insurance, reside too far from many places of employment, hindering their employability. Not only has the number of Wisconsin cities with passenger airports shrunk over the past half century, but so have those with passenger rail and bus service. Taxi services are unavailable in many locations.

### Passenger Rail Service

Passenger rail service in early 2020 was limited to seven trains a day each direction between Milwaukee and Chicago, plus one train a day each way between Milwaukee and La Crosse, continuing to Minneapolis and points west. Kenosha is the only Wisconsin city with commuter railroad service to Chicago via Metra, with nine trains running each direction during weekdays. While Wisconsin had an opportunity to acquire high-speed rail between Milwaukee and Madison using federal dollars, politically that was unacceptable to prominent state politicians, who campaigned against such transportation, and the process of canceling the contract cost the state approximately $200 million. Thus today, if one wants to take Amtrak to get to the state's capital, the nearest train station is 27 miles away in Columbus, from which Amtrak serves Madison by bus. Amtrak Thruway I-41 service operates buses between Milwaukee and Green Bay, with stops in Fond du Lac, Oshkosh, and Appleton. Given the bankruptcy of several national bus lines, much of Wisconsin is no longer served by any scheduled bus service.

### Bus Service

Wisconsin had no scheduled passenger bus service in 2009 north of the line connecting Green Bay, Wausau, Eau Claire, Menomonie, and Hudson, with the exceptions of a bus line that ran from Green Bay to Michigan's Upper Peninsula, with stops in Oconto, Peshtigo, and Marinette, and another that connected Superior with Duluth. Likewise, southwestern Wisconsin had no service, south or west of the line that ran from Madison to La Crosse. Several new lines were added since then, including one that runs from Hurley through Ashland to Superior and another that goes from Madison through Dodgeville and Platteville to Dubuque, Iowa, yet northern Wisconsin remains largely unserved (WDOT 2018c). In its discussion of "Intercity bus challenges," the Wisconsin Department of Transportation (2009, 2–10) lists these concerns, among others:

> Extremely long intercity bus travel times between some of Wisconsin's largest metropolitan areas (for example, a trip between Wausau and Madison, which is a 143-mile distance, takes almost seven hours and requires a transfer in Milwaukee). . . . Infrequent service (some routes have only one daily round trip or less). . . . Lack of affordable, convenient, alternative modes between key destinations (for example, no alternative is available between Janesville or Beloit and Milwaukee without transfer through Madison or Chicago).

Many issues remain a decade later. Lack of public transportation hinders economic development within many parts of Wisconsin. Anybody who has traveled across western Europe, where frequent high-speed train service is abundant, the transportation networks are highly developed, and the airports are modern with easy intermodal transit into not only city centers but to outlying communities, can imagine how much improvement Wisconsin's public transportation network needs.

## UTILITIES INFRASTRUCTURE AND PIPELINES

Pipelines and electrical transmission lines are important parts of the transportation grid. While petroleum products can be transported by rail and truck, and now crude oil is shipped across Wisconsin in great quantities by rail, pipelines are often used to transport these fluids. Pipelines to transport and distribute natural gas are far more extensive.

## Pipelines

Wisconsin had 72,377 miles of natural gas pipelines in 2016, of which 4,482 miles were the main transmission pipelines (Figure 10.15), with the remainder being mainline and service distribution pipelines (WDOT 2018a). Seventy percent of Wisconsin's homes are served by natural gas distribution lines. Large quantities of natural gas also fuel electrical power plants and industries. As with other types of transportation, rural areas of Wisconsin are less likely to have access to gas distribution lines. Thus, propane tanks are frequently visible by many farmhouses. In addition, much of northern and central Wisconsin lacks access to main transmission pipelines, although the area just south of Lake Superior is crossed by lines leading to Michigan's Upper Peninsula.

Several petroleum products pipelines lead to terminals from which gasoline and petroleum products are transported by truck to service stations and other users. About a third of Wisconsin's gasoline and diesel is provided by a pipeline operated by Flint Hills Resources LP, which maintains terminals near Waupun, Madison, Junction City, and Milwaukee. A petroleum pipeline extending to Green Bay from Milwaukee was taken out of service in 2016 due to its deteriorated condition, with the capacity of terminals in Waupun and Junction City increased, enabling greater amounts of fuel to be trucked to northeast Wisconsin.

## Electrical Transmission Grid

The electrical grid within Wisconsin displays the influence of the locations of generating facilities, of residential and industrial energy consumers, and the heritage locations of transmission lines, given strong opposition of local landowners to the expansion of new lines across or near their properties (Lenehan 2018). Historically, the first commercial generation of electricity came from hydro-electric dams, with Appleton leading the nation. While steam-powered electrical generation facilities soon came to dominate production, with coal-fired power plants still contributing 55 percent of the state's electrical power in 2017 (USEIA 2018), many newer generating stations now utilize natural gas, which now provides 17.8 percent of the state's electrical generation. Both of these types, as well as nuclear power plants, of which only the Kewaunee Power Station remains in operation, require massive amounts of water for both steam production and cooling. Thus, power plants are typically located along lakes or rivers. Because electrical resistance consumes over 4 percent of the current for every 100 miles of transmission, power must be generated relatively close to the region of consumption.

Over the past two decades Wisconsin has witnessed a rapid growth in use of wind turbines. By 2018 there were 466 wind turbines scattered across 18 wind farms (Figure 10.16). Fond du Lac and Dodge Counties have the greatest number, with the largest wind farm operated by WE Energies having 88 300-foot-high turbines (Gotter 2018). Most wind turbines in Wisconsin are atop the Niagara Escarpment, known for its steady winds. Although wind power remains below the output of hydroelectricity, renewable energy now provides 12.5 percent of Wisconsin's electrical generation (USEIA 2018). However, it requires different connections to the electrical grid than hydroelectric power and

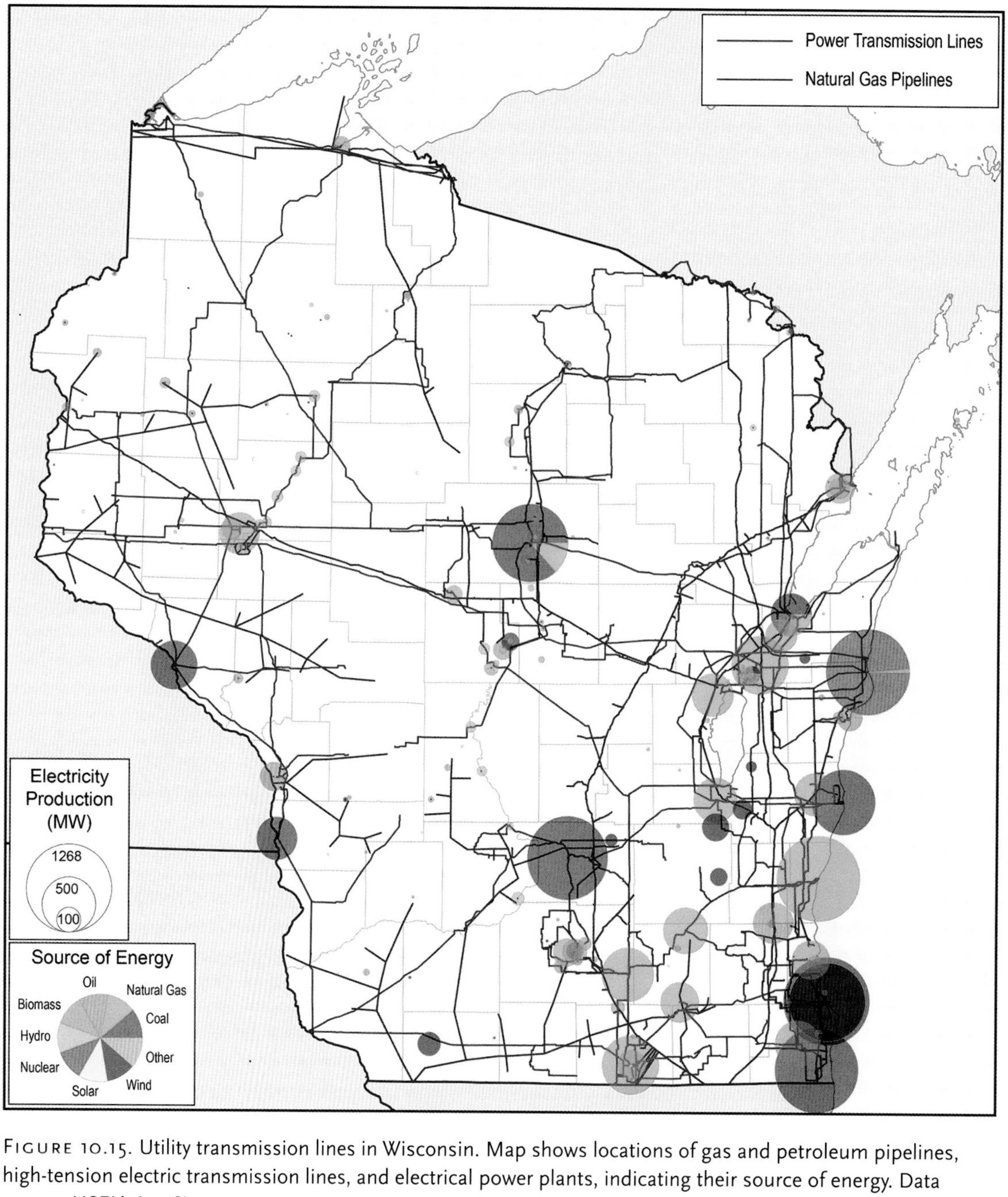

Power Transmission Lines
Natural Gas Pipelines
Electricity Production (MW)
1268
500
100
Source of Energy
Oil
Natural Gas
Coal
Other
Wind
Solar
Nuclear
Hydro
Biomass


FIGURE 10.15. Utility transmission lines in Wisconsin. Map shows locations of gas and petroleum pipelines, high-tension electric transmission lines, and electrical power plants, indicating their source of energy. Data source: USEIA (2018).

FIGURE 10.16. Wind turbine atop drumlin in Cedar Ridge Wind Farm east of Eden, Fond du Lac County.

coal and natural gas facilities, given that its location atop the ridges is different from sites along waterways. Nevertheless, most wind power is produced where nearby communities and industry can consume the power without constructing hundreds of miles of new high-tension power lines.

## Broadband Access

Just as the provision of electrification was critical in the development of the rural hinterland in the 1930s and 1940s, today cellular network coverage and high-speed Wi-Fi service are necessary for these areas to fully engage in the state's economy (Gordon 2017). While high-speed internet coverage in urban areas by both cable and Wi-Fi is both extensive and rapid, such is not the case within many of Wisconsin's rural areas. For example, in 2017 most of Kenosha, Milwaukee, and Calumet Counties had access to wireline and fixed wireless technologies that provided download speeds of 25+ megabits per second, yet well over half of Ashland, Clark, Douglas, Forest, Iron, Jackson, and Menominee Counties, just to name a few, had coverage at speeds of less than 3 megabits per second or no service at all (WBO 2018). Although broadband expansion grants are being promoted, the lack of adequate coverage inhibits the integration of these regions into the broader economy and limits educational opportunities. Indeed, the provision of online educational services during the pandemic

FIGURE 10.17. Large 1.1 million-square-foot distribution center under construction in Chippewa Falls. This facility, from which semitrailer trucks are dispatched, serves Mills Fleet Farm stores across a four-state area.

in spring 2020 proved impossible in many rural areas of Wisconsin, as it is "estimated 43% of Wisconsin's rural residents lack access to high-speed internet, compared to about 31% of rural residents nationwide" (Cameron 2020).

## PRÉCIS

This chapter describes transportation access and constraints that have influenced industrial development (discussed in the next chapter) within various communities of Wisconsin, shown by the contemporary density of roads, the still operating railroad lines, airline service, and the decline of shipping in the state's largest city. At the same time, evolving transportation access explains the growing pattern of warehouse distribution centers in parts of Wisconsin (Figure 10.17), discussed in the chapter describing urban development (chapter 12).

*Chapter Eleven*

# Manufacturing and Industry in Wisconsin

M ANUFACTURING OF ALL VARIETIES comprises secondary economic activity. Simple processing of primary goods, including lumber, metals, and agricultural products, is discussed in chapters 5 to 9. Because producers must have access to both raw materials and markets, Wisconsin's transportation network is reviewed in chapter 10. Tertiary economic activities, which include a wide variety of services, are primarily discussed in chapters 12 and 14, which describe urbanization and the recreation and tourism industries, respectively. Before reviewing the contemporary pattern of manufacturing in Wisconsin, we explore those geographic factors that were critical in determining the initial location of manufacturing facilities and the role of industrial inertia in explaining the contemporary pattern. First, let's review the factors.

## GEOGRAPHIC CONTROLS UPON LOCATION OF INDUSTRY

Five geographic factors are critical in determining the best location for a manufacturing facility: (1) location with respect to raw materials; (2) location with respect to energy; (3) location with respect to market; (4) location with respect to labor; and (5) location with respect to transportation. Ideally, the best location for a manufacturing facility will be favorably situated with respect to all five factors. Adverse location with respect to any of the first four factors may be aggravated or ameliorated by the type of transportation that serves the manufacturing facility. Let's consider several examples.

Location close to raw materials minimizes the cost of inputs, with transportation costs adding to the cost of production the greater the distance that the raw materials must travel and the greater the likelihood that these materials may be degraded during shipment. Thus, as discussed earlier, a century ago Wisconsin's cheese producers located their factories near the milk cows to minimize the cost of transporting bulky milk cans and the risk of milk spoilage. Sawmills in Oshkosh and Eau Claire were optimally located with respect to logs being floated downriver. Over time, changes in transportation may alter the importance of being located close to raw materials. Now milk is transported greater distances to cheese factories, but close proximity remains important, particularly when materials are bulky, heavy, or perishable, but of relatively low value. Likewise, after the cutting of the forests, the woodworking mills at Oshkosh and Eau Claire lost their competitive advantages.

Locations near sources of power became important when manufacturing had evolved from handicrafts, with its reliance upon human muscles and skill to make items one at a time. At one time wool and linen fibers were spun at home to make thread and then woven into cloth in the same homes. Likewise, a cobbler made shoes one at a time from leather, or a blacksmith hammered out knives or cut nails. Factories became reliant upon locations that could provide power to turn grindstones to make flour, to spin saws that cut lumber, and to operate shuttles that wove cloth. Initially, water power provided the needed energy to factories, which were located along rivers where it could be easily harnessed. Locations where mill dams could be erected, such as along the Lower Fox River or the Upper Wisconsin River, became favored for erecting flour mills, sawmills, and paper mills. Later, sites where coal could be easily and cheaply shipped, such as near docks or along railroads, were advantaged. With changes in technology and transportation, locations that could readily receive large and inexpensive supplies of natural gas and electricity became desirable for manufacturing facilities.

Location with respect to markets is important, as nearness minimizes the cost of transporting the item to consumers. Although French traders knew of the presence of lead ore in the 1700s, they were not in a position to capitalize upon its mining, given that there was no economical way to ship heavy lead ingots to distant markets. Being close to a market means that the producer can receive a larger share of the sale price than a competitor at a greater distance, whose items must bear increased transportation cost. For example, given the demand for a wide variety of logging equipment at the time Wisconsin's forests were being clear-cut, manufacturing of such equipment in several Wisconsin cities gave their producers a major advantage. They were close to the market, knew the needs of their customers, and had minimal cost to convey their logging and sawing equipment to where it was needed.

Location with respect to labor is critical. Not only do many industries require large numbers of workers, but certain industries require labor with highly specialized skills. Initially, Wisconsin relied heavily upon immigrant labor, as noted in the discussion of the number of Germans and other Europeans who worked in sawmills a century ago. Wisconsin's cheese industry benefited from the skills that immigrants brought from Switzerland and other European locations. Locating a manufacturing facility in a location without an adequate labor supply will necessitate the recruitment of workers, who may require higher pay. It may involve hiring groups that might have been overlooked previously, such as the hiring of African Americans during the two world wars. Recruitment of workers is far easier in a community that already has a large labor pool, from which workers in related industries may be hired. Furthermore, certain communities provide a variety of urban amenities desired by recruits.

Transportation does not equally serve all locations. While many manufacturing facilities require direct access to a railroad or shipping dock, which will determine their location within a community, there are many communities that lack such services. Thus, no manufacturer of bulky items, such as those exceeding the size or weight limits for a semitrailer load, which is typically under 45,000 pounds, would locate in a community that lacked a railroad or port facility. Not all highway access is equal either. As mentioned in chapter 10, weight limits for many rural bridges leave many sites along rural roads unsuitable for manufacturing, or even places where one might want to truck logs or crops. Narrow and winding highways may be subject to congestion, reducing the speed of travel, limiting the viability of many locations for either manufacturing or warehousing facilities. With just-in-time delivery of many items demanded by customers, certain manufacturing facilities need to be close to commercial airports. As discussed in the previous chapter, this limits the ability of manufacturers in much of northern Wisconsin to serve such customers.

## INDUSTRIAL INERTIA AND MANUFACTURING

Decisions made by manufacturing companies a century or more ago continue to strongly influence the location of many industries, even industries that did not exist at that time. Construction of manufacturing facilities represents a major financial investment. Many companies continue operations in the same quarters for decades and upgrade the old facility, if necessary, rather than invest in a new facility. Start-up manufacturers often acquire an existing structure, even if it is not what they would build if they built new, simply because structures that have been outgrown by earlier occupants are often less expensive. Thus, flour mills along the Lower Fox River evolved into that area's early paper mills a century and a half ago (Weichelt 2016). Today in Winnebago County's hamlet of Pickett we see a company processing the mash from a nearby ethanol plant, using a building that was once occupied by a vegetable cannery. In Eau Claire buildings erected in 1917 and occupied by a tire company that once employed 1,300 workers along a now abandoned railroad line have become Banbury Place, which houses apartments plus many small businesses, including one that deals in rubber. In Milwaukee the century-old brick buildings of several former breweries have been converted into warehouses, a variety of small businesses, and loft apartments.

Transportation links that were established to serve one industry are often utilized by other manufacturers. Indeed, the presence of a railroad or a transportation node is often exploited by later developing businesses. New railroad lines have not been established in Wisconsin for nearly a century, and those industries that require rail access by necessity must locate near an existing line, or sufficiently close that a short spur line can be built. Thus, communities that are well connected continue to attract new industry often long after their original employers are gone. Many types of industries, such as those manufacturing transportation or agricultural equipment, spawn a large number of local suppliers. Their presence can help attract other manufacturing concerns that benefit from the presence of these ancillary establishments.

Employees at one type of manufacturing plant may prove crucial to the successful establishment or expansion of other factories. For example, manufacturers of logging equipment evolved into makers of trucks in Clintonville and Oshkosh. When Leach Motors, then the nation's largest manufacturer of garbage trucks, closed its factory in Oshkosh, some of its workers found ready employment at Oshkosh Truck Corporation, when it was still primarily a manufacturer of cement mixers, fire engines, and construction equipment, but expanding into military vehicles. Reductions in employment at several paper makers in the Fox Valley have provided manufacturing workers to packaging companies that have expanded their employment.

Because of the significant capital and human investment in specific types of manufacturing, geographers have long noted industrial inertia. Put simply, once a particular type of industry develops in a region, it tends to persist, even when some of the factors that were responsible for its development are no longer present. Thus, papermaking has long been associated with the Fox Valley, brewing with Milwaukee, and cheese making with Wisconsin, just to name a few. It is impossible to explain the contemporary pattern of manufacturing and industry without considering its historical geography.

## MANUFACTURING AND INDUSTRY IN WISCONSIN IN 1910

The geography of manufacturing in Wisconsin was summarized by Ray Whitbeck (1913) in his *Geography and Industries of Wisconsin*, which included many maps showing the location of woodworking, milk-processing, brickmaking, and canning and pickling factories, plus other manufacturing facilities in 1910. At that time nine Wisconsin cities had "one or more manufacturing establishments employing 1000 or more persons: Beloit, Cudahy, Kenosha, Menasha, Milwaukee, Oshkosh, Racine, South Milwaukee, and West Allis" (Whitbeck 1913, 70). Milwaukee was by far the largest industrial center in the state, employing 32.6 percent of all manufacturing wage earners in Wisconsin, 59,502 individuals (USBC 1912, 1353).

### Types of Manufacturing in 1910

The U.S. Census of Manufactures report indicated, "In 1909 the state of Wisconsin had 9,721 manufacturing establishments, which gave employment to an average of 213,426 persons" (USBC 1912, 1331). Of these individuals, 182,583 were considered wage earners, with the others being proprietors, firm members, and salaried employees. Of the wage earners, 18.7 percent were employed in one of the state's 1,020 lumber and timber products mills, 6.1 percent worked at making furniture or refrigerators—at that time ice boxes, of which there were 114 establishments—and 4.1 percent were employees of one of the state's 57 paper mills. Altogether, including workers at cooperage shops and those making other wooden goods, forest-related industries employed 29.6 percent of Wisconsin's manufacturing work force. Wisconsin remained a major lumber producer in 1910, yet its output had fallen by half since 1890. Nevertheless, Wisconsin had the nation's third-largest woodworking industry. As we saw in chapter 6, large concentrations of sawmills and woodworking mills were located in Oshkosh, Eau Claire, Chippewa Falls, Sheboygan, and Milwaukee, among other cities, while paper mills were concentrated upon the Lower Fox River and the Wisconsin River north from Nekoosa.

The product of the paper mills was reported by the census: "In 1909 news paper represented 33.7 per cent of the total value of paper products; book paper, 18.2 per cent; writing paper, 15.3 percent; wrapping paper, 23.8 per cent; tissues, 4.6 per cent; and other paper products, 4.3 per cent" (USBC 1912, 1349). Printing and publishing employed 5,300 persons, another 2.9 percent of all wage earners.

The second leading manufacturing sector in Wisconsin included foundry and machine shop products. As explained in the census report:

> This industry embraces, in addition to the foundries and machine shops, the establishments engaged in the manufacture of gas machines and gas and water meters; hardware; cast-iron and cast-steel pipe; plumbers' supplies; steam fittings and heating apparatus; and structural ironwork. The industry is really of greater importance in the state than is indicated by the statistics, as some machine shops manufactured distinctive

products, and, as a result, were assigned to other classifications. In 1909 the industry gave employment to an average of 24,219 wage earners, or 13.3 per cent of the total for all manufacturing industries in Wisconsin. (USBC 1912, 1333)

These "distinctive products" included railroad cars; automobiles; agricultural implements; iron and steel; copper, tin, and sheet-iron products; electrical machinery; and stoves and furnaces. These together employed 24,870 wage earners, more than those working in foundries and machine shops. In addition, 3,437 persons made carriages and wagons, which at the time typically involved far more wooden than metal components. Thus, "distinctive products" were not only a prominent part of Wisconsin's manufacturing heritage of a century ago, but several provided the foundation of important industries of today.

The 1910 Census of Manufactures noted, "While Wisconsin is largely an agricultural state, the advance in the relative importance of its manufacturing industries, as measured by value of products, has been marked" (USBC 1912, 1331). Much of this manufacturing relied upon raw materials that were raised or grown by Wisconsin's farmers. By their value, butter, cheese, and condensed milk making constituted the state's third most important industry, involving 2,630 establishments. More than 3,000 persons worked at Wisconsin's 370 flour mills, grist mills, and slaughterhouses. Brewing had become an even larger industry, with 5,061 wage earners employed by the 136 establishments producing malt liquors.

The state's fourth-largest industry, as measured by its product value, was the tanning of leather. Drawing upon the bark of hemlocks cut from the state's forests and the hides obtained from the slaughterhouses that processed its livestock, plus large numbers of hides shipped into the state, this industry directly employed 7,548 wage earners. It indirectly employed far more. Indeed, 9,172 workers made shoes, leather gloves, and other leather goods.

## Milwaukee as Largest Manufacturing Center

Milwaukee was the largest manufacturing center in Wisconsin, and it was ranked tenth nationally in the value of its goods manufactured. As the Census of Manufactures noted,

> More than one-third of the total value of the foundry and machine-shop products of the state; about three-fifths of the value of products for the leather, brewery, and printing and publishing industries; nine-tenths of that for the fur-goods and the paint and varnish industries; and the entire output of the millinery and lace-goods industry were reported from Milwaukee. (USBC 1912, 1339)

Although its manufacturing was relatively diversified, Milwaukee led the nation in leather production (Figure 11.1). The Milwaukee area, and particularly Cudahy, just south of Milwaukee, was "the only important meat packing center in Wisconsin" (Whitbeck 1921, 103), slaughtering approximately 1 million hogs a year. Yet this represented only 2 percent of the nation's wholesale slaughtering, compared to 29 percent in Illinois, where the Chicago stockyards dominated the industry. Hides from these facilities, as well as from throughout the Midwest, were processed at Milwaukee's 12 tanneries, which employed 5,166 individuals in 1909 (USBC 1912, 1354). This spawned related industries, such that by "1918 there were 31 factories, large and small, engaged in making boots and shoes, 12 engaged in making leather gloves and mittens, and 25 others in making other leather goods; these do not include industries which use leather only as a part of their material, such as furniture, carriages, trunks, automobile parts, etc." (Whitbeck 1921, 111). The most prominent tannery was the Pfister and Vogel Leather Company, which operated four tanneries in Milwaukee.

Milwaukee was engaged in manufacturing a wide variety of iron and steel products. Its involvement with iron production dated from Wisconsin's territorial days, and Milwaukee was ideally situated to bring together the three essential resources for making steel. Iron ore came both by railroad from mines in Dodge County at Mayville and Iron Ridge and by ship from Minnesota's Mesabi Range and mines near Hurley. Coal came by ship from Pennsylvania, while limestone came from nearby Racine County. Milwaukee's 125 foundries and machine shops employed 9,018 persons in 1909, and the number of employees making metal products grew considerably during the years encompassing World War I.

In his summary of the more than 200 Milwaukee area establishments making iron and steel products in 1920, a geographer at the time noted:

FIGURE 11.1. Hide House Tannery Complex in Milwaukee. These structures now provide space for artist studios and small businesses, which occupy buildings erected in 1898 that had accommodated a tannery and patent leather shoe manufacturer.

Some of these are of great size. For example, the Allis-Chalmers Company employs over 6,000 persons. . . . The shops of the Chicago, Milwaukee and St. Paul Railway employ more than 5000 persons, making and repairing cars and locomotives. The International Harvester Company employs about 4,000 persons, and there are nearly a dozen plants engaged in making iron and steel products, which employ from 1,000 to 2,000 persons each. . . . The Bucyrus Company of South Milwaukee employs 1,500 people making of the powerful steam shovels, dredges, excavators, etc. used in such operations as the digging of the Panama Canal and the great open-pit iron mines of Minnesota. The National Enameling & Stamping Company—makers of enameled ware—is one of the largest in the world. The Harley-Davidson motor cycle company has grown into a great plant in a few years, employing 2,500 persons. (Whitbeck 1921, 115)

Several of these companies remain in business today under the same names.

## Manufacturing of Specific Products in Many Wisconsin Cities

The manufacturing of specific products dominated the economies of many Wisconsin cities, as explained in the 1910 census report.

The paper and pulp mills were by far the most important manufacturing industry of Appleton, contributing 58.7 per cent of the value of all manufactured products of the city. Of the total value of manufactured products for Beloit, the foundries and machine shops contributed 80.4 per cent. The blast furnaces constituted the most important industry of Ashland, while in Eau Claire the lumber and timber products and the paper and wood-pulp industries predominated. The chief industry of Fond du Lac was the tanning, currying, and finishing of leather; in Green Bay the flour mills and gristmills and the paper and pulp mills are most important; and in Janesville the manufacture of agricultural implements is the leading industry. The principal industries in Kenosha were the tanning, currying, and finishing of leather and the manufacture of furniture and refrigerators and of automobiles, including bodies and parts; in La Crosse, flour mills and gristmills and breweries; in Madison, the foundry and machine-shop and the printing and publishing industries and the manufacture of agricultural implements and of electrical machinery, apparatus, and supplies; in Manitowoc, the malt industry; in Marinette, the lumber and timber products industry; in Oshkosh, the lumber and timber products and the match industries; in Racine, the manufacture of agricultural implements, automobiles, including bodies and parts, and carriages and wagons and materials; in Sheboygan, the manufacture of furniture and refrigerators and the tanning, currying, and finishing of leather; in Superior, flour mills and gristmills; and in Wausau, the lumber and timber products and flour-mill and gristmill industries. (USBC 1912, 1339)

Today, most of those cities listed as being dominated by the lumber industry in 1909 have turned to other enterprises, yet those with paper mills have persisted. Some of the industries, such as tanning, have not only largely disappeared from Wisconsin, but from the nation. Within several cities, such as Kenosha and Racine, their involvement with the automobile industry persisted into the twenty-first century before their factories closed.

We trace the evolution of manufacturing in Wisconsin over the past century by looking at four groupings of industries: (1) those that have persisted, such as paper, printing, and food processing;

(2) those that have evolved into related industries, even though the products being manufactured have dramatically changed, such as transportation equipment and metal working; (3) those that have largely disappeared from the scene, such as tanning and apparel; and (4) those new industries, involving products and processes inconceivable a century ago, such as supercomputers, biomedical devices, and software. Wisconsin remains a manufacturing state, with 16.2 percent of its civilian workforce directly employed in manufacturing in December 2019 (USBLS 2020). Wisconsin had the second-highest percentage among the states and nearly twice the national average, which was 8.5 percent. Many additional workers are employed in transporting and marketing the numerous products and in servicing the manufacturing facilities.

## PAPERMAKING IN WISCONSIN

Wisconsin's first paper mill began operating in Milwaukee in 1848, using rags for pulp. The initial mill along the Fox River opened five years later in Appleton. Mill locations were influenced by both market and availability of water, both for power and use in the papermaking process. The first wood pulp mill in Appleton was established in 1871. By the 1880s, when mills were operating along the Fox,

FIGURE 11.2. Aerial view of pulp and paper mill along Fox River at Kaukauna, Outagamie County, showing its railroad access. A mill was erected at this site in 1883. Today's mill is operated by Expera Specialty Solutions, owned by Ahlstrom-Munksjö of Finland, employing approximately 700 persons.

Wisconsin, and Chippewa Rivers and at eight other locations scattered across the state, utilization of wood pulp was well established. When the 1910 Census of Manufactures was conducted, reporting data for 1909, Wisconsin had 57 establishments making paper and wood pulp, collectively employing 7,548 wage earners. Ten years later the number of establishments remained the same, but employment had jumped to 12,789 (USBC 1912, 1923).

Pulp and paper mills were concentrated in three river valleys a century ago. The greatest concentration was in the Lower Fox River Valley, where 32 mills were located between Neenah and Green Bay (Figure 11.2). Another dozen mills were strung along the Wisconsin River from Nekoosa to Rhinelander (Whitbeck 1913). Five paper mills were located along the Chippewa River from Eau Claire northward, plus a mill along the tributary Flambeau River at Park Falls.

### Fox River Valley Paper Mills

Within the Fox River Valley the Kimberly-Clark Company dominated the industry, operating 8 mills. A century ago the "Kimberly-Clark mills in the valley produce[d] nearly one-half million pounds of paper and 200,000 pounds of pulp a day" (Whitbeck 1915, 56). Yet, it was not the only major paper producer. Neenah had 6 paper mills, 3 operated by Kimberly-Clark, and Menasha had 5. Of these

FIGURE 11.3. Former paper mill by dam along Fox River in Appleton. This area has undergone redevelopment, with many of the mills converted to apartments, while other mill buildings have become shops and offices.

11 mills, 9 relied upon water power, being located along the Fox River or canals drawing water from it. Appleton (Figure 11.3), just downriver from Menasha, had "eleven paper and pulp mills, nine of which manufacture pulp or paper, and two of which use paper in further manufacturing, such as coating or enameling" (62).

The presence of these paper mills, which themselves produced a huge variety of paper, ranging from high-quality book and ledger paper to newsprint, kraft wrapping paper, and tissues, spawned other local industries. Some used the paper in their manufacturing of printed wrappers for soap, gum, and other products, the making of envelopes and boxes, or in the printing of books and catalogs. Other local manufacturing concerns supplied the paper mills with equipment. "A majority of the factories and mills in Appleton which do not themselves make pulp or paper make special machinery or supplies for the use of paper mills, both in the Valley and elsewhere. The Appleton Wire Works makes fourdrinier wires, cylinder moulds, etc., for the paper mills. The Valley Iron Works' chief line is the manufacture of beating engines for the paper mills, and the Appleton Woolen Mills makes paper makers' felts and jackets" (Whitbeck 1915, 64). Factories that supplied the mills were not just limited to Appleton. Menasha's largest metal-working factory also supplied the paper mills.

### Paper Mills along the Wisconsin River

The paper industry developed a little later within the Wisconsin River Valley, with its first paper mill opening in what is now Wisconsin Rapids. Additional mills were established in nearby Nekoosa and Port Edwards later in the 1890s. By 1921 there were 15 mills in the valley, with the largest numbers in Wisconsin Rapids and Stevens Point, as the paper mills replaced lumber mills as the region's primary employer (Weichelt 2016). In her dissertation describing the historical geography of the paper industry in the Wisconsin River Valley, Katie L. Weichelt noted, "Investors, pulp mills, paper mills, timber producers, and other related firms worked to fashion networks that would create a new economic region" (90). This process of clustering explained the growth of the paper industry in both the Fox and Wisconsin River Valleys. While separate, expertise and investors from the Fox Valley assisted in the development of paper mills in the Wisconsin Valley.

Both valleys saw the establishment of company towns, where the paper companies built housing, community buildings, retail facilities, and parks for their employees. In the Fox Valley both Kimberly and Combined Locks are examples, the former town named after a partner in the Kimberly-Clark Company and the latter named after the mill. In the Wisconsin River Valley company towns developed at Nekoosa, Port Edwards, and Biron.

### Paper Mills Face Challenges as Twenty-first Century Begins

Paper manufacturing within Wisconsin flourished throughout the twentieth century, with employment at the paper mills peaking at 21,500 in 1967, at which time most of the mills that were operating a half century earlier in the Lower Fox River and Wisconsin River Valleys remained in business (Schmid 2012). Four remained in business within the Chippewa and Flambeau River Valleys, along with single mills in Ashland, Niagara, Marinette, Peshtigo, Oconto Falls, and Shawano, plus several

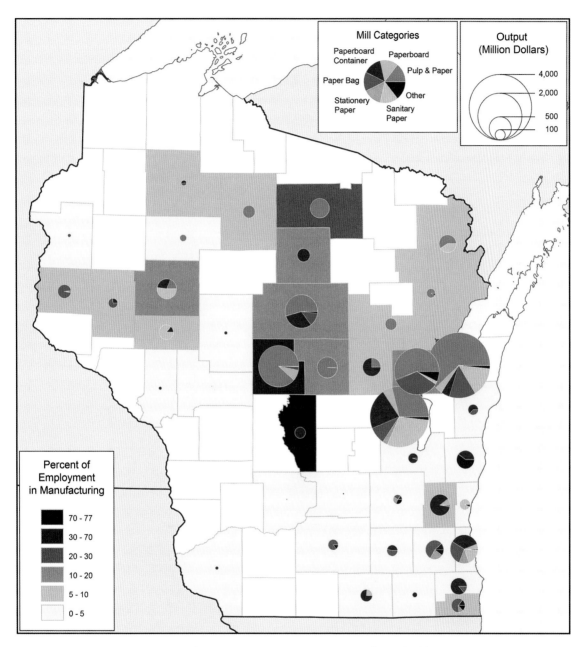

FIGURE 11.4. Wisconsin's paper industry in 2019. Wisconsin's paper mills are highly concentrated in the Lower Fox and Upper Wisconsin River Valleys, accounting for a large share of manufacturing employment and often highly focused upon specific fiber products. Data source: WIST (2019).

smaller operations in southeast Wisconsin. Employment was to remain near this level through the remainder of the century.

In 2002 Wisconsin still had 18,824 workers employed at 45 pulp or paper mills, along with an additional 19,184 persons working at one of the state's 201 plants engaged in converted paper products, largely making cardboard or paperboard boxes, paper bags, sanitary paper products, and stationery, including envelopes. The Appleton-Oshkosh-Neenah Combined Metropolitan Statistical Area, which includes Calumet, Outagamie, and Winnebago Counties, accounted for 31.9 percent of the state's employment in paper manufacturing. The Green Bay Metropolitan Statistical Area, which includes the remainder of the Lower Fox River Valley, had 19.2 percent of Wisconsin's paper workers. Thus, the Lower Fox River Valley had 10,092 employees working in its 22 pulp and paper mills, over half of the state's total, plus 9,574 workers producing converted paper products (USCB 2005).

## Paper Manufacturing Today

Paper manufacturers faced multiple challenges in the current century, yet Wisconsin leads the nation in both paper employment and productivity (WIST 2019). The dramatic increase in the use of email and the internet with the rise of the digital economy substantially decreased the demand for stationery, envelopes, and newsprint. Imports of cheap paper and paper products from China undercut many Wisconsin producers. The economic recession that began in late 2007 provided additional challenges, and paper mills in Niagara and Port Edwards closed in 2008. The mill in Kimberly ceased operating in 2009. In 2011 paper mills in Whiting and Brokaw closed (Schmid 2012). Several paper companies were sold to foreign owners, while others, such as Kimberly-Clark, saw their headquarters moved to another state (Schmid 2013, 2014). Yet other mills have successfully reoriented their product type, moving from fine printing papers and expanding into packaging (B. Adams 2018).

Wisconsin had 10,251 employees working in its 36 remaining pulp and paper mills in 2017 (Figure 11.4), according to the Census Bureau's County Business Patterns data (USCB 2019b). Of these workers, 49.7 percent were within the Lower Fox River Valley, while 20.2 percent were in Wood County, including the Wisconsin Rapids area (Figure 11.5). Specialty papers, which range widely from candy wrappers, medical packaging, and filtration papers to wine bottle labels, just to name a few, now account for 16 percent of Wisconsin's paper tonnage and "are frequently pointed to as a sustaining force in Wisconsin's paper industry" (WIST 2019, 13).

Statewide, an additional 18,394 workers were employed by converted paper product manufacturers. These 169 establishments made paperboard, boxes, bags, stationery, envelopes, and sanitary paper products, among other items. Paper bag and coated paper manufacturing was concentrated in Outagamie and Brown Counties. Winnebago County led in the manufacture of paperboard and sanitary paper products, such as tissues, paper towels, and toilet paper. Stationery was primarily produced within Winnebago, Waukesha, and Milwaukee Counties (WIST 2019). Considering both paper mill and paper converting employment, Brown and Winnebago Counties led the state, employing 6,009 and 5,287 workers, respectively, in 2018 (WIST 2019).

Printing has long been a prominent Wisconsin industry, benefiting from its proximity to the paper mills (Figure 11.6). Its workforce nearly equals total employment in pulp and paper mills and

FIGURE 11.5. Unloading pulp logs from railroad cars at Domtar pulp and paper mill along Wisconsin River in Nekoosa, Wood County.

in converted paper product manufacturing. Its 700 establishments had 27,931 employees in printing and related support jobs in 2017, as reported in the County Business Patterns (USCB 2019b). The leading centers of such printing employment are Brown County, with 2,387 employees; Dane County, 1,964; La Crosse County, 1,494; Milwaukee County, 2,871; Outagamie County, 1,103; Washington County, 1,456; Waukesha County, 5,523; and Winnebago County, 2,238. In addition, Dodge and Sauk Counties each had between 1,000 and 2,499 printing employees, yet disclosure rules preclude reporting exact statistics. The largest printing company in Wisconsin is Quad/Graphics with 6,700 employees, with facilities in Sussex, Hartford, Lomira, West Allis, Pewaukee, Burlington, and New Berlin.

## FOOD PROCESSING AND CHEESE MAKING

Processing of agricultural products included key Wisconsin industries over a century ago. Its prominence rose between 1909 and 1919, as the quantity of lumber produced plummeted and the value of dairy and meat products dramatically increased with postwar food shortages in Europe. Conversely, with the onset of Prohibition, malt liquors, the fifth leading manufactured product of 1909, ceased legal production by the end of 1919. In 1919 slaughtering and meat packing was Wisconsin's number

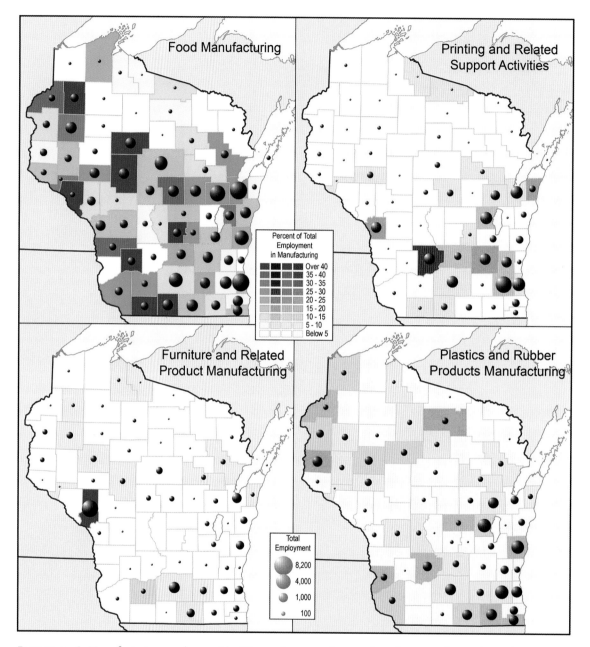

FIGURE 11.6. Manufacturing employment by Wisconsin county in 2017, focusing upon four industrial sectors: food manufacturing; printing and related services; furniture and related products; and plastics and rubber products. Maps indicate percentage of workforce employed in manufacturing and the number of employees working in each industrial sector. Data source: USBLS (2019).

one industry, as measured by the value of the product, leathering and tanning was third, cheese was fourth, condensed milk was eighth, and flour-mill and grist-mill products were tenth. Yet, measured by the size of the work force, these five activities employed 16,047 wage earners, 6.1 percent of the state's total manufacturing employees (USBC 1923).

Dairy and meat products facilities clearly involved the largest number of establishments. Wisconsin had 2,323 cheese factories, 498 creameries making butter, and 51 establishments making condensed milk in 1919. Altogether, they accounted for 27.6 percent of the state's industrial establishments. Wisconsin had 742 bakeries and 103 facilities for canning and preserving fruits and vegetables (USBC 1923). A commonality regarding most of these industries was that they were typically small, with relatively few employees. The average bakery had 4 employees, the creamery fewer than 3, and the cheese factory fewer than 1 wage earner, as the proprietor and family provided most of the labor. In contrast, the mean-sized condensery averaged 54 employees. This is very misleading given that over half of the state's condenseries were located in Portage County (Emery 1913) where they were in small rural communities, in contrast to the large national dairy producers' condenseries in Jefferson and Walworth Counties. Slaughterhouse and meat-packing facilities averaged 153 employees. Over 45 percent of these employees were working in Milwaukee (USBC 1912, 1923).

A century ago, as noted in chapter 9, cheese and butter were made close to where the cows were milked, typically in rural settings. A similar pattern was discernible regarding canning factories, although they were far more concentrated in the eastern third of the state than were cheese factories. In 1910 the greatest concentration of canneries was from Brown, Door, and Kewaunee Counties south through Ozaukee County, with a scattering of canneries extending west from Ozaukee County into eastern Columbia County. The only county elsewhere in the state to have as many as three canneries was Barron (Whitbeck 1913).

## Cheese and Other Dairy Products

As the twentieth century progressed, the number of cheese factories and creameries dramatically dropped, even though the state's output of cheese rose. Improved transportation and the shift from use of milk cans to bulk tanks permitted farmers to have their milk transported greater distances to larger cheese factories and condenseries that benefited from economies of scale. Thus, across the rural landscape many hundreds of former cheese factories have been converted to other businesses or into family homes (Cross 2011a). Others have simply disappeared.

Wisconsin has led the nation in cheese production for the past century. It accounted for 71.5 percent of the nation's output in 1923 and 1927 (Ebling, Bormuth, and Graham 1939), yet because of increased production in other states, particularly California and Idaho, its share of production had fallen to 26.3 percent in 2018, even though its output had increased over ninefold since 1935 (Cross 2012; WASS 2019). Only two foreign countries, Germany and France, produce more cheese than Wisconsin. Manufacturing of dairy products engaged 23,448 Wisconsin workers in 2017 working at 229 facilities. Of these workers, 17,386 were employed by cheese manufacturers (USCB 2019b).

Cheese factories are now more focused upon specialty and artisanal cheeses than they were several decades ago (Figure 11.7). The number of operating cheese factories has risen slightly from 114, its low

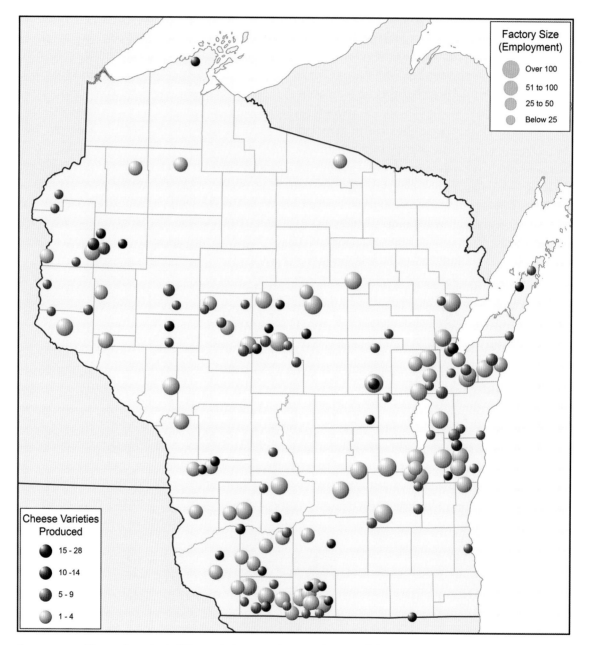

FIGURE 11.7. Cheese factories in Wisconsin in 2019, showing the size of the operations and the number of cheese varieties that each produces. Data from WDATCP (2019c) and MNI (2019).

reached in 2005 and 2006, rising to 128 in 2016 before falling to 118 in 2018 (WASS 2017, 2019). Thirty-five cheese factories produce under a million pounds of cheese annually, while 37 factories each manufacture over 25 million pounds (WASS 2019). There has been growth in numbers of both the smallest and largest cheese factories.

Wisconsin's cheese factories (Figure 11.8), regardless of size, are primarily clustered in three areas of the state: east-central Wisconsin, southwest Wisconsin, and from western Marathon County to the Minnesota border. The larger operators are slightly more prominent in east-central Wisconsin and portions of central Wisconsin, as well as Green County (Cross 2012). The rural location of most cheese factories is quite conspicuous. As noted a decade ago, "Of the state's largest cities, only two—Green Bay and Appleton—have commercial cheese factories, with the small cheese factory in Madison being the Babcock Hall facility operating on the campus of the University of Wisconsin" (530). This has changed only a little, but two small artisanal cheese factories were operating at the same address in Milwaukee in 2019, making French, cheddar, gouda, and mozzarella cheeses (WDATCP 2019c).

FIGURE 11.8. S & R Cheese Corporation's historic buildings were originally built for a brewery in Plymouth, Sheboygan County. Now named Sartori Company, its manufacturing facility has expanded across the street and railroad tracks, including the building and storage tanks on the left.

Over the past three decades shifts in the types of cheese being manufactured occurred. As recently as 1989 American or cheddar cheese accounted for 39.9 percent of Wisconsin's cheese production, yet it comprised 30.0 percent in 2018. Mozzarella constitutes 33.3 percent of the entire state output. If we add Asiago, Parmesan, Provolone, Romano, and other Italian cheeses, altogether Italian cheeses account for 50.5 percent of the state's cheese output. Hispanic cheeses have grown to 2.7 percent of production, while Swiss cheese has fallen to 0.5 percent (WASS 2019).

Producers of certain cheeses, such as cheddar, are distributed across the state in a manner akin to all the cheese factories. Mozzarella is made statewide, but more so in east-central Wisconsin. In contrast, the manufacture of "foreign cheeses," which initially did not include Italian cheeses, has long been focused upon Green County (Swiss cheese) and Dodge County (Muenster and brick cheese). Swiss cheese production remains focused upon Green and Lafayette Counties, where 11 cheese factories making Swiss cheese are listed in the state's dairy plant directory for 2019–2020. Muenster and brick cheeses are produced in those same two counties, but much less so in the Dodge County area. Mexican and Hispanic cheese producers are also concentrated in the Green and Lafayette County region (Cross 2012; WDATCP 2019c).

## Meat Packing

Slaughtering and meat packing employed 17,087 Wisconsin workers in 2002 (USCB 2005), over seven times as many as in 1909. Except for the handling of poultry, nearly half of these employees processed meat from carcasses that were slaughtered elsewhere. Given that Wisconsin is not a leading hog producer but is famous for its production of sausages, it is clear that much of the pork that it processes is raised in Iowa and Minnesota. Of the 135 Wisconsin establishments involved with animal slaughtering and processing in 2002, 13 were in Milwaukee County, where over a quarter of the state's meat-processing workers were employed. Other counties where over 1,000 workers were employed included Barron, Brown, Dane, Jefferson, and Waupaca, with over 500 employed in Marquette and Sheboygan Counties (USCB 2005). These locations correspond with the location of Jennie-O Turkeys in Barron, Oscar Mayer in Madison, and Hillshire Farms in New London.

Corporate mergers and resulting decisions have resulted in recent changes to Wisconsin's slaughtering and meat-packing industries—too recent to be shown in the census of manufacturing statistics. Oscar Mayer, which employed 4,000 workers at its Madison meat-packing plant and corporate headquarters in the 1970s, saw its workforce fall to 1,300 by 2013. Following the acquisition of Kraft Foods, its owner, by a conglomerate in 2015, Madison's meat-packing facilities and corporate offices were closed in 2017, with the work redistributed to other states. Hillshire Farms, long established in New London, became a subsidiary of Tyson Foods in 2014, but it continues its operations in Wisconsin. Other major Wisconsin meat-processing companies include Johnsonville Sausage, which employs approximately 1,200 workers at its plant near Sheboygan Falls (Figure 11.9), plus another couple hundred in Watertown, and Abbyland Foods, whose total workforce is nearly 1,000 workers, with its headquarters and processing facilities at Abbottsford along the Clark-Marathon County line. In total 17,692 workers were employed at Wisconsin's 145 animal slaughtering and processing facilities in 2017 (USCB 2019b).

FIGURE 11.9. Headquarters building and manufacturing facilities of Johnsonville Sausage, located amid farm fields near the small hamlet of Johnsonville in the Town of Sheboygan Falls in Sheboygan County. A giant 20-by-20-foot sculpture of the word *BRAT* is located across the highway from the headquarters building.

## Canning and Freezing of Fruits and Vegetables

Canning and freezing facilities to process fruit and vegetables operated at 88 Wisconsin locations in 2002. An additional 6 establishments produced dried and dehydrated food. Total employment in fruit and vegetable preserving, including specialty food manufacturing, was 11,656 (USCB 2005). By 2017, employment had slightly fallen to 10,997. Wisconsin had 10 establishments freezing fruits, juice, and vegetables, employing 2,340 workers. Almost the exact same number of workers, 2,383, were engaged in fruit and vegetable canning at one of the state's 46 canning plants (USCB 2019b).

The largest concentrations of these establishments are in central and east-central Wisconsin (Figure 11.10), with Portage County alone accounting for over a third of the state's employment at frozen fruit and vegetable plants. Prominent processors of fruits and vegetables in Wisconsin include Seneca Foods, which has processing facilities in Cambria, Clyman, Cumberland, Gillett, Janesville, Mayville, Oakfield, and Ripon; Lakeside Foods with plants in Belgium, Manitowoc, Random Lake, and Reedsburg; and Del Monte Foods, which operates canneries in Markesan and Plover. Employment in these facilities is highly seasonal. The website for Del Monte's former Cambria plant, now operated by

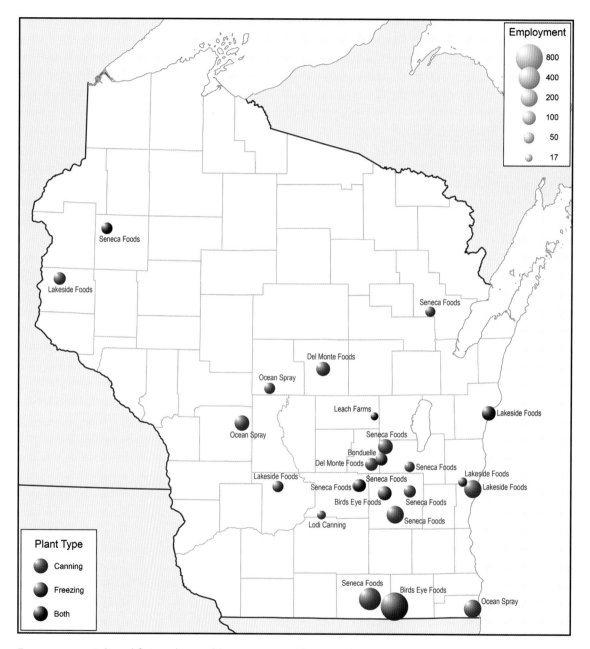

FIGURE 11.10. Selected fruit and vegetable canneries and freezing plants in Wisconsin in 2019, showing the size of the operations and the name of their operators. Data from MNI (2019).

Seneca, indicates it employed 46 salaried employees and maintenance mechanics, plus 280 seasonal employees.

## Brewing

The brewing industry of today is dramatically different from what existed just over a century ago, as Prohibition brought an abrupt cessation of the industry. With the end of Prohibition in 1933, brewing resumed, but new laws governed its sales and distribution. For example, breweries could not operate their own taverns. While many breweries resumed operations following Prohibition, small local breweries had difficulty competing with large brewers that distributed their product nationally (Apps 2005). By 1983 Wisconsin had just seven breweries remaining in business, with Schlitz Brewing, which had held the position as the nation's largest brewery several times during the twentieth century, ceasing its Milwaukee operations the previous year (Shears 2014). Remaining brewers included such brewing giants as Miller, whose Wisconsin roots date to 1855, and Pabst, whose roots as Philip Best Brewery were even deeper, extending to 1842. Other Wisconsin breweries in 1983 included G. Heileman of La Crosse, Point Brewery of Stevens Point, Huber Brewing of Monroe, Leinenkugel of Chippewa Falls, and Walters of Eau Claire, which became Hibernia. Things were soon to change.

Heileman and Hibernia were to also disappear. Yet the opening of small breweries, focused upon producing craft brews using high-quality ingredients in contrast with the cheaper lower-quality product that doomed Schlitz, resulted in over 100 microbreweries being established in Wisconsin (Figure 11.11). These and other, larger breweries focused upon the state, rather than a national market. Sprecher Brewing Company started operations in Milwaukee in 1985, "the first of the new wave" (Shears 2014, 46) of breweries, and subsequently relocated to suburban Glendale in 1993, where it remains. New Glarus Brewery, established in its namesake in 1993, has "found great success, despite limited distribution . . . [and] was the 17th largest American brewery in 2012 despite not distributing outside of the state of Wisconsin" (55).

Wisconsin's largest brewery is now operated by Miller Brewing (Figure 11.12), a subsidiary of Molson Coors, as is Jacob Leinenkugel of Chippewa Falls. Pabst has become a holding company, whose many brands are produced by other brewers. It opened a brewpub on the site of its former Milwaukee brewery in 2017. G. Heileman, following financial difficulties, sold its La Crosse brewery in 1999 to City Brewing Company, which produces a variety of brands, including Old Style. Yet, the bigger story is the expansion of the smaller breweries, particularly New Glarus.

The Census Bureau enumerated 77 breweries in Wisconsin in 2017 (USCB 2019b), up from 62 the previous census, yet still undoubtedly excluding many of the smaller brewpubs that only sold their brew on premises. Total employment was 2,399. The greatest numbers of brewery workers were in Milwaukee County, whose 13 breweries employed an aggregate of 500 to 1,000 workers, and in La Crosse County, whose 4 breweries employed an equivalent range (USCB 2019b). Dane County had 11 breweries, employing 158 workers. The only other counties with over a hundred brewing workers are Chippewa, Green, and Portage, where Leinenkugel, Minhas—the successor of Huber—and Point Brewery, respectively, remain in operation.

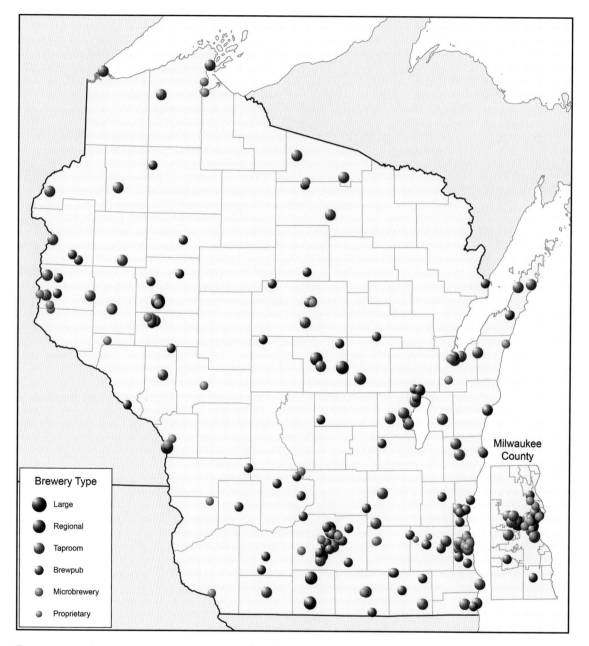

FIGURE 11.11. Breweries in Wisconsin, ranging from large and regional breweries to taprooms, microbreweries, and brewpubs. Data sources: Flanigan (2020) and Brewers Association (2020).

FIGURE 11.12. Miller brewery complex, as viewed from West Wisconsin Avenue bridge in Milwaukee.

Breweries require supporting industries, besides the farms that grow the hops and barley. Barley must be malted before it is used in brewing, a process that involves its germination and drying, producing the sugars required during fermentation. Manitowoc has long been a center of malting (Figure 11.13). The owners of its landmark malting facility, Briess Malt and Ingredients Company, currently have additional malthouses in Chilton and Waterloo, although most activities in Chilton are being phased out.

## TRANSPORTATION EQUIPMENT MANUFACTURING

A century ago Wisconsin workers made a wide variety of transportation equipment. Besides the 2,500 employees making motorcycles in Milwaukee, in Racine the J. I. Case Threshing Machine Company employed 4,500 persons manufacturing tractors and automobiles, Mitchell Motors Company of Racine employed 2,500 workers assembling automobiles, and Racine Manufacturing Company had nearly 1,000 workers making automobile bodies. The Nash Motor Company, formerly the Thomas B. Jeffery Company, employed 5,000 workers in Kenosha manufacturing passenger cars and trucks (Whitbeck 1921). Milwaukee companies also made railroad cars, locomotives, and excavating equipment. Farther north, shipbuilding was taking place at Manitowoc, Sturgeon Bay, and Superior. During World

FIGURE 11.13. Malt plant in Manitowoc with mural on malting silos displaying uses of Briess Malting and Ingredients' malted grains, including brewing and baking.

War I shipbuilding employed nearly 2,500 workers in Manitowoc (WPA 1954). Statewide, there were 7 steel ship builders in 1919, employing 6,658 wage earners. In addition, 28 facilities were engaged in building wooden ships and boats, with 1,107 wage earners (USBC 1923). Things changed over the century.

## Automobiles

During the first quarter of the 1900s four Wisconsin companies were nationally known for their automobiles. The Case automobile, of which over 25,000 were made, ceased production in 1927, as its manufacturer devoted its attention to farm tractors and other equipment. The Kissel Company of Hartford manufactured about 35,000 automobiles before succumbing in 1931. Mitchell Motors remained in business until 1923, when it "was perhaps the largest auto company failure up to that time" (Albert 1993, 6). Its facilities were purchased by Nash Motors of Kenosha, which came to operate assembly plants in Kenosha, Racine, and Milwaukee that produced the Rambler series of automobiles as well as the four-wheel drive Quad truck. Ramblers were marketed into the mid-1960s after Nash Motors had become American Motors Corporation (AMC), following a merger with Hudson Motors of Detroit, whose operations were moved to Wisconsin. By the time Chrysler Corporation acquired AMC in 1987, AMC and its predecessors had shipped "more than ten million cars" from its plants in Wisconsin (6). Chrysler ceased automobile assembly in Kenosha in 1988 but continued to manufacture engines there until 2010, employing 1,300 workers as recently as 2006.

General Motors Corporation operated an assembly plant at Janesville from 1919 until 2009. It was initially constructed to manufacture tractors, but during most of its operation it made pickup trucks, automobiles, and sports utility vehicles. Employment peaked at 7,000 workers in 1970, falling to 1,200 by the end of 2008. With the closure of this facility and the one in Kenosha, both victims of the recession that began in 2007, Wisconsin's engagement with automobile assembly had ended. However, Wisconsin factories still make many automobile parts, and the state manufactures trucks, fire engines, and heavy construction equipment, along with farm tractors.

Only one sizeable Wisconsin establishment was engaged in automobile and light-duty motor vehicle manufacturing in 2002, and it was in Janesville, employing between 2,500 and 5,000 workers (USCB 2005). Yet there were 285 establishments, of which 136 had at least 20 workers, employing 37,131 workers in transportation equipment manufacturing. Two facilities were involved in heavy-duty truck manufacturing, each employing between 1,000 and 2,500 employees. Twenty-three establishments each had more than 20 employees involved in motor vehicle body and trailer manufacturing, totaling 4,603 employees. Far more workers, 17,091 in total, worked at one of the 141 establishments making motor vehicle parts (USCB 2005). Now Wisconsin's manufacturing of automobiles is history, but that of trucks and motorcycles continue. Furthermore, over 5,000 workers make engines, electrical components, and other parts for motor vehicles.

## Trucks and Motorcycles

The largest manufacturer of heavy-duty trucks in Wisconsin today is the Oshkosh Corporation, which has expanded tremendously into supplying military vehicles. Ranging from its light-combat tactical all-terrain vehicles to a variety of heavy tactical vehicles, Oshkosh Corporation is the provider of mine-resistant patrol vehicles. Headquartered in Oshkosh, where it operates two primary manufacturing facilities, plus smaller support sites, Oshkosh Corporation provided most of the vehicles used by the American military forces in Iraq and Afghanistan. The company also manufactures fire trucks, aircraft rescue and firefighting vehicles, cement mixers, and other specialized equipment. Pierce Manufacturing, a subsidiary of Oshkosh Corporation, manufactures fire equipment in Menasha. Thus, nearly 5,000 workers make heavy-duty trucks in Winnebago County. To the north, FWD Corporation of Clintonville, which acquired the Seagraves brand, also makes fire engines.

Motorcycle, bicycle, and parts manufacturers employed 2,508 Wisconsin workers in 2017, down over 500 since the census five years earlier. Two companies dominate this manufacturing, Harley-Davidson for motorcycles and Trek for bicycles. Harley-Davidson located its museum upon a redevelopment site in Milwaukee's industrial Menominee Valley and its corporate headquarters is on the site of its first factory in Milwaukee. Most of its manufacturing occurs in Menomonee Falls in adjacent Waukesha County, where over 1,000 workers made motorcycles in 2017 (USCB 2019b). Its product development center is in Milwaukee County, at Wauwatosa. A satellite Harley-Davidson plant making motorcycle components is located in Tomahawk, with over 300 employees. Trek remains headquartered in Waterloo. Railroad car manufacturing employed 316 workers in 2017, less than 5 percent of its figure a century earlier.

## Shipbuilding

During the century since the construction of ships dramatically increased with the need for Great Lakes iron ore carriers during World War I, Wisconsin's shipbuilding industry has largely focused upon two types of craft, those desired by the U.S. Navy and those to serve Great Lakes ports that were and remain too large to build elsewhere.

Before the construction of the St. Lawrence Seaway, oceangoing vessels could not travel into the Great Lakes, although barges connected the Great Lakes to the Gulf of Mexico via the Illinois and Mississippi Rivers. While many ships were constructed to transport grain from ports, such as Superior, to locations along Lakes Erie and Ontario, others were ore carriers. Superior was one of the first shipbuilding locations, building the famous whalebacks, but shipbuilding there ceased during the 1930s. Manitowoc was another shipbuilding center, yet a sharp curve on the Manitowoc River near its yards limited the size of ships that could be launched. In 1968 the Manitowoc Company's shipyard was closed, and its owner acquired a long-established shipyard at Sturgeon Bay and moved its operations there. Manitowoc remains the home of Burger Boats, which employs about 350 persons who build three custom yachts annually.

Huge ships were built at Sturgeon Bay to transport iron ore to steel mills near Chicago, Gary, Detroit, and Cleveland. The largest ore carriers, up to 1,080 feet in length and weighing over 35,000 tons, were constructed at the shipyards in Sturgeon Bay. While no new ships of this size have been constructed there or anywhere else in the Great Lakes since the 1980s, Bay Shipyards at Sturgeon Bay has drydocks and other facilities to undertake major retrofitting, servicing, and repair of ore carriers during the winter when the shipping season is inactive. Fourteen ships were at the shipyard for work during the winter of 2017–18. Building new vessels continues, with work on a 740-foot articulated barge underway in 2018. In 2019 work began on a 639-foot self-unloading bulk carrier that Fincantieri Marine Group states "is believed to be the first ship for the U.S. Great Lakes service built on the Great Lakes since 1983" (FMG 2019). Large ocean-going yachts have also been built in Sturgeon Bay.

Naval shipbuilding was a major activity at Manitowoc during World War II, employing 7,000 workers. Twenty-eight submarines were constructed and floated through the Chicago Ship and Sanitary Canal to reach the Gulf of Mexico. Most saw action during the war. A surviving submarine is docked by the Wisconsin Maritime Museum in Manitowoc. For the past two decades naval vessels, for both the U.S. Navy and the Saudi Navy, have been built at the shipyards at Marinette. Operated by Fincantieri Marine Group, the Italian shipbuilding giant that also acquired Sturgeon Bay's shipyards from the Manitowoc Company, the Marinette shipyard has built over a dozen 394-foot littoral combat ships (Figure 11.14). It also built Staten Island Ferries. Currently, 1,400 workers are employed at the Marinette shipyards. Shipbuilding, statewide, employed 3,344 workers in 2017, nearly half as many as were employed in 1919.

FIGURE 11.14. Two ships being built for the U.S. Navy at Fincantieri Marine Group's shipyard along the Menominee River in Marinette. The photograph was taken the day before the USS *Minneapolis-St. Paul*, the littoral combat ship festooned with flags, was launched.

## FOUNDRIES AND FABRICATED METAL PRODUCTS

Manufacturing of automobiles, trucks, motorcycles, and ships requires numerous metal components, and their making and the preparation of their constituent metals keeps foundries busy. In 1919, 18,635 wage earners were employed by foundries and machine shops, with the greatest concentration of these in southeast Wisconsin (USBC 1923). By 2002 Wisconsin had 23,515 persons employed in primary metal manufacturing, of which the largest number, 19,721, were employed at one of the state's 138 foundries (USCB 2005). Following the Great Recession, Wisconsin's employment in foundries had fallen to 12,721 by 2017, and 105 foundries remained in business. At that time only four counties—Milwaukee, Sheboygan, Washington, and Waupaca—had over 1,000 workers employed at foundries. Over 500 persons were employed in Dane, Manitowoc, Ozaukee, Sauk, Waukesha, and Winnebago County foundries.

Machine shops, which included turned product and screw, nut, and bolt manufacturing, and establishments making architectural and structural metals are the largest employers in the fabricated metal products industries. Together, they employed 28,659 workers in 2002. In total, including forging and stamping, boiler and tank manufacturing, hardware making, and a wide variety of other metal products, all fabricated metal product manufacturing employed 63,476 Wisconsin workers in 2002 (USCB 2005).

Following recovery from the economic downturn that began in 2007, employment in this sector reached 66,674 by 2017, involving 1,935 establishments. This included 16,486 employees in machine shops, 15,743 workers making architectural and structural metals, 7,486 forging and stamping metals, and 4,898 making boilers, tanks, and shipping containers (USCB 2019b). Milwaukee County led the state with a total of 8,182 fabricated metal products employees in 2017 (Figure 11.15), nearly matched by Waukesha County's 7,638 workers, followed by Marathon County with 5,270 employees. Both Washington County and Brown County had over 3,000 such workers, while Racine, Dodge, Outagamie, Manitowoc, and Ozaukee Counties all exceeded 2,000 fabricated metal products employees (USCB 2019b). Furthermore, more than 1,000 employees made fabricated metal products in nine additional counties.

Statewide, 59,947 Wisconsin employees were engaged in machinery manufacturing in 2017, working at one of 1,035 establishments. This manufacturing is highly diverse, including making agricultural implements; construction machinery; industrial and metalworking machinery; manufacturing heating, air conditioning, and ventilation systems; making engines and power transmission equipment; and manufacturing other types of machinery. Specific types of machinery manufacturing are regionally concentrated. Agricultural implement making and, in particular, building lawn mowing and gardening equipment and snowblowers employ more than 1,000 workers in both Calumet and Dodge Counties, home of Ariens Company and John Deere Company, respectively. Green, Monroe, and Racine Counties also have many workers similarly employed.

Manufacturing of construction machinery employed 1,518 workers in Milwaukee County and 735 in Waukesha County in 2017. Industrial machinery manufacturing engaged 1,374 workers in Brown County, 1,039 in Outagamie County, and 637 in Waukesha County. Manufacturing of ventilation, heating, and air conditioning equipment employed more than 1,000 workers in La Crosse County, where Trane Inc. has major facilities. Production of metalworking machinery involved 1,751 persons in Waukesha County, and Milwaukee and Washington Counties both had 695 workers. Engine, turbine, and power transmission equipment manufacturing employed 1,012 workers in Milwaukee County, 976 in Waukesha County, 1,017 in Racine County, and 769 in Rock County (USCB 2019b).

## FALLEN FLAGS: INDUSTRIES AND COMPANIES LARGELY OF THE PAST

Some industries and companies that were well established in Wisconsin a century ago remain vibrant today, such as the Kohler Company. Another metamorphosed into an international conglomerate whose activities are considerably different from what was seen a century ago. Indeed, Wisconsin Axle in Oshkosh spawned Rockwell International, which swallowed Milwaukee's Allen-Bradley (Figure 11.16), before Axletech was split off. It now operates the original Oshkosh factory, parts of which date to 1912. In contrast, many prominent industries of a century ago have either disappeared or only remain as tokens.

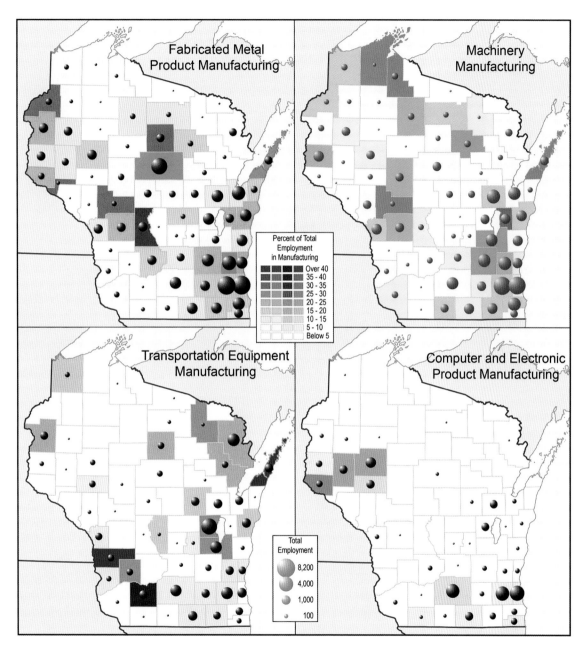

FIGURE 11.15. Manufacturing employment by Wisconsin county in 2017, focusing upon four industrial sectors: fabricated metal products; machinery manufacturing; transportation equipment; and computer and electronic products. Maps indicate percentage of workforce employed in manufacturing and the number of employees working in each industrial sector. Data source: USBLS (2019).

FIGURE 11.16. Allen-Bradley Building, famous for its clock towers, one of which is shown, in Milwaukee. The Allen-Bradley Company, founded in Milwaukee early in the twentieth century, was acquired by Rockwell International in 1985. This building is now occupied by Rockwell Automation, headquartered in Milwaukee. Their business is industrial automation and information technology.

### Leather Making and Clothing Production Fade

Leather and leather goods manufacturing employed 16,592 wage earners in 1919, down slightly from 1909, when 16,720 were employed, as demand for saddles and harnesses had begun to fall (USCB 1912, 1923). Not only is there little demand for saddles today, but most shoes, gloves, and purses are manufactured overseas, often using synthetic materials. Thus, by 2017 employment in all aspects of manufacturing leather totaled 1,743 persons in Wisconsin, of which 60.5 percent were making footwear. Milwaukee County, with 391 employees, had the largest number of workers manufacturing leather (USCB 2019b).

Textile manufacturing and clothing manufacturing have similarly faded from the scene. In 1919 Wisconsin had 72 establishments making knit goods, employing 8,736 wage earners, and 24 plants whose 1,070 wage earners made woolen and worsted goods. In addition, 3,385 workers at 59 facilities labored making men's clothing, while 844 wage earners made women's clothing (USCB 1923). By 2017 combined employment in textile mills, textile product mills, and apparel manufacturing totaled 2,677.

Only Brown, Dane, Outagamie, Rock, and Winnebago Counties had more than 100 persons employed at textile mills the previous year, and several of these mills were making textiles used in paper manufacturing. Apparel manufacturers employed more than 100 workers only in Green Lake, Monroe, and Sheboygan Counties, for a total of 664 statewide (USCB 2019b). Thus, while the Oshkosh B'Gosh brand still remains, its manufacturing of overalls and children's clothes left Wisconsin several decades ago, first for the southern states and then overseas. The company itself was acquired by another corporation, which is headquartered in Georgia.

### Kitchenware and Aluminum Products Manufacturing Declines

Kitchenware and aluminum products have long been associated with Manitowoc. The Mirro Aluminum Company, headquartered there, was long known for its kitchen cookware, and it was the world's largest such manufacturer well over a half century ago. It had expanded into aluminum watercraft. Competition led to the closing of the company in 2003, the sale of its manufacturing facilities, and the demolition of its large factory building in Manitowoc (Figure 11.17). Nevertheless, a successor company, headquartered out-of-state, continues to manufacture some aluminum cookware in Manitowoc.

FIGURE 11.17. Remains of entrance doorway of the Mirro Aluminum Company Plant in Manitowoc, which was demolished in 2017. The seven-floor brick facility, built in 1911, had occupied a full city block. When photographed in early 2020, the site was still awaiting redevelopment.

In 2017 Wisconsin had 11 establishments manufacturing metal kitchen cookware, utensils, cutlery, and flatware, employing a total of 1,253 workers (USCB 2019b). During the previous year, the largest number—over half of the total—worked in Sheboygan County, followed by Washington County, which employed about a quarter. Manitowoc County's employment in kitchenware was reported as being between 100 and 249 workers (USCB 2018).

### Furniture Manufacturing Survives with New Companies and Locations

Other industries have survived, but not the companies nor establishments that were prominent in the past. Furniture manufacturing employs nearly 50 percent more workers today than it did a century ago, but there have been major shifts in its location. In 1919, furniture was manufactured at 107 Wisconsin establishments, who employed 10,463 wage earners (USBC 1923). At that time the major centers of furniture manufacturing were Milwaukee, Kenosha, and Sheboygan. In 2017, furniture and related products were made at 371 Wisconsin facilities by 17,071 employees. Over 5,000 workers made furniture in Trempealeau County (USCB 2019b), where three plants are operated by the same company, Ashley Furniture Incorporated. While its headquarters and largest Wisconsin facility are in Arcadia, its others are in Independence and Whitehall. Ashley Furniture, founded in 1945, had 35 employees in 1976. It is now a conspicuous exception to the disappearance of other furniture manufacturers.

Furniture manufacturing employment in Brown and Dane Counties totaled 1,210 and 1,336 workers, respectively. Yet, the counties of the main furniture manufacturing centers of a century ago all now employ far fewer furniture makers than Trempealeau, Dane, and Brown Counties. Sheboygan County had 761 furniture manufacturing employees in 2017, Milwaukee County had 741, while Kenosha County only had 48 (USCB 2019b). The industry survived and expanded, but with new companies in different locations.

### Loss of Famous Companies and Products

Wisconsin has lost many companies whose products were once well known in both the United States and overseas. For example, the Parker Pen Company was founded in 1888 in Janesville, where pens were manufactured for over a century, until the factory closed in 2009 and manufacturing was moved to Mexico. The company headquarters had moved to England two decades earlier. Horlicks Malted Milk Company of Racine marketed its malt and dried milk products worldwide, initially for infants and then as a high-energy health food. Its availability spawned malt shops serving milk shakes containing its malted milk. While Horlicks products continue to be made overseas by a British conglomerate, their manufacture is no longer connected with Wisconsin. Hamilton Manufacturing in Two Rivers was the nation's largest manufacturer of wooden printing type used to print most of the nation's leading newspapers a century ago. It then evolved into Thermo Fisher Scientific, which manufactured a wide variety of office and laboratory fixtures, cabinets, and laboratory equipment. Yet, in 2013 it closed. Its plant that once employed 1,100 persons was razed over the following two years.

Just as employment burgeoned at the Manitowoc and Sturgeon Bay shipyards during World War II, the Badger Ordnance Works (Figure 11.18) was established just north of Prairie du Sac to manufacture smokeless explosives. Ten thousand acres from over 80 farms were acquired, 1,640 buildings

FIGURE 11.18. A small section of the huge Badger Ordnance Works, established during World War II, which had occupied over 10 square miles of land north of Prairie du Sac in Sauk County. This photo was taken in June 1990.

and bunkers were built by 12,000 construction workers in 1942, and 7,500 persons were employed at the world's largest munitions factory for the duration of the war (Thompson 1988; Meine 2015). While employment dramatically fell with the ending of that war, portions of the facility remained open to provide explosives during both the Korean and Vietnam Wars. Operations ceased in 1997. Today, the lands, of which many have undergone environmental cleanup with the industrial buildings razed, are largely devoted to agricultural research and environmental restoration, under the auspices of the U.S. Dairy Research Center, the Wisconsin Department of Natural Resources, and the Ho-Chunk Nation (Meine 2015). Little visible evidence remains of the massive industrial complex.

## HIGH-TECH INDUSTRIES

Several regions of Wisconsin are highly involved in cutting-edge high-tech industries. The expressways that extend from Chicago through Madison leading to Minneapolis–St. Paul and through Milwaukee leading to Appleton and Green Bay have been referred to as the "I-Q Corridor," given the large number of computer- and biotechnology-related industries that have located in those cities, along with La Crosse, Eau Claire, Chippewa Falls, Marshfield, Oshkosh, Manitowoc, Sheboygan, Racine, Kenosha, as well as Janesville and Beloit (WTC 2015). Biotechnology employed 24,000 Wisconsin workers, spread among 647 companies. Wisconsin's biohealth industry involves about 4,000 concerns, which collectively support 107,600 jobs (Newman 2018). While many of the jobs involve provision of health care and health testing, this sector also includes development and production of pharmaceuticals, medical devices, bioscience products, and related software. Employment in medicine manufacturing employs 3,875 workers in Wisconsin, yet professional and scientific research and development services employ nearly twice as many persons (USCB 2019b). For example, Exact Sciences employs a thousand workers in Madison, and Promega employs nearly as many in Fitchburg (Newman 2018).

In 2017, 246 establishments employed 19,602 workers in computer and electronic product manufacturing (USCB 2019b). Although the state's agreement to provide nearly $4 billion in incentives to Taiwan-based Foxconn to establish its factory in Racine County's Mount Pleasant remains highly controversial, its potential to directly employ 13,000 workers making glass for flat-screen televisions and smart phones, plus large numbers of additional workers at other factories in its supply chain, should elevate Wisconsin's status as a state engaged in cutting-edge industries. We begin by looking first at computing.

### Supercomputers

The earliest and fastest supercomputers manufactured in the nation came from Cray Computers, which was founded in 1972 in Chippewa Falls, the hometown of the computer's developer and the site of its manufacturing. By the end of the decade it was making the world's fastest computers, and it has continued making some of the world's fastest and largest supercomputers well into the twenty-first century. Now a subsidiary of Hewlett Packard, the company's business headquarters is on the West Coast, yet two of its manufacturing facilities remain in the Chippewa Falls–Eau Claire area. It has major research and development facilities in the Minneapolis–St. Paul area. Wisconsin had 1,700 persons employed in computer and peripheral equipment manufacturing in 2018 (USBLS 2019).

### Software Development

Computers cannot work effectively without software. According to the U.S. Bureau of Labor Statistics (USBLS 2019), in May 2018 there were 20,730 Wisconsin workers engaged in software development, including both software systems and applications. An additional 3,030 persons worked in web development, and 5,530 computer programmers were employed. Furthermore, 15,150 persons were computer systems analysts, whose duties, as described by the Bureau of Labor Statistics, are to "analyze science, engineering, business, and other data processing problems to implement and improve computer systems." Software development is concentrated within the Madison and Milwaukee metropolitan areas. In the Madison area, software development alone employed 8,770 persons in 2018, with 24,490 persons working in the broader computer and mathematical occupations. Within the Milwaukee-Waukesha metropolitan area, 26,940 persons worked in computer and mathematical occupations, of which 6,710 were writing systems and applications software (USBLS 2019).

Wisconsin's largest software company is Epic Systems, headquartered in Verona, in Dane County west of Madison, where 9,800 persons, including more than 5,000 programmers, were employed

FIGURE 11.19. Portion of the office campus of Epic Systems, a leading producer of medical records software, as viewed from Military Ridge State Trail just west of Verona, Dane County.

in 2018 (Kelly 2018). Its campus of 25 modern office buildings is sprawled across 1,100 acres (Figure 11.19). Epic specializes in healthcare software, with international sales and branch offices located in Europe, Asia, and Australia. Other major software companies in the Madison area include Single-Wire Software, which creates emergency communication software, and Widen, which creates image database software in Monona, nestled between Madison and Middleton. Overall, Dane County had 4,699 persons working in computer systems design and related services in 2018 (USBLS 2019), while these software and computer systems development companies employed many other persons in support capacity.

In the Milwaukee-Waukesha metropolitan area several major software developers operate, including Zywave in Milwaukee, which prepares personal financial software, and ReServe Interactive in Delafield, which focuses upon event management and hospitality software. Relying upon both their own software and that developed by others, Fiserv, a Fortune 500 company headquartered in Brookfield, a Milwaukee suburb, is a major provider of financial services technology used by many banks, credit unions, and insurance companies. In total the Milwaukee-Waukesha metropolitan area had nearly 6,000 persons working as computer network and users support specialists in 2018. Wisconsin's third-largest center for software writing includes the Green Bay and Appleton metropolitan areas, where 1,150 and 660 software writers were employed in 2018, respectively. Other metropolitan areas that had at least 250 software development employees included Oshkosh-Neenah and Wausau (USBLS 2019).

## WISCONSIN'S LARGEST COMPANIES: THEIR INDUSTRIES AND HEADQUARTERS

Wisconsin has the second-largest share of its population employed in manufacturing among the nation's states. Johnson Controls, headquartered in Glendale within Milwaukee County, led Wisconsin's firms included on the Fortune 500 list in 2016 (Weiland 2016). In that year Wisconsin had 10 companies on the list, of which Johnson Controls, Rockwell Automation, Oshkosh Corporation, and Harley-Davidson were engaged in traditional manufacturing, while Fiserv created software that underpins financial services technology. Three years later 9 Wisconsin-headquartered companies were on the Fortune 500 list (Weiland 2019), as Johnson Controls had been dropped, given its headquarters was relocated to Ireland, following its merger with Tyco. Nevertheless, its operational headquarters remains in Milwaukee.

The Fortune 1000 list for 2019 included 25 Wisconsin-based companies. Of the additional 16 companies, 8 are engaged in manufacturing in the state. These included Spectrum Brands (which includes Rayovac), Quad/Graphics, Bemis, Snap-On, Regal Beloit, A. O. Smith, Plexus, Modine Manufacturing, Rexnord, and Generac Holdings. Products of these firms include printed materials, packaging, tools, batteries, electric motors, generators, heat transfer equipment, air compressors, and water heaters.

### Insurance Carriers

Northwestern Mutual Life Insurance, Wisconsin's top-ranked Fortune 500 company, and American Family Insurance Group are joined by two companies within the Fortune 1,000 companies that are insurance providers. Insurance carriers are big business in Wisconsin, employing 61,858 persons in 2017. Not surprising, given the location of the two Fortune 500 Companies, Milwaukee and Dane Counties had the greatest number of persons employed by insurance carriers, 19,511 and 14,258, respectively. Brown County with 6,024 employees and Portage County with 3,587 workers are the homes of Humana and United Healthcare and Sentry Insurance, respectively. Five additional counties had more than 1,000 individuals working for insurance carriers: Marathon, Outagamie, Sheboygan, Waukesha, and Winnebago (USCB 2019b). These counties contain the headquarters of Wausau Insurance, Thrivent, Acuity, Secura, and Jewelers Mutual, among other companies.

Two of the top Fortune companies are utilities, while one company provides employment services, and another is Schneider National, a trucking giant. Firms that are privately owned, such as S. C. Johnson and Son, Inc. of Racine and Kohler Company of Sheboygan County, are not included in the listing.

### Role of Business Headquarters

In examining the Fortune 500 and Fortune 1000 lists, which fluctuate yearly depending upon corporate revenues, mergers, spin-offs, and relocation of headquarters, several observations can be made:

- Many of the largest employers in Wisconsin, be they retail giants like Walmart or manufacturing companies such as Kimberly-Clark, are headquartered out of state. Thus, they have a far different impact upon the local employment mix than do companies headquartered in the state.
- Companies headquartered in Wisconsin provide a much greater variety of job opportunities, particularly at the professional and management levels, to local residents and are frequently more invested in their communities, shown by donations to fund public parks, community centers, and other facilities. Their corporate officers are part of the Wisconsin community, interacting with other local residents and businesses, becoming invested in the welfare of the community in which they live and are raising their families.
- Mergers and acquisitions have transitioned many firms with Wisconsin roots to firms headquartered in other states or nations, resulting in the loss of management employment at many levels, with many of the operational decisions that affect these companies being made by individuals who lack connections with the workers and their community.
- Those cities that have retained or gained corporate headquarters display a greater diversity of employment opportunities and economic vitality than communities that lack such corporate management staffs. More of the corporate administrative salaries and tax dollars remain in Wisconsin.

As of 2019, of Wisconsin's 9 Fortune 500 companies, 7 were headquartered in the Milwaukee-Waukesha metropolitan area—5 within the city of Milwaukee proper, while both Madison and Oshkosh hosted 1 such company (Weiland 2019). When the 16 Fortune 1000 companies are added, 6 are in the Milwaukee-Waukesha metropolitan area, 3 are in the Madison area, and 2 are in the Oshkosh-Neenah metropolitan area. The following cities each have 1: Beloit, Green Bay, Kenosha, Racine, and Stevens Point.

While most of this chapter is devoted to manufacturing companies, the top four firms in Wisconsin are focused upon other activities, including insurance and retailing. Activities of these large companies underpin the economies of several major Wisconsin cities, the focus of the following chapter, which looks at the urban geography of Wisconsin.

*Chapter Twelve*

# Urbanization in Wisconsin

MILWAUKEE HAS BEEN Wisconsin's largest urban center since territorial days. The second largest city in Wisconsin has always been much smaller than Milwaukee. Today's second city, Madison, has a population that is under half of that of Milwaukee. In 1900 Milwaukee was over nine times larger than Wisconsin's second-largest city, which was Superior. Wisconsin's second-largest city position has been held by three additional places: Racine, Fond du Lac, and Oshkosh. Such urban history prompts several questions: (1) why has Milwaukee so long dominated Wisconsin's urban hierarchy, (2) what factors influenced the development of Wisconsin's larger cities, while other communities remained much smaller cities or villages, (3) why were certain cities ranked highly in the 1800s, but much lower the following century, (4) how might dramatic changes in ranking have influenced the character of the cities that we see today, and (5) what activities characterize Wisconsin's larger cities? In answering these questions we need to consider those forces that propelled many cities to rise to the top ten rank in Wisconsin, and the factors that prompted the rise of these and other larger communities to populations well above those displayed by smaller cities, villages, and unincorporated hamlets.

## BEGINNINGS OF URBAN DEVELOPMENT

Urban centers began developing in the land that was to become Wisconsin over two centuries ago. Even during the colonial period several locales became known for their gathering of fur traders, trading posts, and the occasional posting of soldiers. While they would not meet today's definition as being "urban places," locations such as Prairie du Chien and De Pere were nevertheless among Wisconsin's best-known places, points where goods were transported, stored, and exchanged. During Wisconsin's territorial days four federal court districts were established, with judges presiding in each existing county twice annually, with the exception of Milwaukee County, where court was held more frequently (A. E. Smith 1973). The focus upon places to conduct business and resolve legal disputes resulted in the establishment of urban centers—albeit often small—in the various counties.

For over a century the U.S. Census Bureau has defined communities with 2,500 persons as being urban (Ratcliffe et al. 2016). As noted in chapter 1, Wisconsin's criteria for incorporating new cities requires only 1,000 persons for communities located in rural areas, and only 150 persons to incorporate a village in such an area. What do most of these urban centers share, and what do the largest communities in the various counties share? Why have some of the urban centers become large cities, and for 12 of these, the central city of a Census Bureau–designated metropolitan area?

## Post Offices

Early clusters of population developed where farmers brought their grain to be milled or miners hauled their lead to be shipped to market. With the presence of potential customers, merchants opened stores near where these individuals gathered. One service that residents needed was offered by the post office. Before rural free delivery existed—and the four routes initiated out of Sun Prairie in 1896 were among the earliest in the nation—people had to visit the post office to send or receive mail (Janik 2017). They often shopped and conducted other business during their visit to the post office, which was often located in a store. Thus, a list of post offices, such as was used in producing Figure 7.3 that illustrates the settlement of Wisconsin, indicates those centers that had at least one compelling reason for nearby residents to visit on a regular basis. In 1853 Wisconsin had 496 post offices, ranging from just 1 in Marathon County (in Wausau) and 2 in Outagamie County (in Ellington and Kaukauna) to 37 in Dane County and 33 in Walworth County (Wisconsin Assembly 1853). For some of these post offices, their presence resulted in the smallest population cluster, a hamlet, while the locales of others developed into villages and cities, as additional functions and activities attracted more customers and settlers. By 1901 Wisconsin had 1,897 post offices, 51 fewer than it had before rural free delivery was initiated (Froehlich 1901).

## Town Halls

Settlements often grew around locales that had administrative functions. While the next section focuses upon communities that contained the county courthouse, far more town halls were erected across Wisconsin. Town officials collected taxes and conducted elections at the town hall, and residents of the town gathered at the hall during town meetings, making decisions regarding roads or schools. While some town halls are solitary structures, a few houses—and sometimes a business or two—are often located nearby. Because towns have the lowest level of government, far more local than the county, it is likely that their presence limited the desires of local residents to incorporate into villages. Indeed, a large number of Wisconsin's post offices were established in hamlets that were never incorporated, and which remain unincorporated, yet were provided services by the town. For example, services that might be otherwise found only in incorporated communities could be provided by "granting the powers of village boards, by vote of the people at the town meeting, to town boards in towns with 500 or more people containing one or more incorporated villages," and "when such powers have been granted, the board is empowered to provide the unincorporated villages with the usual urban improvements and conveniences" (Wehrwein 1935, 99–100).

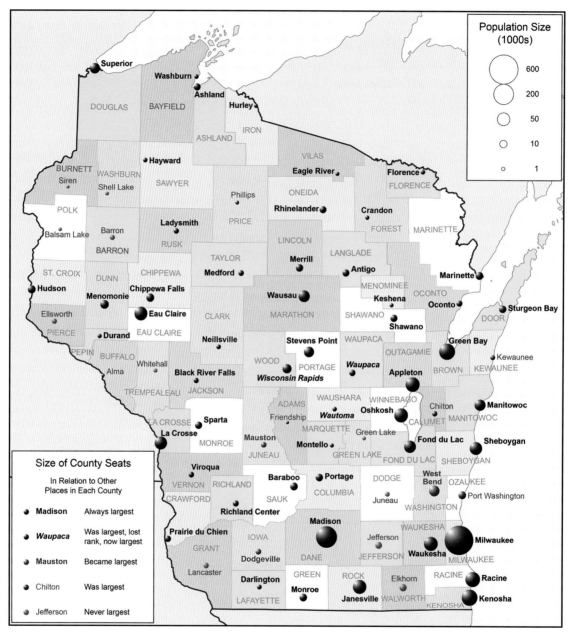

FIGURE 12.1. Wisconsin counties and county seats, indicating those county seats that are, or formerly were, the county's largest community. County seats that are currently the largest community and have been so since their first or second census are shown in bold black; those that were initially the largest and lost that status before regaining the largest place are in bold black italics. County seats that became the largest are in blue and those that were formerly the largest are labeled in red. County seats that neither previously nor currently are the largest community are shown in green. Map compiled from data in Mulhern (2009).

## Transport Centers

Communities that functioned as a transportation center attracted population growth. As noted in chapter 10, communities grew around railroad stations, where farmers brought their harvest and obtained shipments of supplies and farm equipment. Similarly, settlements sprang up along rivers and canals where wharfs could be established and along Lakes Michigan and Superior where boats and ships could find safe harbor. Where two railroads intersected, or where a railroad connected to a port, growth was enhanced, and this often led to the construction of warehouses and employment in unloading and loading the transshipped freight. Both railroad locomotives and steamboats required frequent servicing, needing places for refueling and obtaining boiler water, and transportation centers required many employees. Indeed, a century ago there were few sizable communities in Wisconsin that lacked railroad service (Hudson 1997). Where a railroad bypassed an existing settlement, its population often moved closer to the newly established station, such as occurred at Belmont. Where railroads established extensive yards and maintenance facilities, such as at Milwaukee, Neenah, North Fond du Lac, and Superior, population growth occurred.

## COUNTY SEATS: THEIR ROLE IN URBAN GROWTH

County seats have distinct advantages over other communities (Figure 12.1). While some counties designated a central locality as their county seat, others selected the largest population center at the time the county was established. Either way, usually county seats were or became and remained the largest population center in their counties. There are many examples of communities that fought (at least politically and legally) over their designation as county seat. For example, Dodgeville wrested the Iowa County seat from Mineral Point in 1859, the same year Alma took the Buffalo County courthouse from Fountain City. The most recent example occurred in Burnett County, where a new courthouse built in 1985 in the rural Town of Meenon near Siren replaced the previous one in downtown Grantsburg.

## Courthouses

The courthouse was frequently the most prominent public building in the county. In many communities it was conspicuously located at the center of downtown in the county seat. Not only were courthouses the first public building erected in the community, but they "reflected the hope communities had of serving as the county seat . . . with each community trying to outdo the other in the creation of a fitting building" (Bernstein 1998, 8). Although the traditional courthouse square is not as commonly seen in Wisconsin as it is in many southern states, examples do exist, ranging from the Brown County Courthouse in Green Bay (Figure 12.2), the Sauk County Courthouse in Baraboo, to the Grant County Courthouse in Lancaster (Figure 12.3). In Lancaster not only do retail buildings face all four sides of the courthouse square (Figure 12.4), but highway traffic is routed in a one-way loop around the courthouse, with vehicles from the north travelling along three sides of the square as they move through the city. In Monroe, Green County's courthouse square, with businesses on its four sides, is the focus of its biennial Cheese Days festivities.

FIGURE 12.2. Brown County Courthouse located on a courthouse square in downtown Green Bay. The courthouse, erected between 1908 and 1910, is one block south of the previous courthouse, with additional county offices housed across the street to the north and a local television station located across the street to the east.

FIGURE 12.3. Grant County Courthouse in Lancaster on a public square platted in 1837 to serve as the site of the courthouse. Two previous courthouses were erected on this site before the present structure was built in 1902. Wisconsin's oldest Civil War monument, dedicated in 1867, occupies the northeast corner of the courthouse square.

FIGURE 12.4. Red brick business buildings along Lancaster's West Maple Street, with the oldest erected in 1848, facing the Grant County Courthouse.

## Businesses and Newspapers

The county jail was typically erected near the courthouse, and sometimes it was attached to or occupied part of the courthouse. Besides the sheriff, the courthouse housed the county judge, the county clerk, the county treasurer, and the register of deeds, among a growing number of officials and employees. County seats attracted those who needed to conduct legal business and employees to run the county offices. Frequently law offices were erected nearby, along with retail businesses, hotels, and taverns to accommodate residents and visitors. Given the need to publish various legal notices, newspapers were also established in most county seats, which through their content and advertisements attracted greater attention to businesses in the county seat. In 1880 Wisconsin had 292 newspapers and periodicals (Warner 1880). Of these, 59.6 percent were published in a county seat. In that year newspapers were published in all but two of Wisconsin's county seats. One exception was the newly forming Langlade County, which lacked a newspaper. The other was in Green Lake County, whose four newspapers were published in Berlin and Princeton, which both served as county seat in the nineteenth century, rather than Dartford. It had acquired the county seat from Berlin and was to lose it to Princeton before regaining it in 1898, with its name becoming Green Lake.

## Population of County Seats

Wisconsin had 63 counties in 1880. Within more than three-quarters of these counties—a total of 48—their courthouses were located in the county's largest community. In addition, within two counties where the courthouse was located in a smaller community, it would be later moved to a larger community. The demographic dominance of county seats has persisted. In 2010 the county seats of 54 of Wisconsin's 72 counties were the largest community in the county, just 1 percent under what was noted 130 years earlier. Of the 18 counties where the county seat was not the largest community, in 9 counties the largest community was located at the edge of the county, or overlapping the county line. Examples of the largest city sitting on a county line include Appleton, which has grown across the Outagamie–Calumet County line such that more of its residents live in Calumet County than the population of Chilton, Calumet County's seat; Watertown, which has more residents living in both Dodge and Jefferson Counties than in their respective county seats; and River Falls, which has more Pierce County residents than Ellsworth.

The largest cities in Wisconsin are typically county seats. In 1860, among Wisconsin's 10 largest cities, only Watertown was not a county seat. By 1885 all 10 were county seats, and that status remained until 1950, when West Allis, a suburban city within Milwaukee County, joined the list of top 10 cities. Wauwatosa became a top 10 city in 1970. By 2010 both Wauwatosa and West Allis had fallen from the

TABLE 12.1 Wisconsin's ten largest cities during Federal and State Censuses, 1850 to 1900

| Rank | 1850 | 1860 | 1870 | 1875 | 1880 | 1885 | 1890 | 1895 | 1900 |
|---|---|---|---|---|---|---|---|---|---|
| 1 | Milwaukee | Milwaukee | Milwaukee | Milwaukee | Milwaukee | Milwaukee | Milwaukee | Milwaukee | Milwaukee |
| 2 | Racine | Racine | Fond du Lac | Oshkosh | Racine | Oshkosh | La Crosse | La Crosse | Superior |
| 3 | Kenosha | Janesville | Oshkosh | Fond du Lac | Oshkosh | La Crosse | Oshkosh | Oshkosh | Racine |
| 4 | Janesville | Madison | Racine | Racine | La Crosse | Eau Claire | Racine | Superior | La Crosse |
| 5 | Beloit | Oshkosh | Madison | La Crosse | Fond du Lac | Racine | Eau Claire | Racine | Oshkosh |
| 6 | Mineral Point | Fond du Lac | Janesville | Janesville | Madison | Fond du Lac | Sheboygan | Sheboygan | Sheboygan |
| 7 | Platteville | Watertown | La Crosse | Madison | Eau Claire | Madison | Madison | Eau Claire | Madison |
| 8 | Dodgeville | Sheboygan | Watertown | Watertown | Janesville | Sheboygan | Fond du Lac | Green Bay | Green Bay |
| 9 | Fond du Lac | Beloit | Sheboygan | Eau Claire | Appleton | Appleton | Superior | Madison | Eau Claire |
| 10 | Madison | Kenosha | Manitowoc | Green Bay | Watertown | Janesville | Appleton | Marinette | Marinette |

Data from reports of the U.S. Census with semi-decennial information compiled from various *Wisconsin Blue Books*. See Crane 1861; Froehlich 1901; Heg 1882; Mulhern 2009

top 10 cities, with all 10 being county seats. In discussing urban growth, the focus is upon the changing composition of the 10 largest cities, as shown in Tables 12.1 to 12.3.

## URBAN GROWTH DURING THE NINETEENTH CENTURY

The designation of county seats took place initially when the counties were established, yet some settlements were favored with far more growth than others. By the end of the nineteenth century, Milwaukee had 285,315 residents, and it was the nation's fourteenth-largest city. In 1900 4 other Wisconsin cities had over 25,000 inhabitants and were ranked within the top 150 cities in the United States: Superior, Racine, La Crosse, and Oshkosh. What were the keys to the growth of Wisconsin's 10 largest cities during the nineteenth century? While a variety of goods and services are provided to residents of all communities who support local merchants and service providers, communities that are most favored for growth are those that can market their products and services to a much wider region, bringing in outside income and promoting the hiring of more employees in those industries that market their products regionally, nationally, or globally. Indeed, for much of the first half of the nineteenth century southwestern Wisconsin produced lead for the national market, while during the second half of the century Wisconsin's sawmill centers shipped their lumber throughout the Midwest and Great Plains.

### Transportation Advantages

Proximity to resources and access to favorable transportation clearly played a major role in Wisconsin's urban development during the nineteenth century. While three of Wisconsin's eight largest communities were located in the lead-mining district in 1850, and they grew rapidly before entering a long period of stagnation, three were much larger port cities along Lake Michigan that served the surrounding agricultural hinterland. Milwaukee had a distinct advantage inasmuch as its harbor was much larger and had better links to its hinterland, first by plank road and later by railroad. So much of Wisconsin's wheat harvest was funneled through Milwaukee during the 1860s that it led the world in its shipping of wheat. Milwaukee had grown from rival settlements founded on opposite sides of the Milwaukee River by Solomon Juneau and Bryon Kilbourn. As John Gurda explains in his insightful history of Milwaukee, "The cross-river rivalries, paradoxically, aided the whole community's cause. . . . Competition gave rise to resources that might otherwise have emerged much later: two newspapers, a commodious courthouse, good hotels, and crude but serviceable street systems, all before 1840. These wilderness amenities helped to make Milwaukee a destination, a focal point for new arrivals from the East as well as aspiring farmers from the countryside" (Gurda 1999, 43).

By 1860, 8 of Wisconsin's 10 largest cities were on navigable lakes or rivers. The Rock River, which received shallow draft boats and steamboats to Janesville, was only useful for water power at Watertown (Figure 12.5), where the river fell 20 feet in just two miles. Madison lacked water transportation, even though it was positioned between two lakes. As discussed in chapters 6 and 10, regularly scheduled steamboat service served cities along Lakes Michigan and Superior and the Mississippi and Fox Rivers. Milwaukee had particularly good service, and on a busy day in the summer of 1873 it saw the arrival of 146 schooners, typically transporting cargo, and 29 steamers, which carried passengers (Gurda 2018, 15). Oshkosh, Eau Claire, Fond du Lac, and Marinette all benefited from log drives down

FIGURE 12.5. The Rock River at Watertown was well suited for providing hydropower. The industrial facilities that once lined the river are now gone or have been remodeled into loft apartments or replaced by modern apartment buildings. The Lower Watertown Dam remains, as does the former powerhouse, from which power was supplied to four industries a century ago.

the Wolf, Chippewa, and Marinette Rivers, highlighting the importance of transportation and proximity to raw materials for growth of many of the top 10 cities. As Robert C. Nesbit explains, "Even before the evolution of the Great Lakes commodity carriers of the latter decades of the century, and the concomitant development of harbors, adequate locks, channel improvements, and navigation aids, the mixed fleet of smaller sail and steam vessels assured Wisconsin's lake ports of a flow of raw materials and coal at a cost that made possible the commitment to heavy industry" (Nesbit 1985, 147). By the end of the century, the rise of mining in Minnesota's Mesabi Range and the shipment of wheat by railroads from the Northern Plains to wharfs along Lake Superior propelled Superior to its second rank among Wisconsin cities.

## Manufacturing Centers

Manufacturing employment brought large numbers of immigrants to Wisconsin's largest cities, where favorably capitalized industries developed. While manufacturing in Oshkosh and Eau Claire initially focused upon cutting of the lumber, making wood products, and manufacturing equipment to operate the mills and obtain the logs, activity in Milwaukee, Racine, and Sheboygan was more diversified, as noted in chapter 10. The magnitude of immigration was illustrated by five countries (Austria-Hungary, Belgium, Denmark, Germany, and Sweden and Norway, then united) posting consuls in Wisconsin by 1880, with Milwaukee, Madison, and Green Bay each having at least one consulate (Warner 1880). The 1880 Census of Occupations showed that 44.1 percent of Milwaukee's employed workers were born in the United States, while those born in Germany comprised 39.1 percent of the workforce. For many occupations, including laborer, shoemaker, brewer, saloon keeper and bartender, musician, baker, blacksmith, cooper, butcher, and leather maker, among others, German-born workers outnumbered native-born Americans (USCO 1883).

Manufacturing jobs underpinned the economies of the state's largest cities. In 1890 Milwaukee had 26,624 adult male industrial workers, with foundries and machine shops employing the largest proportion, 10.5 percent. Its manufacturing activities were the most diversified, with six industries each employing more than 1,000 workers in making carpets, railroad cars, leather, malt liquors, lumber, and brick. Oshkosh had 3,826 industrial employees, with 48.2 percent working in lumber mills. Similarly, La Crosse had 3,295 industrial workers, of which 52.9 worked in lumber mills. Racine was focused upon making agricultural implements and carriages and wagons, which accounted for 27.1 and 24.7 percent, respectively, of its 2,959 adult male manufacturing employees (USCO 1895).

## Business Services

The larger cities provided banking and financial services that were largely lacking in smaller communities. Wisconsin had 36 national banks in 1880, which at the time issued national currency bearing their names. Of these banks, 15 were located in the state's largest 10 cities, of which all but one (Eau Claire) had a national bank. Three were in cities that were to join the top 10 list. Eau Claire had a state-chartered bank, and 43.3 percent of Wisconsin's state banks in 1879 were located in the state's 10 largest cities (Warner 1880). In 1890 Milwaukee was the headquarters of Northwestern Mutual Life Insurance Company, Wisconsin's only class A life insurance company. It began business in 1858 and over a century and half later is the largest business headquartered in Wisconsin. Class B insurance companies were headquartered in Milwaukee, Madison, and Manitowoc, while Wisconsin had 52 class C life insurance providers, representing fraternal beneficiary orders or independent beneficiary societies. Of these, 17 were located in Milwaukee, and 6 were headquartered in Racine; all of Wisconsin's largest 8 cities had at least 1 in 1890 (Jenney 1895).

The largest cities were also the best connected for business interactions, whether one considers railroad access, streetcars, or postal service. Milwaukee, which was served by five railroads in 1890, had 98 trains to and from the city daily. Superior had 72 trains daily, La Crosse had 34 trains, Madison 32, Fond du Lac 28, Oshkosh 24, Appleton 22, Eau Claire 18, Racine 16, and Sheboygan 12 (Billings 1895). Not only were these cities well connected with their hinterland, but they also had streetcar service. All of Wisconsin's 10 largest cities in 1890 had street railways, but no other city in the state did. The greatest number and mileage were in Milwaukee, where five companies operated 52.75 miles, using a combination of animal (horse or mule), electric, and steam power. The second-largest mileage was in La Crosse (9.39 miles), followed by Oshkosh (8.0 miles of track) (H. C. Adams 1895). A decade later all 10 largest cities, along with 13 other Wisconsin cities, had post offices providing free delivery to homes and businesses, whereas residents in other communities needed to visit their post offices to retrieve their mail.

### Education and Medical Services

Before the end of the 1800s, free public high schools offering four years of courses were operating in 151 Wisconsin communities (Figure 12.6), and other communities operated three-year high schools or schools that did not meet the free criteria for listing (Froehlich 1899). In addition, most of the state's largest cities had institutions offering higher education. Private colleges, universities, or seminaries were operating in Appleton, Milwaukee (which had five), Racine (which had two), Watertown, and Waukesha, plus five communities that were never top 10 cities. The University of Wisconsin, whose first building in Madison was erected in 1851, had by far the largest faculty and number of students of the institutions of higher education at the end of the century (Figure 12.7), better than five times larger than both Beloit College and Lawrence University, and two and a half times the enrollment of

FIGURE 12.6. White Limestone School operated in Mayville, Dodge County, from 1858 until 1981, providing instruction from kindergarten through high school. It is currently a museum.

Oshkosh Normal School. State-financed normal schools, for the training of elementary and high school teachers, were operating in Milwaukee, Oshkosh, and Superior, along with four cities that were not in the top ten: Platteville, Whitewater, River Falls, and Stevens Point (Froehlich 1899).

By the end of the century, hospitals, mostly affiliated with religious orders, were operating in scattered locations. Wisconsin's first hospital that provided care supervised by a physician was St. John's Infirmary, established in Milwaukee in 1848. By 1885 there were "three denominational hospitals in Milwaukee, two in Racine, one at La Crosse, one in Ashland, two at Chippewa Falls, and one (short lived) at Madison" (Bardeen 1925, 253). Secular hospital care was focused upon Milwaukee. Its Milwaukee County Hospital, dating from 1870, was the "first general hospital in the state under secular control" (256), and a second secular hospital was established in Milwaukee in 1888. Milwaukee was also the site of the National Home for Disabled Soldiers, opened in 1867 as just one of three facilities in the nation to provide care to disabled soldiers following the Civil War. Over time it was expanded to become today's Zablocki Veterans Affairs Medical Center, the largest veterans' hospital in Wisconsin. Special purpose health facilities, such as tuberculosis sanitariums, developed largely in the early twentieth century.

FIGURE 12.7. Science Hall, erected in 1887 at the University of Wisconsin in Madison, has long been home to that university's Geography Department.

## Urban Amenities

Wisconsin's largest cities offered many amenities in the late nineteenth century that were unavailable in the smaller communities, similar to today's situation. These include full-time police and fire departments, public parks, paved streets, street lighting, waterworks, and sewers. For example, in 1890 Milwaukee had 3,682 street lights, most of which were gas or oil. Oshkosh had 168, mostly electric arc, La Crosse had 128, and Racine had 133, all of which were electric arc. Milwaukee had 249 miles of paved streets, of which 41 miles were paved with block, 29 miles were paved with wood plank, and 178 miles were surfaced with gravel, which at that time was considered being paved. La Crosse had 15 miles of streets paved with macadam, while Appleton had 14 miles of paved streets, mostly with gravel (H. C. Adams 1895).

Most larger cities had parks and venues for theatrical performances. For example, in 1880 Milwaukee had three theaters that together could accommodate 3,000 persons, plus twelve concert and lecture halls, with a combined seating capacity of 8,400, and one of its beer gardens had a concert hall that could accommodate 5,000 persons. That same year Fond du Lac had an opera hall that could seat 800, a music hall that could seat 600, and an armory that could accommodate 1,500. La Crosse had

FIGURE 12.8. Grand Opera House, erected in 1894, in New London, Waupaca County. This theater has been renovated and is used for both movies and theatrical performances.

1,000 seats in its opera house, with another 700 in its Germania hall. Madison had an opera house that could seat 600, plus seating at the University of Wisconsin. Oshkosh had two halls used for theatrical productions, together seating 1,300 (Waring 1887). Even more cities gained opera houses and theaters by the end of the century (Figure 12.8). Of course, given Wisconsin's ethnic population, Milwaukee led the nation in seating capacity of its beer gardens. In its account on the "Social Statistics of Cities," the 1890 Census reported that "Milwaukee had 5 beer gardens, with a reported seating capacity of 105,000, or 514 per each 1,000 of population" (Billings 1895, 41). That same year Milwaukee had 1,315 saloons.

## Legacy of Nineteenth-Century Urban Growth

The legacy of urban growth during the nineteenth century remains strong within many Wisconsin cities, as shown by the large number of downtown business buildings (Figure 12.9), churches, schools, and houses that remain from that period. Six of Wisconsin's 10 largest cities in 1860 were among the 10 largest cities 150 years later.

Today's housing stock in several cities is disproportionately old. For example, 37.8 percent of Milwaukee's housing units, according to the 2012–2016 American Community Survey (USCB 2019a)

FIGURE 12.9. Businesses continue occupying the Taylor-Goss Block, erected in 1880, along Second Street in Hudson, St. Croix County.

estimates, were built before World War II. Given that very few structures were erected during the Great Depression, most of these houses are over a century old. The preponderance of old housing stock, both working-class (Figure 12.10) and majestic houses of the wealthy, define the character of many Milwaukee neighborhoods today (Gurda 2016). In Racine 39.1 percent of the housing predates 1940, notably in its Southside Historic District (Figure 12.11). For Sheboygan it is 33.1 percent, and in Oshkosh it is 29.5 percent. In several cities of northern Wisconsin that rapidly grew in the late 1800s, the numbers are similarly high. Superior's percentage built in 1939 or earlier is 35.5. Ashland's figure is 41.6 percent, and Marinette's percentage is 36.3. Several cities in southwestern Wisconsin that blossomed during the lead-mining period have even larger shares of historic housing. For example, 51.2 percent of Mineral Point's housing units were erected before 1939, as were 44.7 percent of Shullsburg's dwellings.

Wisconsin's settlements clearly corresponded to Walter Christaller's central place model during the nineteenth century, and that spatial arrangement persists to the present. Small settlements and neighborhoods typically supported a general store, church, and post office, while residents needed to visit larger centers to obtain more specialized goods or professional services. The mix of services offered by the smaller centers changed during the twentieth century as transportation improved, with

FIGURE 12.10. Century-old two-story frame houses remain within many neighborhoods in Milwaukee. This photograph is along South Fourth Street within several blocks of St. Stanislaus Catholic Church.

FIGURE 12.11. Greek Revival house erected in 1842 in Racine's Southside Historic District. Many Wisconsin cities have fashionable residential districts dominated by houses erected during the nineteenth century.

gas pumps often added to the general store or small communities getting gasoline stations. Full department stores required a much larger customer base than small communities could provide, and thus were located in larger centers that included a number of surrounding smaller communities within their hinterland.

## URBAN GROWTH IN THE FIRST HALF OF THE TWENTIETH CENTURY

Multiple forces influenced Wisconsin's urban development during the twentieth century, of which changes in transportation technology played a major role. The first half of the century was dominated by railroads, interurbans, and streetcars, while the second half saw automobiles rising to prominence and the almost complete abandonment of passenger rail service. Improvements in medical care, particularly the establishment of general hospitals, focused the provision of health care upon larger urban centers, while the development of commercial radio and then television were likewise urban centered. Changes in immigration also occurred, with rapid immigration from Europe flooding the state's largest cities with industrial workers during the first 15 years of the century, while the last 15 years saw a wave of immigrants from Southeast Asia along with a steady flow from Latin America. During the century's middle six decades, little immigration occurred from abroad, but a massive

TABLE 12.2 Wisconsin's ten largest cities during Federal Censuses, 1900 to 1970

| Rank | 1900 | 1905 | 1910 | 1920 | 1930 | 1940 | 1950 | 1960 | 1970 |
|---|---|---|---|---|---|---|---|---|---|
| 1 | Milwaukee | Milwaukee | Milwaukee | Milwaukee | Milwaukee | Milwaukee | Milwaukee | Milwaukee | Milwaukee |
| 2 | Superior | Superior | Superior | Racine | Racine | Madison | Madison | Madison | Madison |
| 3 | Racine | Racine | Racine | Kenosha | Madison | Racine | Racine | Racine | Racine |
| 4 | La Crosse | Oshkosh | Oshkosh | Superior | Kenosha | Kenosha | Kenosha | West Allis | Green Bay |
| 5 | Oshkosh | La Crosse | La Crosse | Madison | Oshkosh | Green Bay | Green Bay | Kenosha | Kenosha |
| 6 | Sheboygan | Madison | Sheboygan | Oshkosh | La Crosse | La Crosse | La Crosse | Green Bay | West Allis |
| 7 | Madison | Sheboygan | Madison | Green Bay | Sheboygan | Sheboygan | West Allis | Appleton | Wauwatosa |
| 8 | Green Bay | Green Bay | Green Bay | Sheboygan | Green Bay | Oshkosh | Sheboygan | La Crosse | Appleton |
| 9 | Eau Claire | Eau Claire | Kenosha | La Crosse | Superior | Superior | Oshkosh | Sheboygan | Oshkosh |
| 10 | Marinette | Fond du Lac | Fond du Lac | Fond du Lac | Fond du Lac | Eau Claire | Eau Claire | Oshkosh | La Crosse |

Data from reports of the U.S. Census and 1905 Wisconsin Census. See Mulhern 2009.

TABLE 12.3 Wisconsin's ten largest cities during Federal Censuses, 1970 to 2020

| Rank | 1970 | 1980 | 1990 | 2000 | 2010 | 2010 Metro (MSA) | 2020 |
|---|---|---|---|---|---|---|---|
| 1 | Milwaukee | Milwaukee | Milwaukee | Milwaukee | Milwaukee | Milwaukee-Waukesha-West Allis | Milwaukee |
| 2 | Madison | Madison | Madison | Madison | Madison | Madison | Madison |
| 3 | Racine | Green Bay | Green Bay | Green Bay | Green Bay | Green Bay | Green Bay |
| 4 | Green Bay | Racine | Racine | Kenosha | Kenosha | Appleton | Kenosha |
| 5 | Kenosha | Kenosha | Kenosha | Racine | Racine | Racine | Racine |
| 6 | West Allis | West Allis | Appleton | Appleton | Appleton | Oshkosh-Neenah | Appleton |
| 7 | Wauwatosa | Appleton | West Allis | Waukesha | Waukesha | Kenosha | Waukesha |
| 8 | Appleton | Eau Claire | Eau Claire | Oshkosh | Oshkosh | Eau Claire | Eau Claire |
| 9 | Oshkosh | Wauwatosa | Oshkosh | Eau Claire | Eau Claire | Janesville | Oshkosh |
| 10 | La Crosse | Janesville | Janesville | West Allis | Janesville | Wausau | Janesville |

Data from reports of the U.S. Census. See Mulhern 2009.

inflow of African Americans from the lower Mississippi River Valley reshaped several of Wisconsin's largest cities. The first half of the century was dominated by two world wars and the Great Depression, while the postwar expansion of the economy and construction of a modern passenger car–based transportation network reshaped Wisconsin's urban geography during the last half century. We first consider the period up to 1950.

## Shifts in Urban Rankings

The list of Wisconsin's 10 largest cities underwent several notable shifts between 1900 and 1950, even though Milwaukee always occupied the top position and Racine held either second or third place. Superior started in the number two position but had disappeared from the list by 1950. Marinette and Fond du Lac, which had been in the tenth position, were off the list by 1950. Oshkosh and La Crosse fell from the top half of the list to the bottom half. These downward shifts can be attributed to the reliance of local economies upon the forestry and mining industries that peaked shortly after the century began, while those cities with more diversified manufacturing, particularly of transportation equipment and machinery, and those functioning as major wholesaling and warehousing centers, grew in prominence. At the same time, the relative position of Madison, buoyed by state employment,

TABLE 12.4 African Americans as percentage of total population in Wisconsin's largest cities

| Cities (1910 ranking) | 1910 (%) | 1950 (%) | 1980 (%) | 2010 (%) |
|---|---|---|---|---|
| Milwaukee | 0.26 | 3.42 | 23.10 | 44.78 |
| Superior | 0.45 | 0.01 | 0.36 | 1.44 |
| Racine | 0.29 | 2.09 | 14.74 | 22.57 |
| Oshkosh | 0.29 | 0.02 | 0.59 | 3.10 |
| La Crosse | 0.19 | 0.06 | 0.29 | 2.25 |
| Sheboygan | 0.03 | 0.02 | 0.12 | 1.80 |
| Madison | 0.56 | 0.67 | 2.70 | 7.26 |
| Green Bay | 0.18 | 0.03 | 0.25 | 3.55 |
| Kenosha | 0.15 | 0.45 | 3.62 | 9.95 |
| Fond du Lac | 0.25 | 0.05 | 0.09 | 2.55 |
| West Allis* | 0.00 | 0.05 | 0.07 | 3.64 |
| Eau Claire* | 0.17 | 0.04 | 0.25 | 1.14 |
| Wauwatosa* | 0.03 | 0.02 | 0.67 | 4.46 |
| Appleton* | 0.05 | 0.01 | 0.08 | 1.67 |
| Janesville* | 0.26 | 0.13 | 0.22 | 2.57 |

* Cities not in top 10 in 1910 but that were to join the listing in later years.
Data from reports of the U.S. Census: USBC 1912, 1952c; USCB 2019a.

the presence of the University of Wisconsin, and several major industries, steadily rose from seventh to second place among Wisconsin's cities.

These shifts were accompanied by profound changes in the racial composition of Wisconsin's cities, as those with growing industrial employment, such as Milwaukee, Racine, and Kenosha, attracted large numbers of African American workers during the two world wars, while those cities with stagnating or declining economies, such as Oshkosh, Fond du Lac, Sheboygan, and La Crosse, lacked jobs to attract newcomers (Table 12.4). At the same time, limitations on immigration that occurred with the beginning of World War I that were then codified into law in 1924 substantially reduced the prominence of the foreign-born population within Wisconsin's cities (Table 12.5). The last census conducted before World War I showed that among Wisconsin's 10 largest cities, the foreign-born share of their populations ranged from 16.1 to 35.8 percent in 1910. Forty years later that proportion was below 10 percent in half of the largest 10 cities. At the beginning of same period, in none of Wisconsin's cities did African Americans comprise as much as 1 percent of the population, yet by 1950 they exceeded 2 percent in two cities; however, their share had fallen in most of the largest cities. In both Oshkosh and Sheboygan, there were fewer than 10 African Americans, while in Milwaukee, the African American population had reached 21,772.

Table 12.5 Foreign-born population as percentage of total population in Wisconsin's largest cities

| Cities (1910 ranking) | 1910 (%)** | 1950 (%) | 1990 (%) | 2000 (%) | 2013–17 (%)*** |
|---|---|---|---|---|---|
| Milwaukee | 29.8 | 10.0 | 4.7 | 7.7 | 9.7 |
| Superior | 34.1 | 11.6 | 1.7 | 2.0 | 2.3 |
| Racine | 32.9 | 12.3 | 3.7 | 5.7 | 7.2 |
| Oshkosh | 22.4 | 5.4 | 2.8 | 3.4 | 3.0 |
| La Crosse | 19.9 | 3.3 | 4.2 | 3.3 | 3.1 |
| Sheboygan | 32.8 | 10.8 | 5.0 | 7.7 | 9.5 |
| Madison | 16.3 | 4.8 | 5.7 | 9.1 | 11.7 |
| Green Bay | 16.1 | 2.9 | 2.4 | 6.8 | 9.1 |
| Kenosha | 35.8 | 15.0 | 5.0 | 5.9 | 7.5 |
| Fond du Lac | 16.3 | 4.5 | 1.5 | 3.3 | 5.5 |
| West Allis* | 43.4 | 10.2 | 2.8 | 3.6 | 6.4 |
| Eau Claire* | 23.2 | 3.8 | 3.5 | 2.7 | 4.3 |
| Wauwatosa* | 24.5 | 6.1 | 3.9 | 3.8 | 5.3 |
| Appleton* | 19.4 | 3.7 | 2.8 | 5.1 | 7.0 |
| Janesville* | 14.4 | 4.5 | 1.9 | 2.4 | 3.5 |

* Cities not in top 10 in 1910 but that were to join the listing in later years.
** Foreign-born whites, as nativity not reported for nonwhites.
*** American Community Survey 5-year estimates only. Data no longer collected as part of regular census.
Data sources: USBC 1912, 1952c; USCB 2019a.

## Urban and Suburban Expansion

Improved passenger rail service, plus expanded interurban service linking many suburban communities to central cities and streetcar lines extending into more distant neighborhoods, facilitated the expansion of cities early in the twentieth century. For example, suburban communities flourished around Milwaukee, focusing upon their industrial facilities and linked to Milwaukee by streetcar, electric interurbans, and railroad lines. West Allis was described at the time as "the leading suburb of Milwaukee and its growth is notable" (Whitbeck 1921, 118), with 12,000 persons employed in its industrial facilities, of which Allis-Chalmers was the largest, not only in the suburb, but in the entire metropolitan area. Besides West Allis, three other suburbs were incorporated cities in 1920: South Milwaukee, Cudahy, and Wauwatosa, with the first two noted for their manufacturing facilities—Bucyrus Company and Cudahy Brothers, respectively—while the latter was more residential. Besides these close-in suburbs, interurban service extended outward from Milwaukee north to Sheboygan, west to Waukesha, and southwest to East Troy and Burlington; two lines went south to Kenosha, with one continuing to Chicago until service was abandoned in 1963 (Cutler 1965).

During the depths of the Great Depression one suburb, Greendale, represented an effort of the U.S. Resettlement Administration to establish "'greenbelt towns' on the outskirts of major cities" (Gurda 1999, 288). Only three such resettlement efforts were made nationally to relocate urban families into a greenbelt community, whose buildings were partly erected by Works Progress Administration labor. While built to accommodate about 2,500 residents, the village "has long since been overwhelmed by more prosaic development, [yet] the original Greendale, with its curving streets, cul-de-sacs, and picturesque town center, remains a much admired-example of Depression-era urban planning" (Gurda 1999, 289). The village of Kohler, located just west of Sheboygan, represents another example of a greenbelt suburb, although founded as a company town three decades earlier. Its carefully designed residential neighborhoods, over which the company still maintains design controls, are near its commercial area, including an upscale resort hotel, located across the street from Kohler Company headquarters and industrial facilities.

## Milwaukee

The city of Milwaukee expanded during this period. In 1902 it had the nation's "third highest number of persons per acre, trailing only Boston and Baltimore" (Gurda 1999, 183). Beginning with densely packed neighborhoods around a relatively small core at the beginning of the century, as shown by the U.S. Geological Survey's 1906 map of Milwaukee (Figure 12.12), developers created new neighborhoods that were annexed to the city, which extended services needed by the developers (Simon 1978; Rast 2007). Yet the resistance of some areas to annexation, preference for annexation into a suburban city or village with a wealthier population, or the desires of industrialists to locate in a community with lower tax rates or fewer environmental regulations, such as happened with Cudahy and South Milwaukee, resulted in much of Milwaukee being ringed by other incorporated communities by mid-century, limiting the city's further territorial growth. By 1957, with the establishment of the city of Greenfield, "the 'iron ring' of suburbs around the city of Milwaukee was complete" (Gurda 1999, 342).

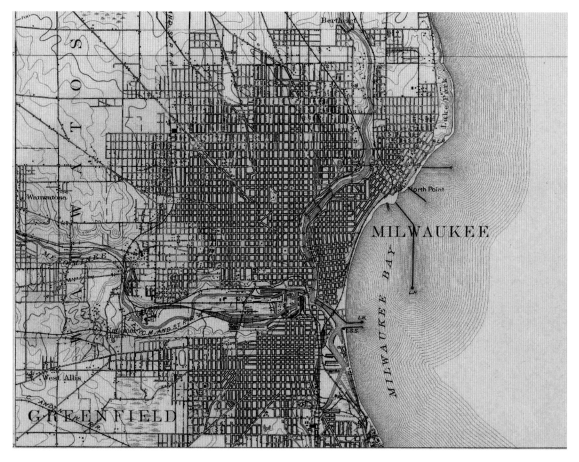

FIGURE 12.12. Milwaukee was a highly compact city in 1906, as shown on the U.S. Geological Survey's Milwaukee 15-minute topographic map.

No land in Milwaukee County remained outside an incorporated city or village. However, that did not limit the suburbanization that was extending south from Milwaukee and north from Chicago that Richard Cutler (1965) was to describe as metropolitan coalescence.

It was during the first half of the twentieth century that Milwaukee reached its greatest prominence as Wisconsin's leading metropolitan center. In 1950 the population within Milwaukee's city limits was 637,392, or 18.6 percent of the state's population. Milwaukee County held over 25 percent of the state's residents. Milwaukee was described just before World War II with many superlatives. "Milwaukee ranks twelfth in population among the cities of the United States and ninth in the value of manufactured goods. For health, safety, and solvency, it resolutely holds a place in the highest brackets among the 13 largest cities in the country. . . . [It has] the lowest homicide rate and the lowest motor death rate among all United States cities of more than 250,000 population" (WPA 1954, 241). By the 1920s Milwaukee County had about 25 percent of Wisconsin's hospital beds and 8 of the state's 32 hospitals with 70 or more beds. Of the 13 Wisconsin hospitals that were approved by the American Medical Association in 1923 to provide internships for newly degreed physicians, 6 were in Milwaukee. Of the other 7 hospitals approved for internships, 4 were located in top 10 cities (La Crosse, with two hospitals, Madison, and Oshkosh), plus one in Fond du Lac, which had just lost its top 10 status (Bardeen 1925). With the other two located in Ashland and Marshfield, the distribution of today's medical care was largely in place a century earlier.

## Most People Lived in Smaller Communities

In contrast with the largest cities that provided their residents with the greatest variety of goods and services, most Wisconsin residents lived in smaller cities, villages, or hamlets, if not on dispersed farmsteads. By the middle of the century, two geographers, Glenn T. Trewartha (1943) and John E. Brush (1953), had intensely studied small hamlets and central places within 12 counties in southwestern Wisconsin, providing a far more comprehensive example of midcentury settlement patterns than available for other parts of the state.

### *Hamlets*

The smallest agglomerations of population are the hamlets, which "serve as trade centers of the most rudimentary sort . . . [with] at least five residential structures or other buildings used for commercial or cultural purposes clustered within one-quarter of a mile" (Brush 1953, 385). Trewartha (1943, 40) reports, "in the 12 Wisconsin counties studied there were in 1940, as determined by field observation, 167 bona fide hamlets with populations of 20 to 150." At that time hamlets included at least one retail or service establishment, but "only grocery stores and elementary schools are typical . . . [yet] taverns, filling stations, and churches are common" (Brush 1953, 385). Indeed, filling stations were the most common business in hamlets, and gasoline pumps could be found in 86 percent of the hamlets, although often in front of some other business establishment rather than a full filling station. General stores were typically present in hamlets, and three-fifths of them had taverns. Nearly half of the hamlets had a cheese factory or creamery, over half had a church, nearly four-tenths had a school, and one-fourth had the town hall (Trewartha 1943).

The vast majority of the unincorporated hamlets within the region included, or had included, a post office, and nearly half had been platted by their developers. While most of the hamlets within southwestern Wisconsin were located at intersections, particularly of roads other than state and U.S. highways, only one-eighth were along railroads (Trewartha 1943). Even when studied in the mid-twentieth century, the hamlets, typically located a little over four miles from another, were experiencing change. Not only was an overall decline in rural population occurring, but "the improvement and rerouting of roads, the common use of motor cars, the decline of the fourth class post-office, and the serious depression of the last decade [the 1930s] . . . forced readjustments upon rural trading communities, and [with] these adjustments some hamlets may suffer and a few become extinct" (Trewartha 1943, 50). Nevertheless, new hamlets were also being formed, "catering chiefly to the transient car population, . . . many of whose residents are non-farmers" (50), providing services such as tourist cabins, filling stations, taverns, but not the general store found in earlier established hamlets.

FIGURE 12.13. Downtown businesses, including a post office and supper club, in the village of Bagley, Grant County. This village had a population of 293 in 1940 but had grown to 379 by 2010, even though it is only reached by county roads, as the nearest state highway is 10 miles distant.

## Villages

Villages, of which many were incorporated, were communities of more than 150 residents, typically of sufficient size "for a business core to develop, a feature that is not conspicuous in most hamlets" (Trewartha 1943, 38). Villages (Figure 12.13) at the time were described as having "in addition to the groceries, taverns, and filling stations found in hamlets, at least four other retail businesses, selling autos, implements, appliances, lumber, hardware, or livestock feed [along with] three other essential services, such as auto repair, banking, and telephone exchange" (Brush 1953, 385). In the mid-twentieth century, 70 percent of the villages in southwestern Wisconsin had public high schools.

## Small Cities

Fourth-class cities, those that typically had at least 1,000 residents, can be viewed as market towns, providing a greater variety of retail establishments and services than villages. In 1950, 19 such cities existed in southwestern Wisconsin, providing "certain specialized types of retail goods, personal services, and recreational facilities related directly to the concentration of town dwellers and not maintained in smaller centers" (Brush 1953, 387). These included meat markets, bakeries, news dealers, gift and sporting goods shops, laundries and dry cleaners, taxi services, parks, libraries, golf courses, and medical clinics and hospitals, along with city water and sewage systems (Brush 1953), among many other retail shops and amenities. Grocery supermarkets, drugstores, department stores, jewelry stores, and weekly newspapers were commonly found in these small cities. These communities provided accommodations to commercial travelers, such as traveling salesmen, having hotels and passenger railroad terminals. In 1950 Wisconsin had 132 fourth-class cities, of which 115 had populations ranging from 319 to 5,000 (WLRL 1950).

## URBAN GROWTH IN THE LAST HALF OF THE TWENTIETH CENTURY

The urban population of Wisconsin outnumbered the rural population by 57.9 to 42.1 percent in 1950. By 2000, urban residents accounted for 68.3 percent of Wisconsin's population. While Milwaukee County had both the greatest urban population, 937,009 persons, and the greatest proportion of its population that the Census Bureau categorized as being urban, 99.7 percent, urban residents accounted for over half of the population of 23 additional counties in 2000. Besides Milwaukee County, urban residents exceeded 80 percent of the population within seven more counties: Brown, Dane, Kenosha, La Crosse, Racine, Waukesha, and Winnebago, which included the cities of Green Bay, Madison, Kenosha, La Crosse, Racine, Waukesha, and Oshkosh, respectively. The city of Milwaukee was home to 16.3 percent of Wisconsin's urban residents, or 11.2 percent of the state's total population (USCB 2019a).

### Shifts in Urban Rankings

Milwaukee was the only top 10 Wisconsin city to have lost population over the previous half century, after peaking at 741,324 in 1960 before losing 144,350 residents over the next 40 years. Several suburban cities similarly peaked and subsequently declined or experienced small gains and thus fell in their rankings. For example, West Allis, which was the state's seventh-largest city in 1950, experienced a 58.7 percent increase in population over the next decade, becoming the state's fourth largest city. While it posted a small increase during the 1960s, reaching 71,723 residents in 1970, its state ranking had fallen to sixth. It fell every decade since then and was Wisconsin's tenth-ranked city in 2000. As a suburban city, surrounded by other incorporated communities, West Allis suffered a population exodus akin to what Milwaukee was experiencing, particularly with the demise of Allis-Chalmers, which had been Wisconsin's largest corporation. A similar trajectory was followed by Wauwatosa, which reached its maximum population of 58,676 in 1970, when it was the state's seventh-largest city, before steadily declining for the rest of the century. As suburban growth from the Milwaukee area expanded westward across the county line, Waukesha, which experienced growth every decade during the twentieth century, was the state's seventh largest city in 2000.

Given the substantial urban growth that occurred during the last decades of the twentieth century, the U.S. Census Bureau began reporting statistics for metropolitan statistical areas. Wisconsin's largest was the Milwaukee-Waukesha primary metropolitan statistical area (PMSA), which included all of Milwaukee, Ozaukee, Washington, and Waukesha Counties in 2000 and had a population of 1,500,741, or 28.0 percent of Wisconsin's residents. To the south was the Racine PMSA, with a population of 188,831. The Census Bureau added the Racine County data to that of the Milwaukee-Waukesha PMSA, reporting the population of the Milwaukee-Racine combined metropolitan statistical area (CMSA), which covered five counties and included 31.5 percent of the state's population. To the south, Kenosha County constituted the Kenosha PMSA, which was included within the Chicago-Gary-Kenosha CMSA. Given that many residents had come to rely upon commuting to work within a metropolitan area, which shared a passenger airport, major newspaper(s), and television stations resulting in a regional media market, Wisconsin's larger urban areas during the twentieth century had largely outgrown their city limits.

### Milwaukee: Wisconsin's Primate City

Milwaukee can be viewed as functioning as primate city, given how it has long dominated the economic activity of the state, even though it lacked the state capitol. Wisconsin's largest banks in the last half of the twentieth century, including First Wisconsin National Bank, the state's largest, along

FIGURE 12.14. Banking and legal office buildings in Milwaukee. The high-rise building on the right is the 411 East Wisconsin Center, housing law offices and investment and insurance offices, among other tenants. The older building in the center is the Milwaukee Federal Building and U.S. Courthouse, while the high-rise building behind it is the US Bank building. The Foxconn NA headquarters are hidden between the courthouse and US Bank.

with Marshall and Ilsley Bank, Marine Bank, and Continental Bank and Trust, among others, were clustered in downtown Milwaukee east of the Milwaukee River (Figure 12.14). While branch banking only came at the end of the century, Milwaukee's largest banks developed bank holding companies, including banks in many other communities that were ostensibly independent, given their separate offices and boards of directors. Milwaukee was also home to several major financial services companies, such as Fiserv and M&I Data Services, world leaders in bank data processing. The city had the corporate headquarters of 11 Fortune 500 companies as recently as 1980, but the recession of the early 1980s and deindustrialization reduced that number to 6 by the end of the decade (Gurda 1999).

Retail businesses, once focused upon department stores in the central business district and in several established neighborhoods, were beginning to relocate from the downtown. While the Grand Avenue Mall, extending over four blocks and opened in 1982 to connect several established department stores and the Plankinton Arcade built in 1916, continued to attract downtown shoppers, suburban shopping centers provided competition. Although the first shopping centers were near Milwaukee's "historic edge," the "next generation of centers—Brookfield Square (1967), Southridge (1970), and Northridge (1972)—reflected the progressive unbundling of the city . . . [being] miles farther removed . . . [and] significantly larger than their predecessors" (Gurda 1999, 384–385). These shopping complexes increasingly attracted customers, given their extensive selection of stores, spacious parking lots, and proximity to newly developed residential areas.

In 1950 Milwaukee had 18.6 percent of the state's population, down from 19.7 percent in 1930. Rapid expansion of highways and expressways and rapid suburbanization, with the conversion of former farmland into tract housing, shopping centers, and industrial parks, contributed to the city's loss of population during the last half of the twentieth century. Even Milwaukee County lost population between 1990 and 2000, the only county in the state to do so, while Waukesha and Washington Counties displayed growth rates exceeding both the state and national average.

*Proximity to Chicago*

Milwaukee's proximity to Chicago has long constrained its functioning as a regional center. A renowned urban geographer noted, "As the primate city of Wisconsin . . . Milwaukee has an important function as a trade center, although with respect to many wholesale and retail functions it lies within the shadow of Chicago, 85 miles to the south" (Mayer 1975b, 101). This is even more true for those cities that lie between Milwaukee and Chicago, which affects their economic base (Mayer 1975a). Chicago was, and remains, the nation's railroad hub, and rail traffic from Milwaukee heading south and east was routed through Chicago. When railroad ferries operated across Lake Michigan, Milwaukee shippers were assured rates comparable to those from Chicago, yet that service to Milwaukee ended in 1979. Furthermore, "in air transportation, too, Milwaukee lies in the shadow of Chicago. . . . Located a short distance from [Chicago] O'Hare, the world's busiest airport, however, Mitchell Field has somewhat less traffic than the size and importance of metropolitan Milwaukee would otherwise indicate" (Mayer 1975b, 102). By the 1990s General Mitchell Airport was the hub of a large regional airline, Midwest Express, providing connecting service to several other Wisconsin cities, as well as nonstop flights to both the East and West Coasts. The airport was considered Chicago's third airport, but it was not to maintain its Wisconsin hub function.

### White Flight and Changing Racial Composition

White flight in response to rapidly changing racial composition of many neighborhoods, including integration of public schools that local officials had long delayed, spurred the growth of suburban communities and changed Milwaukee into a highly segregated city. A legacy of the Depression era, redlining had defined much of the city west of the Milwaukee River as "'hazardous' for investment . . . [as] these neighborhoods were old and in poor condition and had African American people living there" (Foltman and Jones 2019). This designation focused federal mortgage assistance upon non-minority neighborhoods, as banks generally declined to make loans within redlined neighborhoods. This, along with deed restrictions prohibiting sales of property to nonwhites, such as in Washington Heights in suburban Wauwatosa, led Milwaukee to be "among the most segregated" cities in nation by the 1960s (Foltman and Jones 2019). With the demise of many of its industries in the 1980s, Black unemployment surged, and median household income in Milwaukee was less than half that within the suburbs in 1990 (Gurda 1999). As Gurda (1999, 389) explains, "the growing concentration of poverty claimed an increasing share of local government's decreasing financial resources, and growing concerns about schools, crime, and taxes gave middle-class whites even more reasons to leave."

By 2000 Milwaukee's population was 37.3 percent African American, but this population was largely crowded into the former German neighborhoods north of downtown and west of the Milwaukee River. In contrast, many suburban communities had few Black residents. For example, Cudahy was 0.9 percent African American, Shorewood was 2.4 percent, St. Francis was 1.0 percent, Wauwatosa was 2.0 percent, and both Waukesha and West Allis were 1.3 percent (USCB 2019a).

### Madison and the Fox Valley

Madison gained 113,369 residents between 1950 and 2000, more than doubling its population. While Milwaukee was clearly the state's most important business center, given the presence of its banks, insurance companies, and many corporate headquarters, Madison had the state's leading university and the state's administrative offices. It was corporate headquarters of Rayovac, Oscar Mayer, and American Family Insurance. The presence of the state's top research university, which had an enrollment of 40,740 students during the 1998–99 academic year, was a strong enticement to a variety of businesses and industries that benefited from proximity to its scientific expertise. Incorporated suburban communities, including Middleton, Fitchburg, Monona, and others had come to surround much of Madison, and Monona and Maple Bluff are themselves surrounded by Madison.

The area along the Fox River from Oshkosh through Green Bay included two metropolitan statistical areas in 2000, the Green Bay PMSA which included Brown County, and the Appleton-Oshkosh-Neenah PMSA, which included Calumet, Outagamie, and Winnebago Counties. Long noted for its paper industry, this four-county region had grown to include 585,143 residents, including 3 of Wisconsin's 10 largest cities. As described in the next section, this region has become one of Wisconsin's growth nodes.

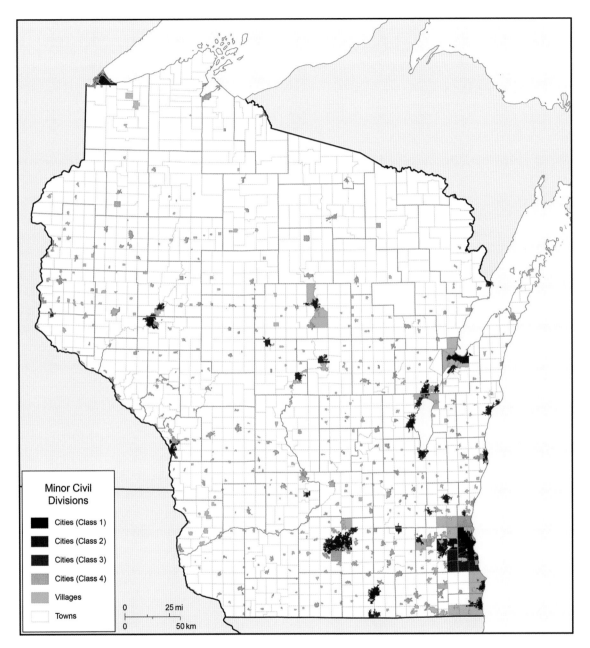

FIGURE 12.15. Incorporated cities and villages in Wisconsin in 2019. Cities are shown by class 1–4. Some villages now occupy the entirety of their former town. Data from U.S. Census Bureau (https://www.census.gov/geographies/mapping-files/time-series/geo/carto-boundary-file.html) and WLRB (2019).

## WISCONSIN'S URBAN CENTERS TODAY

Much of what we see today within Wisconsin's cities and villages is a legacy of the past, which necessitated our discussion of the historical geography of the state's cities. For many communities, trends that were well underway during the last decades of the twentieth century have continued into the present, whether we look at locations of the metropolitan areas or the distribution of Wisconsin's villages and cities (Figure 12.15). As of April 2, 2019, Wisconsin had 190 cities and 412 villages (WLRB 2019). At the beginning of 1999 the number of cities was the same, yet Wisconsin then had 395 villages.

### Metropolitan Areas of Wisconsin

All but one of the villages established in Wisconsin between 2000 and 2019 are within U.S. Census Bureau–defined metropolitan statistical areas. Thus, while metropolitan areas increasingly dominate Wisconsin, their central cities are increasingly surrounded by incorporated suburbs, such as Milwaukee began experiencing a century earlier. For example, since 2010 portions of Winnebago and Calumet Counties have been incorporated into the villages of Fox Crossing and Harrison, respectively, both now part of the Appleton-Oshkosh-Neenah combined statistical area. Similarly, to the north and south of Wausau, respectively, the villages of Maine and Kronenwetter incorporated, with the latter being the largest village in the state, with an area of 52 square miles. Five new villages incorporated within the Racine and Kenosha MSAs, including Mount Pleasant, the site of the giant Foxconn manufacturing facility, and Caledonia, immediately to its north. Both villages have populations of about 25,000.

Metropolitan areas now cover all of Wisconsin's Lake Michigan shoreline south from the Door-Kewaunee County line with the exception of Manitowoc County, which is considered a micropolitan statistical area. The Green Bay MSA includes both Kewaunee and Oconto Counties as well as Brown County, and the Fox Valley lies within three additional MSAs: Appleton, Oshkosh-Neenah, and Fond du Lac. A continuous swath of metropolitan and micropolitan areas extends west from Lake Michigan to the Mississippi River, including the state's two largest combined metropolitan statistical areas, the Milwaukee-Racine-Waukesha CMSA and the Madison-Janesville-Beloit CMSA. Sprawling east from Minnesota, the Minneapolis–St. Paul-Bloomington MSA includes Pierce and St. Croix Counties, while smaller MSAs are focused on La Crosse, Wausau, Eau Claire, and Duluth, which includes Wisconsin's Douglas County. While much land within the metropolitan statistical areas is indeed urban, the extent of the built-up areas is somewhat deceptive, given that large portions of these areas are rural and include small cities and villages, surrounded by farm fields or forests. Let's first consider the metropolitan impacts upon these rural areas.

### Metropolitan Services and Hinterlands

Television market areas are typically focused upon their surrounding metropolitan areas. There are no commercial television stations, except for ones in Crandon, Antigo, and Rhinelander and several repeater or translator stations, outside of the state's metropolitan areas. In addition, there are several metropolitan areas—Appleton, Oshkosh-Neenah, Sheboygan, Racine, and Janesville-Beloit MSAs—that must rely upon broadcasts from stations in Green Bay, Milwaukee, or Madison, respectively, for television service. Kenosha borders the Chicago television market. In contrast, several much smaller metropolitan areas, such as Eau Claire, La Crosse-Onalaska, and Wausau, have two or more commercial network stations.

Cities with commercial television stations, as well as advertisers whose businesses are focused upon the primary urban center of the broadcast area, dominate the local news and business coverage, giving them commercial and political advantages within their hinterlands. Furthermore, several metropolitan areas are split between different congressional districts, and one of the congressional districts (Wisconsin's sixth district in the 2010s) was spread over four television markets (Green Bay, Milwaukee, Madison, and Wausau) with no station broadcasting from within the district, diminishing the political discourse of residents in metropolitan areas without television stations.

Airports, as described in chapter 10, provide commercial passenger service to eight Wisconsin destinations, seven of which are in metropolitan areas, with the other one offering the least service. As with television market areas, six Wisconsin metropolitan areas lack commercial air service, with those living in Kenosha, Racine, Janesville, Sheboygan, and Fond du Lac having the greatest distance to travel to an airport, yet many are within an hour's drive to one of the state's two busiest airports. Downtown Oshkosh is 20 miles from Appleton International Airport, the state's third-busiest airport. In contrast, downtown Sheboygan is 67 miles from Milwaukee's General Mitchell International Airport, the state's busiest.

Wisconsin's metropolitan cities provide many specialized services and shopping opportunities that are not available in smaller cities and villages. For example, we already reviewed how the state's university system and major hospitals were focused upon the state's larger communities even a century ago. All of the state's four-year public universities are located in either a metropolitan or micropolitan area (Figure 12.16). Eleven Wisconsin hospitals have the capacity to provide "initial definitive trauma care regardless of the severity of injury" (WDHS 2019c), operating Level II trauma care facilities or Level I facilities, which are also engaged in research and teaching.

Wisconsin's three Level I trauma centers are Froedtert Hospital and Children's Hospital of Wisconsin in Milwaukee and University Hospital in Madison. Hospitals with Level II trauma centers are located in Green Bay, Neenah, Marshfield, La Crosse, Eau Claire, Wausau, Janesville, and Summit (Waukesha County). All of these except Marshfield, which is micropolitan, are in metropolitan areas (Figure 12.17). Of Wisconsin's Level III trauma centers, which provide "assessment, resuscitation, stabilization, and emergency surgery and arrange . . . transfer to a Level I or II facility for definitive surgical and intensive care as necessary" (WDHS 2019c), all but three (Mauston, Medford, and Rhinelander) are located in Wisconsin's metropolitan or micropolitan areas. Rhinelander has the northernmost Level III trauma center in Wisconsin, and even Level IV centers, which can offer little other than patient stabilization and advanced life support preparatory to transit to a better-prepared hospital, are sparsely distributed across the northern third of the state. Before we explore the orientation of retail marketing toward metropolitan centers, we need to consider that the largest cities also face serious challenges.

## MILWAUKEE: CHALLENGES AND OPPORTUNITIES

Milwaukee remains the state's largest city, its only first-class city. It is ranked first in many things, not all of which are desired attributes. Its population and that of the county are highly segregated by race and ethnicity (Figure 12.18). Milwaukee's "urban landscape and housing markets have long been

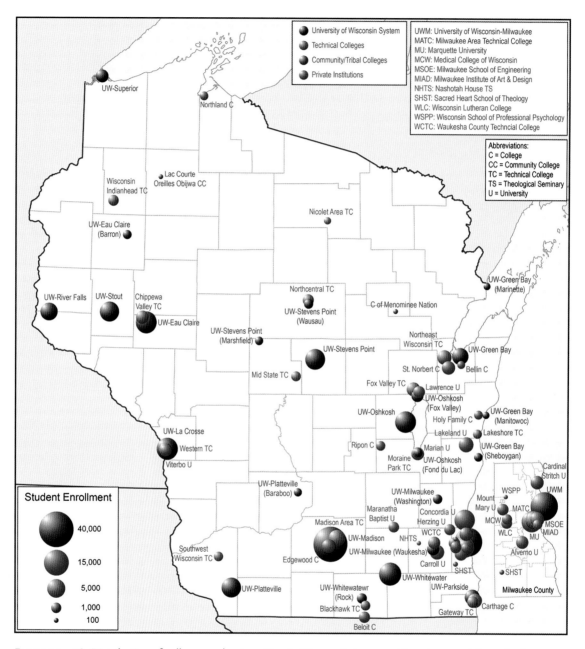

FIGURE 12.16. Distribution of colleges and universities in Wisconsin in 2019. Map shows public and private four-year colleges and universities, two-year colleges, seminaries, and technical schools. Data from Wikipedia (https://en.wikipedia.org/wiki/List_of_colleges_and_universities_in_Wisconsin).

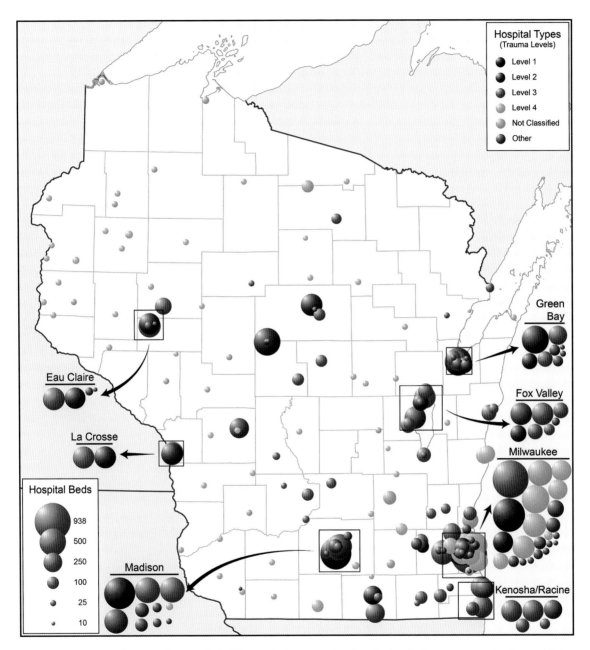

FIGURE 12.17. Distribution of hospitals in Wisconsin in 2019, showing the level of trauma care that is provided. Level I trauma hospitals provide the most sophisticated care, while Level IV trauma centers prepare patients for transfer to higher level centers. Data from WDHS (2019c).

marked by racial segregation, a process deepened by economic restructuring, White flight, and suburbanization. Indeed, 2010 census data identify the city as the most racially segregated city in the nation" (Bonds, Kenny, and Wolfe 2015, 1070). Black suburbanization in Milwaukee's suburbs was the smallest among the nation's 100 largest metropolitan areas (Ward 2007). Failure of external nonprofit groups to involve local residents and engage in their issues, rather than those conceptualized by outsiders, hinders neighborhood revitalization in certain minority neighborhoods, given "the significance of race in structuring inequality" (Bonds, Kenny, and Wolfe 2015, 1080). As Ward (2007, 786) explains, "between 1990 and 2004, the income of residents in inner-city neighborhoods, adjusted for inflation, declined by between 2.8% and 8.6% in Milwaukee's near northwest side."

## Inner-City Poverty and Segregation

Income in Milwaukee's inner-city neighborhoods averages less than half of that in the suburbs of Milwaukee, Waukesha, Ozaukee, and Washington Counties. Crime rates are disproportionately high within these neighborhoods, while the provision of a variety of urban amenities is lacking. The area between Interstate Highway 43 and 26th Street, extending from North Avenue to Capitol Drive, has been a focus of longitudinal study of Milwaukee's inner-city woes. "Sprawling across the city's north side, Milwaukee's zip code 53206 has come to epitomize the social and economic distress facing inner city neighborhoods in this hypersegregated metropolitan area. Over the past decade, [there were] enormous challenges facing residents of 53206—concentrated poverty, pervasive joblessness, plummeting incomes, segregated schools, violence and mass incarceration" (Levine 2019, 8). Statistics for this zip code from the American Community Survey for 2013–2017 illustrate the continuing socioeconomic inequities suffered by the neighborhood that is 95 percent African American, as Marc Levine explains:

- the poverty rate in 53206 was six times greater than in the Milwaukee suburbs;
- over half of the zip code's children lived in the poverty;
- fewer than half of prime working-age males (ages 25–54) in the neighborhood were employed;
- household incomes in 53206 hit new lows while residents continued to abandon the zip code in droves and the neighborhood experienced massive population loss;
- one-quarter of housing units in the zip code were vacant;
- black children born in 53206—especially black males—have experienced virtually no upward intergenerational economic mobility over the past 35 years;
- over 15 percent of black males in their late 20s and early 30s, born and raised in low income households in census tracts across 53206, were incarcerated in jail or prison. (Levine 2019, 11)

Environmental shortcomings are also apparent when comparing the inner city with other parts of Milwaukee and its metropolitan area. Incidents of lead poisoning of children, which can cause developmental problems, are excessively high among Milwaukee's African American schoolchildren (Amato et al. 2013). The availability of green space and urban tree canopy cover within Milwaukee's African American inner city is demonstrably less than elsewhere in the city and county, and "there is a socially inequitable distribution of urban trees within Milwaukee" (Heynen, Perkins, and Roy 2006, 6).

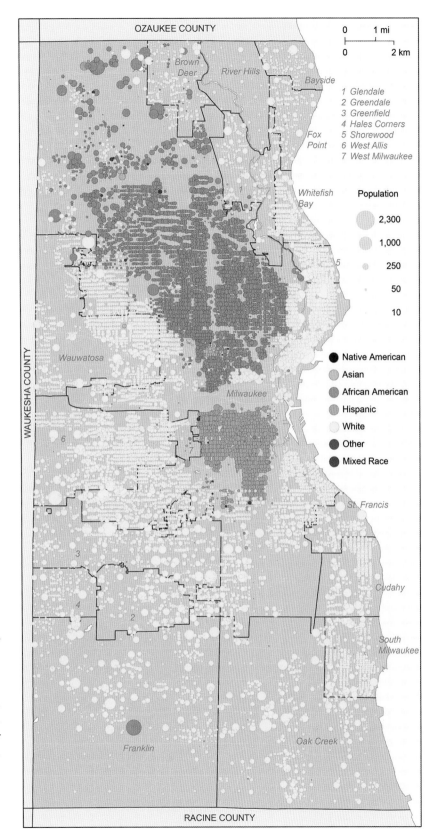

OZAUKEE COUNTY

Brown Deer    River Hills    Bayside

1  Glendale
2  Greendale
3  Greenfield
4  Hales Corners
5  Shorewood
6  West Allis
7  West Milwaukee

Fox Point

Whitefish Bay

Population

2,300
1,000
250
50
10

Native American
Asian
African American
Hispanic
White
Other
Mixed Race

Wauwatosa

Milwaukee

St. Francis

Cudahy

South Milwaukee

Franklin

Oak Creek

RACINE COUNTY

WAUKESHA COUNTY

0    1 mi
0    2 km

FIGURE 12.18.
Distribution of racial
groups and Hispanic
population in Milwaukee
County in 2010. The dots
show the population of
the dominant group
(African American,
Asian, Hispanic, and
non-Hispanic white) per
census block, illustrating
the high level of
segregation of the
population. Data from
2010 U.S. Census.

### Economic Developments in Milwaukee

While racism has clearly influenced the development of Milwaukee's segregated and marginalized neighborhoods, the loss of many of the manufacturing jobs that only a generation or two earlier brought African Americans to Milwaukee also played a major role. Yet employment opportunities are often not easily accessible to these residents. Even a half century ago poor minority neighborhoods suffered "the greatest declines in accessibility to employment opportunities" (Ottensmann 1980, 428). Now, "cuts in public transportation in Greater Milwaukee have significantly eroded the ability of residents in inner city neighborhoods such as 53206 to access employment in suburban locations and have aggravated the region's spatial mismatch" (Levine 2019, 26). Milwaukee is responding to a variety of these issues, capitalizing upon its ethnic heritage, its legacy of urban parks and greenways (Erickson 2004), building a new streetcar system, and establishing several dozen business improvement districts, both downtown and in a variety of neighborhoods and derelict industrial areas.

A visitor to downtown Milwaukee today cannot help but notice the 550-foot Northwestern Mutual Tower (Figure 12.19), built between 2014 and 2017 at the eastern end of Wisconsin Avenue. Construction, which began in 2018, of the 25-story BMO Tower just south of Milwaukee's City Hall on Water

FIGURE 12.19. The lower floors of Northwestern Mutual Life Insurance Company's new 32-story high-rise office tower in Milwaukee. On the left is *The Calling*, an abstract steel sculpture. Partly hidden behind the sculpture is the company's older building.

Street is additional evidence of investments being made by the financial services industry within the city. The city's "downtown and surrounding neighborhoods . . . [are the focus of the mayor's] economic development program" (Ward 2007, 782). Besides new construction, "boutique retail centers, gentrified warehouses, new condominiums, convention centers, refurbished downtown hotels, revitalized walkways, and parking lots have all been built" (786). A mixture of old and new establishments, such as the Milwaukee Public Market and the Wisconsin Center, a large convention center whose architecture hints at the German Renaissance style (Figure 12.20), are designed to attract both visitors and new residents to the downtown.

As part of their effort to rehabilitate and revitalize downtown Milwaukee, "the city's civic and business leaders hope to . . . [marginalize] the contemporary issues of class and race, nostalgically celebrating ethnicity and a working class heritage" (Kenny and Zimmerman 2004, 80). As explored further in chapter 14, Milwaukee's efforts to celebrate its ethnicity and culture underpin a huge tourist industry. Its Summerfest has been described as the world's largest music festival, and a lively nightlife entertainment zone has developed in a two-block area along North Water Street at the edge of the central business district (Campo and Ryan 2008). The city has an extensive acreage of

FIGURE 12.20. Wisconsin Center, which hosted the 2020 Democratic National Convention, along Milwaukee's West Wisconsin Avenue. A modern glass-clad multitenant office building, formerly named the Henry S. Reuss Federal Plaza and now 310W, is on the right.

lakefront parkland and greenways that circle the region, even though portions of the inner city are underserved.

## Challenges in Other Wisconsin Cities

No other cities in Wisconsin face the variety and intensity of challenges that Milwaukee faces, yet less-privileged, economically deprived neighborhoods can be found in most cities that have minority populations, including Kenosha, Racine, Beloit, Janesville, and Green Bay, among others. Racine has been ranked as the second-worst city nationally for African Americans, given their poverty, low educational attainment, and unemployment (Mauk 2019). Just as we have seen how redevelopment efforts, involving business improvement districts or tax incremental districts, have aided in the regeneration of parts of Milwaukee, similar strategies have converted former industrial brownfields along the Fox River in Oshkosh, Neenah, Menasha, Appleton, Kaukauna, De Pere, Ashwaubenon, and Green Bay.

These sites, associated with the former sawmill and lumber industry in Oshkosh and paper manufacturing downriver, have been redeveloped with new apartment buildings, gentrified industrial buildings providing apartments and condominiums, restaurants and shops, and recreational facilities. The site of a furniture manufacturer that had operated in Oshkosh for nearly 160 years became the Menominee Nation Arena, home to two minor league basketball teams, while at the lower end of the Fox River, new housing and shops have been erected in Green Bay along a riverfront trail. Oshkosh has also established a riverfront trail, and small trail segments can be seen in several communities between the two cities. Eau Claire, Wisconsin's other "Sawdust City," has similarly redeveloped its former industrial downtown at the confluence of the Chippewa and Eau Claire Rivers, resulting in establishment of Phoenix Park (Figure 12.21).

Milwaukee's Menomonee Valley saw its manufacturing facilities and rail yards become abandoned, creating "one of the largest undeveloped sections of central city land in America" (Erickson 2004, 213). This became the site of Miller Field, the Potawatomi Casino and Resort, the Harley-Davidson Museum, and the Hank Aaron State Trail, yet considerable space remains (Figure 12.22). This is true in other cities. One of the biggest industrial redevelopment sites in Wisconsin today involves Kenosha's 109-acre former Chrysler Engine Plant (Urban Land Institute 2015), which was razed and its contaminated soils remediated. Although the ultimate use of the land has yet to be finalized, a portion may be utilized for a technology and innovation center. The site of the former American Motors assembly plant in Kenosha, closed in 1988, has been converted into a marina, park, and 350 condominiums (Figure 12.23). Just as deindustrialization has changed the downtowns of many Wisconsin's cities, so have changes in retailing.

### METROPOLITAN DEPARTMENT STORES AND SHOPPING MALLS

Downtown business districts were once the focus of large department stores and other specialized retailers, yet such facilities are now typically located in large suburban malls (Figure 12.24). For example, in early 2020 Macy's Department Stores operated in three Wisconsin metropolitan areas: Appleton, Madison, and Milwaukee, with the stores actually located in the Town of Grand Chute,

FIGURE 12.21. Phoenix Park was established on a brownfield site where the Eau Claire River flows into the Chippewa River in downtown Eau Claire. The Phoenix Manufacturing Company, later replaced by the Phoenix Steel Company, manufactured equipment used in sawmills and logging, including steam tractor log haulers, as displayed in Figure 6.1.

FIGURE 12.22. Redevelopment site along former ship docks within Menomonee Valley in Milwaukee. The 11-story building in the background was built to house a seed company in 1912 and is being converted into loft apartments. Along the foreground a portion of the Hank Aaron State Trail parallels the canal.

FIGURE 12.23. Streetcar runs along Kenosha Harbor's Promenade and Navy Memorial Park. Long an industrial site with coal-handling facilities, American Motors manufactured cars at this location between 1960 and 1988. The site has been redeveloped for recreation and housing, and Kenosha's streetcar line links the harbor with Kenosha's METRA station, from which commuter trains run to Chicago.

Madison, Greendale, and Wauwatosa. Nordstrom operated department stores in Madison and Wauwatosa. Menomonee Falls–based Kohl's had 38 department stores in Wisconsin, with the greatest number in southeast Wisconsin. These middle-market department stores are concentrated in Wisconsin's metropolitan areas, where 30 are located, with another 6 within its micropolitan areas. JC Penney had 10 stores in Wisconsin in early 2019, all of which were located in metropolitan areas. Almost all of Wisconsin's 39 Target discount department stores are located within metropolitan counties, although there is one in Sturgeon Bay.

Both upscale and midscale department stores typically anchor the state's largest shopping malls, which again are often located in suburban communities of metropolitan areas. Wisconsin's four largest malls in early 2020, all of which occupy at least one million square feet, are Southridge Mall in Greendale with 106 stores, Brookfield Square in Brookfield with 112 stores, Mayfair in Wauwatosa with more than 200 stores, and the Fox River Mall in Grand Chute with 136 stores. The first three serve greater Milwaukee, while the latter serves the Appleton and Oshkosh-Neenah metropolitan areas. All of the state's 16 largest shopping malls are in metropolitan counties.

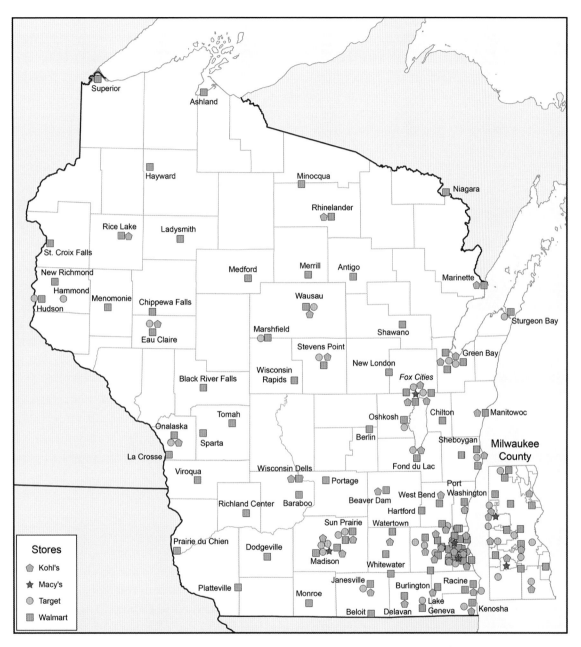

FIGURE 12.24. Selected department stores in Wisconsin in 2019. Map shows location of Macy's, Kohl's, Target, and Walmart department stores, illustrating differences in the distribution of upscale and midscale department stores and discount department stores. Data from Caliper (https://www2.caliper.com/store/product/buy-business-location-data/).

### Retailing in Wisconsin's Smaller Nonmetropolitan Cities and Villages

Wisconsin has considerably more discount department stores than mid- and up-market retailers. While many are located in larger communities, they display a greater presence in smaller cities. Although not providing the same variety nor quality of goods found in the middle-market and upscale department stores, Walmart and Target dominate this retail sector in Wisconsin. Shopko, founded and headquartered in Green Bay, operated 74 stores in Wisconsin until its bankruptcy in early 2019. It provided an alternative to the out-of-state major discount retailers. Shopko's initial closure of 38 stores left several smaller communities, such as Seymour and Winneconne, without an operating pharmacy or hardware vendor, demonstrating the critical role such stores play in the local retail market, even when they are within the outer margins of metropolitan areas. By summer 2019 Shopko closed all of its stores. The demise and departure of big-box retailers from communities is not the only consequence that such stores have upon many rural communities. Their arrivals played a major role in the restructuring of retail in many locations, and for many that was the loss of smaller, locally owned stores (Vias 2004).

Walmart operated 99 stores in Wisconsin in late 2018, of which 83 were Walmart Supercenters and 10 were Sam's Clubs. Its stores are located in 84 communities, ranging in size from Milwaukee to Hayward and unincorporated Minocqua. Of its locations, 10 percent are located in communities with populations under 5,000. An additional 14 percent are in places with under 10,000 residents in counties that are neither micropolitan nor metropolitan. Thus, nearly a quarter of Wisconsin's Walmarts are in small rural cities or villages. An additional 6 percent are in communities with fewer than 10,000 residents within micropolitan counties.

In contrast to the large discount stores such as Walmart and Target, Dollar General operates its smaller discount variety stores in 120 Wisconsin locations, many of which are too small to support the larger stores. Their orientation toward small communities is also shown by the fact that the central cities of 6 of the state's metropolitan areas lack a Dollar General Store, even though some do exist within smaller, more rural cities within their metropolitan counties.

### Metropolitan Supermarkets and Rural Convenience Stores

Wisconsin's largest supermarkets, such as Woodman's Markets, headquartered in Janesville, and Festival Foods, founded in Onalaska but now based in De Pere, are focused upon serving metropolitan customers. All of Woodman's Markets are in metropolitan areas, and all but 5 of the Festival Foods supermarkets are in metropolitan counties, while 3 (Baraboo, Fort Atkinson, and Marshfield) are in micropolitan counties, and 2 (Mauston and Tomah) are in less urbanized counties. Slightly smaller supermarkets, but more of them, are operated in Wisconsin by Kroger's Roundy's Pick-'n-Save subsidiary. Roundy's had 106 stores in 2018, serving many midsized cities as well as metropolitan areas, including underserved parts of Milwaukee (Taschler 2018). Nationally, there are 1.14 supermarkets per 10,000 persons (Ellickson and Grieco 2013), significantly limiting the establishment of large supermarkets in smaller communities. Given that Walmart Supercenters include grocery stores, consistent with national findings, these "supercenters tend to locate in more remote locations than other large supermarket chains . . . [and] are more likely to locate outside a designated MSA while, within MSAs, they choose locations that are far less dense" (Ellickson and Grieco 2013, 5). Fourth-class cities and villages typically have either smaller supermarkets, often independently owned and operated, although they may be affiliated with IGA (Independent Grocers of America) or franchises of Piggly Wiggly, or are only served by a convenience store.

La Crosse–based Kwik Trip has rapidly expanded its network of gasoline stations combined with convenience stores that stock not only drinks and snacks for travelers, but a variety of groceries including fresh breads, milk, meats, vegetables, and fruits. It fulfills the role that was once provided by general stores, but on a much grander scale. While many of their 376 Wisconsin stores, as of 2019, are located in metropolitan cities—with the exception of Milwaukee, where they operate PDQ stores that they acquired in 2016—their presence has been most felt in many smaller rural communities. Given their size and "backward integrated business . . . [that owns] their own milk production facilities" (McCoy 2019), bakery, La Crosse distribution center, and delivery trucks, the family-owned store chain increases its efficiency and keeps costs low, providing a huge challenge for many franchised gasoline stations. Indeed, Kwik Trip is estimated to "control about 15 percent of the convenience store market in Wisconsin" (McCoy 2019).

Today few hamlets have any gasoline stations, unlike the situation a half century ago. Even in cities the presence of a nearby Kwik Trip store has led to the closing of nearby independent grocery stores and gasoline stations (Tetzlaff 2018). Unlike the nation as a whole, where the number of grocery stores exceeds taverns by 13 percent, Wisconsin has 2.7 taverns per grocery store, by far the nation's highest ratio (Phillip 2014). Indeed, within many small Wisconsin communities, it is far easier to obtain a beer than to buy a gallon of milk.

Fast food restaurants are also missing from Wisconsin's smallest cities and villages. For example, McDonald's had 341 restaurants in Wisconsin in 2016, slightly more than one restaurant for every two incorporated communities in the state. Of course, given that the larger cities have multiple restaurants, most of Wisconsin's smaller villages and cities lack a McDonald's restaurant. Culver's, founded and headquartered in Prairie du Sac, operated 138 restaurants in Wisconsin in 2019. Nevertheless, while it had 4 restaurants along State Highway 64 in Marinette, Antigo, Merrill, and New Richmond, it had only 5 restaurants farther north—in Ashland, Minocqua, Rhinelander, Rice Lake, and Superior. Thus, away from major highway corridors and larger cities, one sees few Culver's. Similar patterns are seen for Burger King, Hardees, and Wendy's.

## Rural School Consolidation and Church Closures

School consolidation has closed many of the small schools that were found last century in many Wisconsin hamlets and villages. During the 1950–1951 school year, Wisconsin had 3,797 one-room schools (Watson 1952). Twenty years later they were gone from the scene. Given the higher costs of operating small rural schools and declining enrollment, school districts were consolidated, and many rural schools closed. This process is continuing and expanding to larger communities. For example, in 2018 the elementary school in Arena closed, just one year after another school in the same district closed in Lone Rock, leaving the school in Spring Green as the only public school in a district that encompasses nearly 300 square miles along the Wisconsin River. With the loss of their only school, residents of these southwest Wisconsin villages are concerned about "their town's future—that the closing will prompt the remaining residents and businesses to drift away and leave the place a ghost town" (Bosman 2018).

Closing of many churches with declining congregations is also occurring, and the shortage of priests has necessitated the consolidation of many Catholic churches in small hamlets and villages into larger parishes. For example, within Wisconsin's "Holy Land" east of Lake Winnebago, four of the churches described two decades ago are now closed, and the one in Dotyville has been razed. Small communities with declining populations are particularly in jeopardy.

## Rural Employment Opportunities

Rural communities provide their residents fewer jobs and job choices than Wisconsin's larger cities and metropolitan areas. Their local jobs typically provide lower salaries. For example, median earnings statewide, according to the American Community Survey (USCB 2019a) for the 2013–2017 five-year period, were $35,437, but within a broad swath of rural northern Wisconsin counties, plus several counties in central Wisconsin and Grant County, the median salaries for jobs were below $30,000. Median earnings (in 2017 dollars) were $27,205 in Burnett County, $27,816 in Grant County, $29,030 in Marquette County, and $26,546 in Sawyer County. In contrast, within five metropolitan area counties, the medians ranged from $36,928 in Outagamie County to $40,818 in Waukesha County.

The lack of an adequate number of well-paying local jobs encourages commuting from many rural communities to larger cities. For example, over half of the residents of five counties in central Wisconsin work outside their county of residence, according to the American Community Survey. Over 10 percent of the workers of Adams and Marquette Counties travel over an hour (one way) to work daily, resulting in a commute time equivalent to over a quarter of a standard eight-hour workday. Even if not commuting as far, over 80 percent of Adams, Marquette, and Waushara County residents work outside the minor civil division of their residence. This percentage is only rivaled among residents of Buffalo, St. Croix, Polk, and Calumet Counties, where many workers travel to the Rochester, Minnesota, or La Crosse; Minneapolis–St. Paul; or Appleton or Oshkosh-Neenah metropolitan areas, respectively. Such distant employment also encourages commuters to do much of their shopping in larger urban centers, with a greater selection of goods and lower prices. This, along with burgeoning e-commerce outlets that provide free or inexpensive, but rapid, shipment of goods to rural addresses, undermines many retailers in small communities, furthering the prominence of more favored metropolitan cities.

Small rural communities lack the variety of employers that larger cities provide, yet such communities in parts of Wisconsin rely upon urban residents to visit as tourists, who come to vacation, shop in their stores, and eat in their restaurants and taverns. Yet, their visits are highly seasonal, with many local businesses closed for part of the year. Thus, their owners and employees rely upon part-year work for most, if not all, of their annual income. Given the importance of recreation and tourism, it is the focus of chapter 14.

*Chapter Thirteen*

# Cultural Landscapes and Demographics of Wisconsin

TWO CENTURIES OF MOVEMENT, involving people from earlier settled states and foreign nations who occupied lands of Native Americans, shaped today's population in Wisconsin. Wisconsin's cultural landscape reflects the heritage of its historical geography of settlement, shaped by those who came seeking to exploit its natural resources, to establish farmsteads, and to work in its cities. Its population density, and its variation across the state, is a function of the capacity of the local economy to sustain these settlers and their descendants and of their demographic characteristics, displayed by their fertility rates and longevity. We see people leaving many locales and relocating, both to growing urban centers, as described in the previous chapter, and to places with desirable amenities, which are explored in the next chapter.

Today's population embodies the descendants of those who settled the land, minus those who departed. In many regions of Wisconsin, the character of the initial settlers remains strong, inasmuch as few outsiders have arrived since their original settlement. In other places, waves of additional settlers have added their cultural baggage to what the earlier colonizers brought, and new arrivals continue to reshape the cultural landscape. We begin this chapter by reviewing the patterns of ancestry displayed across Wisconsin, then describe the spatial distribution of population growth and decline, and finally consider spatial patterns of many demographic and cultural characteristics, including age distribution, fertility, socioeconomic standing, religion, political affiliation, and health.

## PATTERNS OF NATIVITY AND ANCESTRY IN WISCONSIN

The Census Bureau's American Community Survey (USCB 2019a) estimates for the 5-year period ending in 2017 show that 71.4 percent of Wisconsin's residents were born in the state, while 4.9 percent were born outside the United States. Persons born in other midwestern states comprised 15.1 percent of Wisconsin's population. Thus, most Wisconsin residents have deeper roots in their state than do residents of most other states (the national average is 58.4 percent). When one considers the adult population (those aged 25 or older), many Wisconsin counties stand out because of their home-grown populations. Several counties in east-central and northeast Wisconsin have among the highest

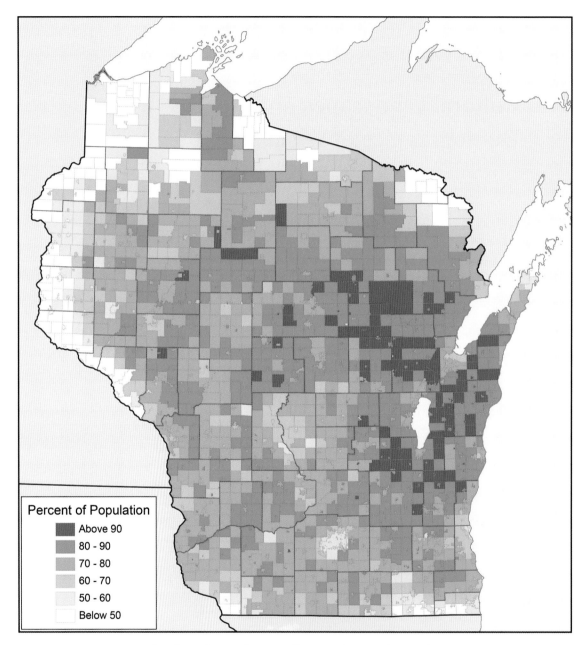

FIGURE 13.1. Percentage of total population of towns, villages, and cities who were born in Wisconsin. Data from U.S. Census Bureau's American Community Survey for 2014–2018.

rates of adult population born within the state of any counties in the nation. In 25 counties at least three-quarters of their adult population was born in Wisconsin. Counties where over 80 percent of the adult population is Wisconsin-born include Menominee (88.8 percent), Shawano (84.1 percent), Kewaunee (83.4 percent), Oconto (82.3 percent), Taylor (81.5 percent), Calumet (81.0 percent), Forest (80.8 percent), Fond du Lac (80.7 percent), and Langlade Counties (80.3 percent). Figure 13.1 displays the percent of the population within minor civil divisions (towns, villages, and cities) who are Wisconsin natives.

## Native American Populations

Menominee County's leading percentage of native-born residents should not be surprising given that the county's borders coincide with the Indian reservation that comprises the Menominee Nation's lands. The 2010 census enumerated 8,388 Menominee statewide, of which 3,085 resided in Menominee County (USCB 2019a). The Menominee comprised 71.5 percent of the county's total population, which was 88.7 percent American Indian when members of other tribes are included. The largest American Indian tribe in Wisconsin, with 19,326 counted, are the Chippewa, who have six reservations scattered across northern Wisconsin. Sawyer County had the most Chippewa, 2,626, yet three other counties each had over 1,800 Chippewa residents, including Ashland, Vilas, and Milwaukee. The Oneida, an Iroquoian people, were relocated in 1838 from central New York State. They are the third-largest Native American group in Wisconsin, with 6,677 enumerated. Although the Oneida Nation reservation straddles the boundary between Brown and Outagamie Counties, their largest number lived in Brown County. Their second-largest number resided in Milwaukee County. The Potawatomi numbered only 1,874 statewide, of which 31.8 percent lived in Forest County. Milwaukee County, where the Potawatomi Casino is located, was home to their second-largest number.

Milwaukee County, with 13,998 American Indians of single or multiple ancestry enumerated in 2010, had the greatest number of any county. Although Milwaukee was established around a trading post, its Native American population was quite small following Indian removal until World War II, when large numbers of Native Americans began taking jobs created by the wartime labor shortage. Given their growing population, a variety of educational and social services became available to support tribal members, and proceeds from the Potawatomi Casino financed the Indian Community School of Milwaukee (Loew 2013).

While 16.0 percent of Wisconsin's Indian population live in Milwaukee County, they comprise 1.4 percent of the county's population. Brown County, with 9,107 American Indians, is 3.6 percent American Indian. In contrast, even though they are home to smaller numbers, five counties in northern Wisconsin range between 11.8 and 19.3 percent American Indian, consequential but still far under Menominee County's proportion. Several northern Wisconsin towns have Native American residents who dominate their population similar to that in Menominee County. For example, the Town of Sanborn in Ashland County is 85.3 percent American Indian, the Town of Russell in Bayfield County is 77.9 percent, and Shawano County's Town of Bartelme is 77.3 percent. These towns comprise the Bad River Indian Reservation, the Red Cliff Indian Reservation, and the Stockbridge-Munsee Indian Reservation, respectively. In addition, another four towns are between 50 and 75 percent American Indian, while seven towns are 25 to 49.9 percent Native American.

Wisconsin currently has 11 federally recognized American Indian nations, plus the Brothertown Nation that has not attained that status. The Brothertown, along with the Stockbridge-Munsee, are Native Americans who were relocated from New England and upstate New York in the nineteenth century, similar to the Oneida, but who are far less numerous. The Ho-Chunk, once called the Winnebago, are federally recognized but lack a compact formal reservation, as their population is scattered among off-reservation trust lands. Nevertheless, they comprise 66.2 percent of the population of Jackson County's Town of Komensky, just north of Ho-Chunk Gaming's Black River Falls Casino. Because of the sovereignty Native American groups are guaranteed by treaties negotiated during the nineteenth century, indigenous nations have established 26 casinos and gaming facilities in Wisconsin, typically on or near their reservations (Figure 13.2).

### Dominant Ancestry

The census of 2000 provides the most recent and detailed data regarding the ancestry of Wisconsin's people. The 2010 census failed to ask ancestry questions except for those individuals of Hispanic and Asian ancestry, and the Census Bureau's American Community Survey sample sizes are so small that no consistently reliable data is available for political units smaller than a county. Even data for

FIGURE 13.2. St. Croix Casino and Hotel in Danbury, Burnett County. This casino is one of three operated by the St. Croix Chippewa Indians of Wisconsin.

smaller counties is suspect for all but the largest ancestry groups. Finally, as anyone who has followed the political news over the past several years already knows, efforts to include citizenship questions on the 2020 census led many demographers to fear that large numbers of minority peoples might avoid submitting information, rendering information from that census as suspect (Liptak 2019). Thus, in exploring ancestry, particularly European ancestry, which defines most of Wisconsin's population, we explore the distribution of various groups at the town and county level, using 2000 census data.

The *Census Atlas of the United States* (Suchan et al. 2007) portrays the prevalent ancestry for each of the nation's counties. The prevalent ancestry in 69 of Wisconsin's 72 counties was German, it was Norwegian in 2 counties, Trempealeau and Vernon, and in Menominee County it was American Indian. Prevalent ancestry indicates which ancestry grouping is most numerous, but it does not necessarily signify that the group comprises a majority of the population. Wisconsin's population of German ancestry comprised over half of the total population within 20 of the state's counties and over half of those persons reporting their ancestry within as many more, given that many individuals did not report their ancestry. Within the two counties where Norwegians were the prevalent group, Norwegians were 39.9 percent of the total population in Trempealeau County and 36.1 percent in Vernon County. Norwegians comprised under half of the population reporting their ancestry in both counties. The population reporting Polish ancestry in Portage County, 33.1 percent, was nearly as great as those reporting Norwegian in Vernon County, yet Germans, with 39.6 percent of the total population, are the prevalent ancestry in Portage County. Looking at ancestry (Figure 13.3) by towns, villages, and cities highlights how Germans dominate the ethnic landscape of much of Wisconsin. It also showcases those areas where other ancestry groups are locally prominent.

## Germans

German ancestry dominates the area north and west of Milwaukee along both sides of Lake Winnebago extending north to the southern border of Menominee County and west across Marathon County into Clark County. Furthermore, "the first areas of German settlement still have a high percentage of the state's German ancestry population" (Zaniewski and Rosen 1998, 74). East of Lake Winnebago settlement of German Catholics is particularly conspicuous. As one scholar notes, "The 'Holyland' is among the most distinctive German settlements in Wisconsin and it . . . consists of several rural German Catholic communities connected by religion and place of origin" (Schlemper 2007, 378). German ethnics account for over half of the population in all of the towns of Washington County and most towns within Dodge and Columbia Counties. Clusters of German predominance extend to the Mississippi River, where German ancestry is strongly displayed in Grant County and portions of La Crosse, Buffalo, and Pierce Counties. German ancestry accounted for over 60 percent of the population in over 150 towns—up to a maximum of 88.5 percent. In larger urban areas it was typically smaller.

Over half of the residents of several sizeable Wisconsin cities reported German ancestry in 2000. German ancestry was indicated by 54.3 percent in Fond du Lac, by 54.1 percent in Neenah, and by 50.4 percent in Sheboygan. Watertown was 56.7 percent German, and West Bend was 58.4 percent.

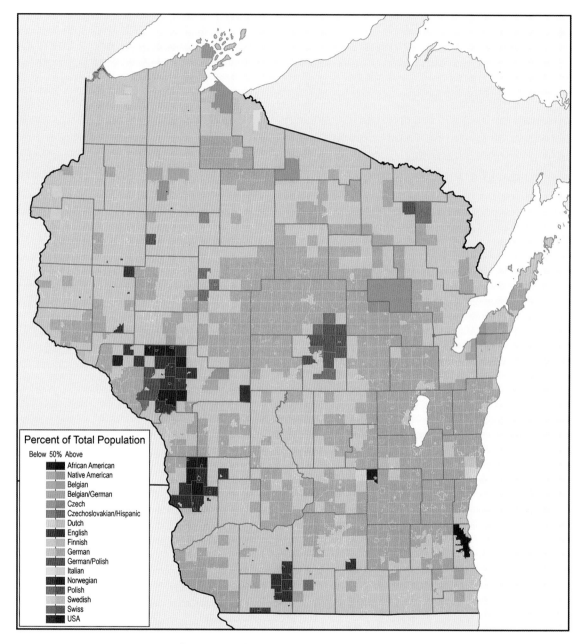

**Percent of Total Population**

Below 50% Above

- African American
- Native American
- Belgian
- Belgian/German
- Czech
- Czechoslovakian/Hispanic
- Dutch
- English
- Finnish
- German
- German/Polish
- Italian
- Norwegian
- Polish
- Swedish
- Swiss
- USA

FIGURE 13.3. Dominant ancestry by minor civil divisions in Wisconsin. Data from 2000 U.S. Census.

Although Milwaukee had the largest German ethnic population, only 20.9 percent of its population reported German ancestry.

Largely a legacy of the early German settlement of agricultural lands, farming in much of Wisconsin remains dominated by German ethnic operators. In 1980 persons of German ancestry accounted for 65.6 percent of Wisconsin's farm population (Salamon 1992). Persons of German or part-German ethnicity comprised 64.6 percent of entrants to dairy farming at the end of the twentieth century (Cross, Barham, and Jackson-Smith 2000).

### Scandinavians

Norwegian ethnics account for most of the descendants of Scandinavians who established farms and cut lumber during the 1800s. In 2000, persons of Norwegian ancestry accounted for over half of the population within 11 towns in western Wisconsin running from western Vernon County into northern Buffalo and Trempealeau Counties. Norwegian ethnics comprised between 35.0 and 49.9 percent of the population within an additional 36 towns within western Wisconsin, plus 5 towns in south-central Wisconsin. In addition, Vernon County's city of Westby was 61.1 percent Norwegian ancestry, while its village of Coon Valley was 56.4 percent.

Wisconsin's Swedish ancestry population is smaller, and only one town, the Town of Stockholm in Pepin County, exceeded 50 percent Swedish ancestry. The much smaller village of Stockholm was only 28.6 percent Swedish. No other towns exceeded 35 percent Swedish ancestry, although 9, of which 8 were in northwest Wisconsin (6 in Burnett County), exceeded 25 percent. In no towns did Finnish ethnics account for at least half of the population. However, Finns exceeded 35 percent within 6 towns in Douglas, Bayfield, Ashland, and Iron Counties. Six additional towns had populations of over 25 percent Finnish ancestry.

### Poles and Czechs

Polish ethnics accounted for over half of the total population within 11 towns in 2000, primarily in Portage County, overlapping into far southern Marathon County, and in Trempealeau County. There the small city of Independence was 51.4 percent Polish. The village of Pulaski in northwest Brown County was 40.8 percent Polish. Persons of Polish ancestry exceeded 35 percent of the population within an additional 19 towns, located near the larger clusters, plus in eastern Shawano County and in Clark and Taylor Counties near Thorp, which was 35.7 percent Polish.

Numerically far more persons of Polish ancestry lived in the city of Milwaukee, which was 9.6 percent Polish, and in southern Milwaukee county, where five of its cities were over 25 percent Polish ancestry. This expansion of the Polish community occurred "as successive generations have moved to adjacent suburbs" (Zaniewski and Rosen 1998, 91). Similarly, Stevens Point in Portage County was 26.7 percent Polish.

Czech ethnic populations are concentrated in Kewaunee County, where Czechs accounted for 41.0 percent of the population in one town and over 25 percent in two additional towns. One town in Grant County was also over one-quarter Czech ancestry.

## Belgians and Dutch

Immigrants from the Low Countries settled several locations in eastern Wisconsin. Belgians focused upon the southern portion of the Door Peninsula, and the 2000 census shows two towns, one in Door County and one immediately to its south in Kewaunee County, whose populations are over half Belgian ancestry. Four nearby towns exceeded 35 percent Belgian ethnicity, and another three were over 25 percent Belgian. This area remains the nation's center of Belgian ethnicity (Cross 2017).

Dutch settlement focused upon the Lower Fox River Valley, where Dutch Catholics predominated, and farther south, where Dutch Protestants prevailed (Schreuder 1997). Persons of Dutch ancestry exceeded half of the population in 2000 in only one town, the Town of Alto in southwest Fond du Lac County. Three towns were 35 to 49 percent Dutch, including the Town of Holland in southern Sheboygan County and the Town of Vandenbroek in Outagamie County. Not only were three additional towns in the Waupun and southern Sheboygan County area over 25 percent Dutch, but so were several towns and incorporated communities along the Fox River. For example, the village of Combined Locks was 27.2 percent Dutch, while the villages of Kimberly and Little Chute were 31.3 and 31.4 percent Dutch, respectively. Sheboygan County's villages of Cedar Grove and Oostburg had even greater Dutch ancestry, 42.5 and 55.0 percent, respectively.

## Irish and English

Irish and English ancestries are well represented in southwest Wisconsin, the legacy of lead mining nearly two centuries ago. While Cornish ethnics are subsumed into English ancestry by the Census Bureau, English ancestry was reported by 17.3 percent of the residents of Iowa County and by 16.4 percent in Lafayette County. Three nearby counties, Grant, Richland, and Rock, were 10.0 to 14.9 percent English.

Irish ancestry is most dominantly displayed in Lafayette County, with 19.0 percent of the population reporting that heritage in 2000. Adjacent Iowa County was 16.4 percent Irish. Both Crawford and Richland Counties were over 15.0 percent Irish. Irish ancestry is widespread within Wisconsin, accounting for 10.0 to 14.9 percent of the population within 40 counties, far exceeding the number reporting that level of English ancestry.

Statewide, persons of Irish heritage are second in number only to the Germans, who outnumber them four to one. Although the Irish had ranked second in 1860, they retook this position only in 2000, following over a century of the Irish "identifying themselves primarily with the United States or Wisconsin. . . . Today, however, there is a resurgence of interest in Irish music, dancing, athletics, history, literature, language, and genealogy throughout the state, throughout the year, culminating in one giant party every August on Milwaukee's lakefront" (D. G. Holmes 2004, 9).

## Other Europeans

Germans, Poles, Czechs, Scandinavians, Irish, Dutch, and Belgians clearly dominated the settlement of Wisconsin, yet other ethnic settlers left a lasting imprint upon local landscapes. For example, Luxemburgers played a major role in settling Port Washington, and that area retains links to the Old Country (Figure 13.4). Ozaukee County was 2.5 percent Luxemburger in 2000, Port Washington was 5.3 percent,

FIGURE 13.4. St. Mary's Roman Catholic Church, built in 1884 at head of Franklin Street in Port Washington, Ozaukee County. This downtown's architecture with its numerous towers and the presence of the meat market advertising ethnic sausages attract visitors to experience its Luxemburger heritage.

while the village of Belgium was 16.2 percent. Italians comprised 19.2 percent of the population in Iron County and 8.5 percent in Florence County, former centers of iron mining. Within the small cities of Hurley and Montreal they comprised 28.7 and 32.6 percent, respectively, of the population. Italians also comprised 10.8 percent of the population in Kenosha County in 2000. While the former Sicilian neighborhood in Milwaukee was largely destroyed by the development of an expressway, 17,499 Milwaukee residents claimed Italian ancestry on the 2000 census.

## Arrivals of New Ethnic and Racial Groups

Wisconsin remains dominated by persons of European ancestry, but its minority populations are increasing. A century ago it had very few non-European residents besides its indigenous population. The arrival of African Americans during the two world wars changed that. So has the use of Mexican American and immigrant labor to harvest Wisconsin's fruits and vegetables and to work in slaughterhouses. Changes in immigration law and an influx of refugees from Indo-China as a consequence of the American involvement in the Vietnam War brought Asian immigrants into Wisconsin. The arrival of these newcomers has had noticeable consequences upon the state's ethnic and racial population, particularly within several of its largest cities.

### African Americans

Wisconsin's African American population is highly urban, focused upon those cities that first attracted industrial workers a century ago. Milwaukee was home to 66.2 percent of the state's Black population in 2010. The city's population was 40.0 percent African American. Given that several suburbs also have large African American populations, 70.7 percent of Wisconsin's Black residents lived in Milwaukee County. Several other cities in southeast Wisconsin have sizeable African American populations. Racine was 22.6 percent Black, and Kenosha was 10.0 percent Black in 2010. Thus, just three counties contained 79.8 percent of Wisconsin's African American population. Yet, Milwaukee and Racine have the dubious distinction as being ranked "as the second and third worst cities in the country for black Americans" (Goldstein 2018), based upon the disparity of incomes between their white and Black populations and their residential, employment, and educational segregation. As noted in chapter 12, Milwaukee's 53206 zip code has come to epitomize the ills of the highly segregated community. Furthermore, the incarceration rate of Wisconsin's African American population is 12 times that of its white population, over twice the national average (Joseph 2016). This has a noticeable impact upon the distribution of the state's Black population.

Although several African American farming settlements were established in Wisconsin during the nineteenth century, very few Blacks reside in rural areas. In only seven towns do African Americans comprise over 5 percent of the population. Four of the five towns with the greatest percentages have prisons (and the inmates are counted as town residents): the Town of Fox Lake in Dodge County, whose population was 23.6 percent Black in 2010, is home of a state prison; the Town of New Chester in Adams County that was 23.5 percent Black contains the state's only federal prison; the Town of Mitchell (21.7 percent Black) in Sheboygan County is the home of Kettle Moraine Correctional Institute; and the Town of Brockway (11.5 percent Black) in Jackson County has a state prison. Forest County's Town of Blackwell, which was 18.7 percent Black, is the home of a federal job corps training center.

Wisconsin has also received immigrants from Subsaharan Africa. The American Community Survey for 2013 to 2017 (USCB 2019a) estimates that they and their descendants number nearly 29,000, of which a third are in Milwaukee County. Both Dane and Racine Counties are also well represented, also having many African Americans. Wisconsin has also received Arabs, who were estimated to number just under 15,000. Sixty percent of these Arabs resided in Milwaukee and Dane Counties. Wisconsin accepted refugees from the Middle East. Twenty-five arrived from Afghanistan, 126 from Iraq, and 80 from Syria in fiscal year 2017, plus 420 from Myanmar, 105 from Congo, and 167 from Somalia, for an overall total of 1,003 refugees. Seventy percent settled in Milwaukee County, 9 percent in Dane County, and 8 percent in Winnebago County (WDHS 2017a).

### Hispanics

Over a third of Wisconsin's Hispanic population resides in Milwaukee County, with 30.7 percent of the state's Hispanics within the city of Milwaukee at the time of the 2010 census. There they comprise

17.3 percent of the population. Hispanics account for over a tenth of the population in several other large Wisconsin cities, including Green Bay (13.4 percent Hispanic in 2010), Kenosha (16.3 percent), and Racine (20.7 percent), and they were nearly that level in Sheboygan (9.9 percent). The growth of Green Bay's Hispanic population largely dates to the 1990s when Hispanic labor was actively recruited in Texas and elsewhere in the West to work in its meat-packing plants (Cruz 1998).

Many rural, non-agricultural areas of Wisconsin have few Hispanics, unless they have food-processing facilities or other industries that provide employment. Several smaller cities and villages with such employers had disproportionately high Hispanic populations in 2010, such as Abbotsford (25.0 percent Hispanic) and Norwalk (35.1 percent), both of which have meat-processing facilities. Arcadia, with its large furniture factories, was 31.2 percent Hispanic. Several resort communities provided employment to large Hispanic populations, including Lake Geneva (17.3 percent Hispanic), Delavan (29.4 percent), and Lake Delton (15.3 percent).

Nearly three-quarters of Wisconsin's Hispanic population is of Mexican ancestry, and persons of Mexican heritage comprised 4.3 percent of the state's population in 2010. Puerto Ricans accounted for 0.8 percent of Wisconsin's population, and 53.3 percent lived in the city of Milwaukee. Cuban ethnics

FIGURE 13.5. Spanish-language signage on restaurant along 5th Street in former German-Polish neighborhood on south side of Milwaukee. Note use of bright colors that are more frequently displayed as the neighborhood has transitioned between ethnic groups.

comprised 0.06 percent of Wisconsin's population, slightly more than those persons of Guatemalan ancestry.

Milwaukee was home to 28.5 percent of Wisconsin's Mexican ancestry population in 2010, comprising 11.7 percent of that city's population and largely living south of the Menomonee River (Figure 13.5). Mexican ethnics are conspicuous in several Milwaukee suburbs, including West Milwaukee (19.1 percent), Waukesha (9.6 percent), West Allis (6.3 percent), and Cudahy (5.5 percent). Several other large cities in southeast Wisconsin also have sizeable Mexican ethnic populations, including Racine (17.4 percent in 2010), Kenosha (12.5 percent), and Sheboygan (8.3 percent). Cities in south-central Wisconsin with sizeable Mexican populations include Fitchburg, a Madison suburb that was 13.9 percent Mexican ancestry, and Beloit (14.9 percent). Even though recent growth of Mexican employment is largest within cities, much of the growth of Mexican ethnic populations in southeast Wisconsin during the 1970s and 1980s involved persons who first came to the state as migrant farm workers before switching to higher-paying factory jobs (González 2017). As their numbers grew, community-building actions in several cities worked to provide appropriate cultural spaces for Mexican ethnics, including bilingual programs in schools and churches and stores providing ethnic foods and wares, which supported rapid growth of this ethnic group.

## Hmong

The 2010 census counted 129,234 Asians (one race) in Wisconsin, of which 47,127 reported Hmong heritage. This number is barely below the 49,240 who reported as being Hmong (either single race or in combination with another ancestry group or race). The Hmong are the largest Asian group in Wisconsin, and one in which little ethnic mixing has occurred. Only two other states (California and Minnesota) have more Hmong residents than Wisconsin. In 2010 Hmong were reported as living within 19 counties, with Milwaukee County having the largest number, 10,917. There, they comprised slightly over 1 percent of the population.

Proportionally, the Hmong are most conspicuous within Marathon County, comprising 4.2 percent of the population. They exceeded 2 percent within three additional counties: Sheboygan (3.5 percent), La Crosse (2.7 percent), and Eau Claire (2.2 percent). Refugees of the Vietnam War or their progeny, 46 percent of Wisconsin's Hmong population in 2010 were foreign-born. Their numbers had increased by 187 percent since 1990. With a birth rate exceeding the state average, Hmong in Wisconsin had a much lower median age than the overall state population and a much broader-based population pyramid. Their numbers grew by 39 percent between 2000 and 2010, even though their immigration had largely ended (APL 2015).

Wisconsin's Hmong, with the exception of those in Milwaukee County, live predominately within the state's midsized cities. One of the key factors in their settlement, except in Milwaukee, was movement into communities that otherwise had no other minority population at the time. Research conducted in La Crosse at the time of their immigration found that nearly half of the local respondents favored continued growth in the Hmong's numbers, although ethnocentrism of the city's residents significantly correlated with their attitudes regarding continued immigration of Southeast Asians. Yet many who opposed their growth evaluated their work ethic positively (Ruefle, Ross, and Mandell 1992).

Overall, by 2010 the Hmong exceeded 1.5 percent of nine counties' total population. Although the Hmong are over twice as likely to labor in manufacturing than the state's typical worker (APL 2015), many Hmong work part-time in their gardens. They are conspicuous vendors (Figure 13.6) at the farmers' markets in communities such as Eau Claire, Oshkosh, and Sheboygan, among other mid-sized cities, as well as in Milwaukee (Steinberger 2018). For many of the Hmong, "selling vegetables at the farmers' market is about more than just growing produce and making money . . . [it shows] satisfaction in farming . . . fills a void in their lives" (Vue 2020, 55), strengthening family bonds and linking with their cultural heritage.

## Ethnicity and Language

A century ago many Wisconsin residents of German ancestry, including those born in the state, and even "some second- and third-generation descendants of immigrants were still monolingual as adults" (Wilkerson and Salmons 2008, 259), having difficulty communicating in English. Today, the vast majority of Wisconsin residents only speak English at home. Yet 8.7 percent speak a language other than English at home. Of these, nearly two-thirds are fully conversant in English, according to the American Community Survey for 2013–2017. Of those speaking a different language, 4.6 percent

FIGURE 13.6. Hmong vendor at farmers' market on Main Street in Oshkosh, displaying vegetables on table and sewn cloth aprons with Hmong needlepoint designs hanging at the back of the shelter.

speak Spanish at home, 1.9 percent speak another Indo-European language, and 1.8 percent speak an Asian language (USCB 2019a).

Recent immigration explains much of the spatial variation in spoken language. Overall, at least 10 percent of the population in eight counties speak a language other than English at home. In five of these, that language is Spanish, with it being the highest in Milwaukee County, where 10.2 percent of the population speak it at home. Yet, the highest percentage (16.9 percent) of residents speaking a language other than English at home is in Clark County, where 12.8 percent use an Indo-European language other than Spanish. The county's large Amish and Mennonite families, who speak German at home and during worship services, account for most of this number, just as it explains why 6.0 percent of Green Lake County's population and 10.1 percent of Vernon County's population speak an Indo-European language other than Spanish.

### Ethnicity and Religion

In many parts of Wisconsin ethnic churches built over a century and half ago remain prominent features, reminding both residents and visitors of the prominent role of religion in promulgating culture

FIGURE 13.7. Evangelische Christus Kirche, erected in 1862, two decades after its founding, in the Dheinsville settlement in the Town of Germantown, Washington County.

(Figure 13.7). Older Protestant churches within German ethnic communities are often engraved with the German word *kirche* for "church," while *kościół* appears within Polish areas. Older gravestones are often inscribed in German, Norwegian, or Polish, illustrating the linkages between ethnicity and religion. Although fewer Wisconsin residents regularly attend church today than in the past, the legacy of ethnicity upon Wisconsin's religious landscape remains strong.

Seventy-one percent of Wisconsin's population identify as being Christian, according to recent surveys by the Pew Foundation. Roman Catholics comprise 25 percent of the state's residents, Evangelical Protestants 22 percent, and Mainline Protestants 18 percent. Among the Evangelical Protestants, Lutherans from an evangelical tradition (primarily Missouri Synod and Wisconsin Synod Lutherans) include 10 percent of the state's population and Baptists and Pentecostals are each 3 percent. Among mainline Protestants, the largest group are Lutherans from a mainline tradition (Evangelical Lutheran Church of America), who account for 9 percent of the state's population, followed by Methodists, who total 4 percent. Orthodox Christians comprise 1 percent of Wisconsin's population, as do Jehovah's Witnesses, Jews, and Muslims. Approximately one-quarter of Wisconsin's adults are unaffiliated, or defined as "religious nones" (Pew Research Center 2019). Yet these individuals experience a local culture influenced by the religious beliefs of their neighbors and ancestors.

## Spatial Patterns of Religion

Clear patterns of religious dominance are evident in Wisconsin (Figure 13.8). Roman Catholics comprise the largest religious group within 64 of Wisconsin's counties, with specific Lutheran synods comprising the largest group in the remaining counties (Figure 13.9). (When Lutheran groups are added together, Lutherans outnumber Catholics in 18 counties.) In 7 counties of west-central and northwest Wisconsin, partly corresponding to areas with numerous Norwegian ethnics, the Evangelical Lutheran Church in America was the most numerous group in 2010. In Shawano County Lutherans of the Missouri Synod are the largest group (ASARB 2019). Within the majority of counties where Catholics are the most numerous group, Lutherans are the largest Protestant group. What has long distinguished Wisconsin is that while both Catholics and Lutherans are numerous and each group is proportionally even more dominant in certain other states, "only North Dakota had a larger proportion of both" (Thompson 1988, 60).

Catholics account for over half of all church adherents in both Menominee and Kewaunee Counties, and over 40 percent in Brown, Manitowoc, Outagamie, Portage, Taylor, and Grant Counties, locations with large numbers of German Catholic, Polish, Dutch Catholic, Belgian, and Czech ethnics. The greatest proportion for any Protestant Christian group in Wisconsin is in Trempealeau County, where 37.6 percent of the adherents are members of the Evangelical Lutheran Church in America. The second-highest proportion of these adherents is in Vernon County, with 22.6 percent (ASARB 2019). Most of the Lutherans in these two counties can trace their heritage to the Norwegian Lutheran Church, yet schisms within the various Lutheran groups resulted in some communities literally having side-by-side Lutheran churches of different affiliations (Legreid 1997). Wisconsin Evangelical Lutheran Synod adherents are most prevalent within Jefferson County, where they comprise 19.1 percent of all

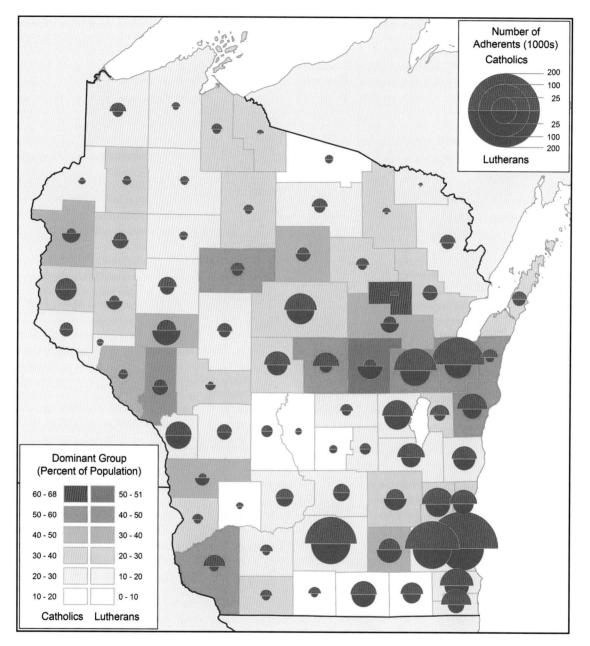

Number of
Adherents (1000s)
Catholics

— 200
— 100
— 25

— 25
— 100
— 200

Lutherans

Dominant Group
(Percent of Population)

60 - 68      50 - 51
50 - 60      40 - 50
40 - 50      30 - 40
30 - 40      20 - 30
20 - 30      10 - 20
10 - 20      0 - 10

Catholics   Lutherans

FIGURE 13.8. Dominance of Roman Catholics and Lutherans by county in Wisconsin. Data from ASARB (2019).

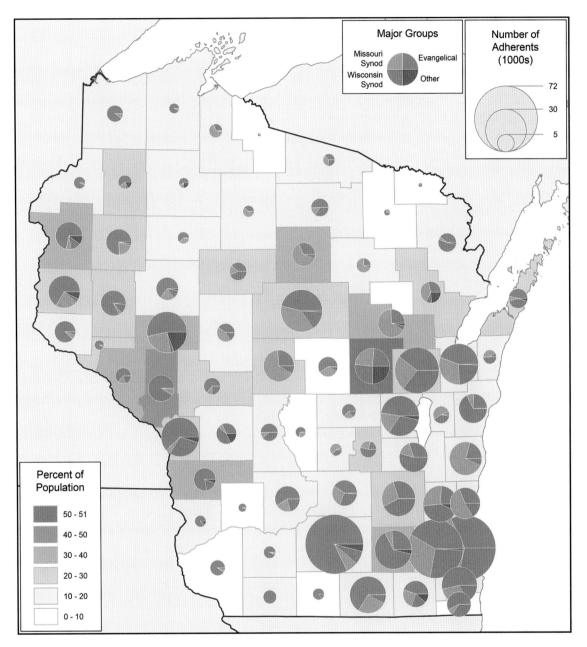

FIGURE 13.9. Lutherans in Wisconsin, showing percentage of adherents per county who are Lutherans and the proportion of Lutherans from the Missouri Synod, the Wisconsin synod, and the Evangelical Lutheran Church. Data from ASARB (2019).

adherents. While members of the Lutheran Church–Missouri Synod are the largest group in Shawano County, they only account for 21.3 percent of the county's adherents.

When the three Lutheran groups are considered together, even though doctrinal differences are substantial, they outnumber other Protestant denominations in most Wisconsin counties. Members of the United Methodist Church account for 7.8 percent of the adherents in Grant County, the county where they had their greatest presence—equal to that of the combined Lutherans, followed by 7.4 percent in Lafayette County, not surprising given the number of early Cornish Methodists who settled that region. Although small in total numbers for the state, the Amish comprise a large proportion of the adherents within several rural counties, including 27.9 percent in Vernon County, 19.9 percent in Clark County, and 16.3 percent in Monroe County (ASARB 2019).

Among the non-Christian religious groups, Jews are the most numerous statewide. Their greatest share of adherents is 2.6 percent within Ozaukee County, followed by Milwaukee County (1.8 percent) and Dane County (1.1 percent). Moslems are most numerous within Milwaukee County, where they comprise just under 1 percent of believers.

*Religious Influence upon Education*

Religion has long influenced many aspects of life among Wisconsin's residents. At one time parochial schools educated most of the children within many parishes, and parochial schools remain important (Figure 13.10). While public schools enrolled 858,833 children in Wisconsin during the 2018–2019 academic year, private schools, many of which are denominationally based, enrolled 122,540 students (WDPI 2019). Catholic parochial schools in the 10 counties of the Milwaukee archdiocese enrolled over 30,000 students within its 107 schools (92 elementary and 15 high schools). The Wisconsin Evangelical Lutheran Synod operated 32 schools in the Milwaukee area. Mennonites, the Amish, and several other Christian groups also run their own schools.

Colleges and universities are operated by several religious groups in Wisconsin, yet enrollment is greatest within the Catholic institutions. The largest Catholic institution is Marquette University in Milwaukee, run by the Jesuits, with 11,600 students and 20 doctoral programs (Figure 13.11). Other Catholic colleges and universities with 2,000 or more students include Alverno College and Cardinal Stritch University, both also in Milwaukee, Edgewood College in Madison, Marian University in Fond du Lac, St. Norbert's College in De Pere, and Viterbo University in La Crosse. Smaller Catholic institutions in the Milwaukee area include Mount Mary University, with about 1,500 women students, and Sacred Heart Seminary and School of Theology. Silver Lake College of the Holy Family, which was recently renamed Holy Family College, operated in Manitowoc until closing in mid-2020.

Presbyterian-affiliated Carroll College has nearly 3,000 students in Waukesha. Carthage College, associated with the Evangelical Lutheran Church of America, has about 2,800 students in Kenosha. The Missouri Synod Lutheran Church's affiliated Concordia University is located in Mequon, enrolling about 4,400 full-time students. Lakeland College in Plymouth, with about 900 full-time students and twice as many part-timers, is affiliated with the United Church of Christ, as is smaller Northland College in Ashland. The Wisconsin Synod Lutherans operate Wisconsin Lutheran College in Milwaukee.

FIGURE 13.10. St. Boniface Parochial School located between St. Boniface Catholic Church and rectory in the small hamlet of Waumandee, Buffalo County. There is no public school in this community, and children not attending this school are bused elsewhere for their education.

FIGURE 13.11. Campus of Marquette University along Milwaukee's West Wisconsin Avenue with office building of Catholic Financial Life in the background. Higher education and life insurance coverage are just several ways that religious affiliated institutions serve their members.

Two small institutions are operated by the Baptists: Maranatha Baptist University in Watertown and Baptist College of Ministry in Menominee Falls. Several of Wisconsin's other private colleges and universities, such as Lawrence University, which was founded with Methodist support, and Beloit College, which was associated with the Congregationalists, no longer have any religious affiliation. Ripon College's ties to two local churches ended a century and a half ago.

### Religious Influence upon Health Care

Hospitals affiliated with various Roman Catholic groups are prominent among Wisconsin's medical service providers. Forty-two Catholic hospitals are located in Wisconsin, which provide 40.7 percent of the state's hospital beds (Kaye et al. 2016). Nearly half of Wisconsin's Catholic hospitals are part of the Ascension medical group (Figure 13.12), which includes hospitals in Milwaukee, Racine, Oshkosh, Appleton, Stevens Point, and Rhinelander, plus nine other locations in eastern and central Wisconsin. The Hospital Sisters of St. Francis operates their Hospital Sisters Health System hospitals in Eau Claire, Chippewa Falls, Green Bay, Sheboygan, and Oconto Falls. Another Catholic group serves the Madison area. Franciscan Healthcare operates hospitals in La Crosse and Sparta.

FIGURE 13.12. Ascension Columbia St. Mary's Hospital in Milwaukee across North Lake Street from an older building of the hospital complex. When relocated to this site in 1858, St. Mary's was named St. John's Infirmary, which had been established a decade earlier as the first public hospital in Wisconsin.

Froedtert Memorial Hospital in Milwaukee and Gunderson Hospital in La Crosse have Lutheran connections. Methodist Hospital is located in Madison. Bellin Hospital in Green Bay, which operates another hospital in Oconto and clinics throughout northeast Wisconsin besides Bellin College of Nursing, has been affiliated with the Methodist Church since 1909. Many nursing homes and assisted living facilities in Wisconsin are also operated in conjunction with various religious groups.

### Other Religious Influences upon Culture

Religious beliefs along with race, ethnicity, and education influence end-of-life decisions and the disposition of the body. The Wisconsin Department of Health Services (WDHS 2019a) reports that 59.1 percent of decedents were cremated in 2017. Burials in cemeteries accommodate the majority of bodies of African American and Asian decedents in Wisconsin and nearly half of Native Americans, but only a third of the non-Hispanic white population. While cemeteries are conspicuous features upon the Wisconsin landscape, with elaborate religious statuary and crucifixes displayed in Catholic cemeteries (Figure 13.13) and far fewer objects in Protestant and secular burial grounds, their use geographically varies. Under a third of decedents are buried within most counties of northwest and north-central Wisconsin, while the majority are buried in the counties of southwest Wisconsin. Other counties with

FIGURE 13.13. Grotto topped with cross displaying corpus flanked by religious statues in cemetery by St. Francis Catholic Church, one mile north of Brussels in Door County.

a large Catholic population, such as Calumet, Kewaunee, and Taylor Counties, also have disproportionate numbers of burials, as does Menominee County, with its Native American population. In contrast, burials are less common within several of the metropolitan counties, including Dane, Winnebago, Brown, and La Crosse (WDHS 2019a).

A mixture of religious beliefs and ethnicity also influence alcoholic beverage sales. Writing about practices in the middle of the twentieth century, William F. Thompson noted that "a number of Scandinavian and Old Yankee settlements . . . prohibited or severely limited the sale of liquor within their communities" (1988, 11). In contrast, the neighborhood or village tavern played a crucial role in the social life of many German and eastern European ethnic communities. A few examples of such restrictions continued into the present century. The city of Sparta and village of Ephraim ended their prohibitions on the retail sale of wine and beer within the past decade (Hernandez 2016), yet liquor still cannot be sold in Sparta.

## SOCIOECONOMIC CHARACTERISTICS OF WISCONSIN'S POPULATION

Besides Wisconsin residents differing culturally, given their ethnic and religious heritage, their social and economic attributes vary geographically, both at the neighborhood and county level. Let's look at these at the county level, for which census and American Community Survey data are most readily available (Figure 13.14).

### Education

The educational attainment of Wisconsin's population is relatively high statewide, particularly the high school graduation rate. Only 10 states have a higher graduation percentage than Wisconsin. Of the population at least 25 years of age, 91.7 percent are high school graduates, 29.0 percent have a bachelor's degree or higher, and 9.9 percent have a graduate or professional degree, according to the American Community Survey for 2013–2017. The four counties with the highest percentage of high school graduates (at least 95 percent) are Dane, Waukesha, Ozaukee, and St. Croix. Sixteen counties have graduation rates of less than 90 percent. The lowest rate is 82 percent in Clark County, related to its large Amish population. Indeed, 11.2 percent of Clark County's adult population lack a ninth-grade education, inasmuch as the Amish do not believe in educating their children beyond the eighth grade. The second-highest percentage (6.2 percent) of adults lacking a ninth-grade education is in Vernon County, which also has a large Amish population.

College graduates, and in particular those with advanced degrees, earn higher salaries than the general population. The county with the largest proportion of its adult population having at least a bachelor's degree is Dane (with 50.0 percent), followed by Waukesha and Ozaukee, which both exceed 40 percent. In 18 additional counties over a quarter of the population have attained a bachelor's degree. These counties are clustered in southeast Wisconsin, the Fox River Valley plus Door County, and in a group from Eau Claire to the St. Croix River. Outliers include La Crosse, Portage, Oneida, Vilas, and Bayfield Counties. The three counties where the largest proportion of their populations aged 25 years or older have graduate or professional degrees include Dane County (20.7 percent), Ozaukee County (17.8 percent), and Waukesha County (14.7 percent).

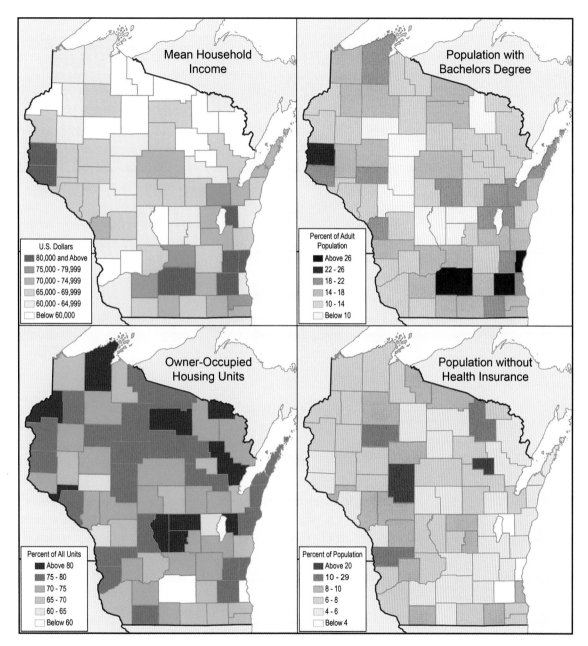

FIGURE 13.14. Socioeconomic characteristics of Wisconsin's counties: mean household income; percentage of adults (25+ years old) with bachelor's degrees; percentage of housing units that are owner-occupied; and percentage of population without health insurance. Map data from the U.S. Census Bureau's five-year American Community Survey for 2014–2018.

### Household Income

Income data from American Community Survey for the five-year period ending in 2017 and expressed in 2017 dollars shows that statewide median household income was $56,759, while mean household income was $74,372 (USCB 2019a). The fact that mean income is much greater than median indicates that very high incomes of a relatively small segment of the population skew the mean incomes upward. Median income shows the midpoint of household incomes—half of the households earn more and half earn less. Mean household income exceeded $100,000 in two counties, Waukesha and Ozaukee, which was highest with a mean of $113,931. One additional county, St. Croix, exceeded $90,000. In contrast, the highest county median income was $81,140 in Waukesha County, and it was $80,526 in Ozaukee County. Three additional counties (Calumet, St. Croix, and Washington) had median household incomes that exceeded $70,000. Another two counties (Dane and Pierce) had median household incomes between $65,000 and $69,999.

Median household incomes in two counties are less than half of that within the top county. Median annual incomes in Iron and Menominee Counties were under $40,000. Ten counties in total had median incomes below $45,000. All of these except Adams County are in the northern third of Wisconsin. Fifty-four Wisconsin counties had median incomes below the statewide median. With the exception of two counties that are part of the Minneapolis–St. Paul metropolitan area, all 18 Wisconsin counties with median household incomes greater than the statewide median are located in southeast, south-central, and east-central Wisconsin, plus Iowa County. Yet even in that area, Milwaukee, Sheboygan, Manitowoc, and Winnebago Counties had median incomes below the statewide median.

### Poverty and Employment Rates

Low-income counties are associated with less participation in the work force, greater proportions of households living below the poverty level, and greater participation in the food stamp or Supplemental Nutrition Assistance Program (SNAP). Statewide, 63.7 percent of Wisconsin's population aged 16 years or older were employed, yet under half of the population was employed in Adams, Menominee, Vilas, and Forest Counties. Part of this low employment reflects the large number of retirees in several counties, but such is not the case in Menominee County. The American Community Survey for 2013–17 estimates that 35.8 percent of the population in Menominee County were below poverty level, while in Milwaukee County it was 20.5 percent. The next highest poverty levels were in Forest, Sawyer, and Vernon Counties, which had rates between 16.0 and 17.9 percent. Statewide, the three counties with the highest participation in the food stamp program were Menominee, Milwaukee, and Iron Counties, where 20.0 to 30.7 percent of the households received food stamps or SNAP.

### Home Ownership and Value

One of the biggest financial assets of many households is the equity in their homes. Yet homeownership and home values vary across Wisconsin. Median values of homes exceed $200,000 (in 2017 dollars) within seven counties, with the median value above $260,000 in both Ozaukee and Waukesha Counties (USCB 2019a). Other counties exceeding $200,000 are Washington, Dane, St. Croix, Door, and Vilas, with upscale retirement homes contributing to the high values in the last two counties.

In contrast, the median value of owner-occupied homes is under $125,000 in nine counties, seven of which are in northern Wisconsin, along with Clark and Adams Counties. Two-thirds (66.96 percent) of Wisconsin's households live in owner-occupied housing. The lowest ownership percentage is in Milwaukee County (49.7 percent), and among its African American population it was 27.6 percent compared with 59.8 percent among its white population. The highest rate is in Florence County (86.5 percent). While poverty partly contributes to higher proportions of individuals renting rather than owning their homes, so do other factors, such as the number of young adults in the community. For example, Dane County, home of the state's largest university, has Wisconsin's second-lowest rate of owner-occupied homes (58.3 percent). The next seven lowest counties all have large University of Wisconsin System campuses.

*Poverty among Elderly*

Demographic data also distinguishes some counties of northern Wisconsin with above average levels of poverty. Ten Wisconsin counties have median ages of at least 50 years. Nine of the 10 counties are in northern Wisconsin. The other is Adams County. Statewide, 15.6 percent of Wisconsin's population is at least 65 years of age, yet within 5 counties of northern Wisconsin, plus Adams County, the proportion exceeds 25.0 percent. The highest proportion, 29.4 percent, is in Vilas County. In that county, 12.5 percent of the population is 75 years of age or older, while in adjacent Iron County it is 13.3 percent. Three percent of Vilas County's population is at least 85 years old.

Counties with large numbers of elderly residents fall within two groupings—those counties with amenities that attract retirees, such as Door County, and those that have an impoverished residual population, who cannot afford to move to a more attractive location upon their retirement. Not only do these counties have smaller proportions of their populations composed of children, but upon reaching adulthood, many of their youth seek employment opportunities elsewhere, often within metropolitan areas either in or out of state.

Because more college-educated persons leave Wisconsin than move into the state, Wisconsin has long been recognized as experiencing a "brain drain." Indeed, the Wisconsin Policy Forum (WPF 2019a) notes that "in 1990, 2000, and 2010, Wisconsin's gap on this measure was the largest in the nation." It remains large, but it was eighth largest in 2017. Persons with college degrees are more likely to seek employment out of state than those without such educational attainment.

## Health Status of Population

Social ills and lower health outcomes often correspond to lower income and minority status. Thus, the health status of Wisconsin's people, as shown by a statewide survey conducted in 2018, is correlated to residence. "Wisconsin's very rural counties also rank poorly for overall health factors and outcomes. Except for Milwaukee and Racine counties, the lowest-ranking counties are also among Wisconsin's most remote and rural, with most in the northern part of the state" (M. Jones and Bourbeau 2018). Data from the 2019 survey showed that all of the counties along the border with Michigan, plus two additional counties in northwest and northeast Wisconsin, three counties in central Wisconsin, three in southeast Wisconsin, and two others ranked in the bottom quarter of the state's counties with respect to health outcomes, "based on an equal weighting of length and quality of life" (UWPHI 2019d, 4).

### Access to Health Care

Access to health care has an impact on health outcomes, with access related to health insurance coverage and to location of health providers. Six percent of Wisconsin residents were uninsured in 2016, ranging from 4 to 18 percent among the counties (UWPHI 2019c). Wealthy suburban counties surrounding Milwaukee, including Ozaukee, Washington, and Waukesha, had the lowest uninsured percentage. Clark County was highest, not surprising given its large Amish and Old Order Mennonite populations, who view insurance as incompatible with their religious beliefs. Yet the uninsured population is disproportionately high in several counties of northern Wisconsin that lack these Anabaptist believers. About 10 or 11 percent of the residents of the following counties were uninsured: Bayfield, Forest, Menominee, Sawyer, and Vilas.

Even for those with insurance, finding a local physician is a challenge. Many rural counties are poorly served by primary care physicians. The 2019 County Health Rankings showed that Marquette County lacks a primary care physician, while four counties—Adams, Buffalo, Burnett, and Menominee—have only two (UWPHI 2019b). Florence and Pepin Counties each have three. While the overall ratio of population to primary care physicians in Wisconsin is 1,250 to 1, within five counties—Adams, Calumet, Burnett, Buffalo, and Douglas Counties—the ratio exceeds 5,000 to 1. The worst ratio is in Adams County, 10,030 to 1, excepting Marquette County, which has none. Although Marquette County has just four dentists, its ratio of population to dentists is fourth highest in the state, with Adams County again the worst, 9,990 to 1 (UWPHI 2019a). Exacerbating this problem is that several of these counties have above average numbers of poor and elderly populations, which, combined with the lack of public transit, even further reduces access to health care.

### Mortality Rates

Mortality rates and rates of death from specific causes also vary across Wisconsin. In 2017 "adjusting for an aging population, mortality rates were highest in Florence, Menominee, Ashland, Forest, and Milwaukee counties" (WDHS 2019a, 8). Heart disease accounted for 22 percent of deaths in Wisconsin during 2017. The "highest age-adjusted heart disease mortality rates were in Oconto (22.3 deaths per 10,000 people), Green Lake (21.4 per 10,000 people), and Vilas (21.0 per 10,000 people) counties. The three counties with the lowest heart disease mortality rates were St. Croix, Rusk, and Kewaunee" (WDHS 2019a, 13), with rates between 11.0 and 11.4 deaths per 10,000. Cancer, which caused 21 percent of deaths statewide, had the highest mortality rates in Florence, Burnett, and Clark Counties and the lowest in Iowa, Pepin, and Calumet Counties. Statewide, the county rates ranged from 9.6 to 31.9 cancer deaths per 10,000. Mortality rates for both cancer and heart disease have steadily fallen over the past decade, yet drug overdose deaths have greatly increased. As the Wisconsin Department of Health Services (WDHS 2019a, 27) reports, "In 2017, the highest drug overdose mortality rates were in St. Croix, Lincoln, and Shawano counties. The lowest mortality rates were seen in Ozaukee, Manitowoc, and Fond du Lac counties."

## POPULATION LOSSES AND GAINS IN WISCONSIN

Wisconsin's population growth has long been under the nation's growth rate. This is displayed by Wisconsin's steady loss of congressional seats. It had 11 seats in 1930 and now has 8. Yet many Wisconsin counties and some cities and villages fall short of slow growth. Population loss has occurred in both Milwaukee (city and county) and many rural counties, particularly within the northern third of the state. The last several censuses document this decline.

### Population Change at the County Level

Milwaukee was the state's only county to lose population between 1990 and 2000. It grew between 2000 and 2010, but it was smaller in 2010 than in 1990. In contrast, 20 counties lost population between 2000 and 2010, of which 14 are in the northern third of the state, 3 in central Wisconsin, plus Manitowoc, Crawford, and Buffalo Counties. The losses during that decade were sufficiently large so that 6 counties lost population between 1990 and 2010. Five of the 6 counties are within the northern third of the state. Estimates from the Census Bureau's American Community Survey indicate that the 2020 census is likely to report a growing number of counties losing population. Indeed, 34 counties had declining populations between 2010 and 2018, with all but three counties (Bayfield, Vilas, and Oconto) within the northernmost three tiers of Wisconsin counties losing population. In contrast, counties with the greatest numerical growth were all metropolitan: Dane County surrounding Madison, Brown County encompassing Green Bay, Outagamie County around Appleton, and Waukesha County just west of Milwaukee. Dane County has grown the most, gaining over 50,000 newcomers in just eight years.

Current populations are less than what they were a century ago in several counties (Figure 13.15). Both Iowa and Lafayette Counties had their largest populations at the time of the 1870 census. Two adjacent counties (Buffalo and Pepin) along the Mississippi River had their greatest population in 1900. Wisconsin's four counties along Lake Superior, plus Price and Clark Counties, had their greatest number in 1920. Four counties reached their maximum population 80 years ago: Forest, Langlade, Richland, and Crawford. Vernon County, whose maximum at the time of the 2010 census was also in 1940, had edged above that number by 2020. All of the above counties are largely rural, although Douglas County is part of the Duluth-Superior metropolitan area. Yet the state's most highly urbanized county, Milwaukee, reached its maximum a half century ago. Manitowoc County's population has fallen since its 1980 maximum.

### Population Density across Wisconsin

Population density, mapped by census blocks in Figures 13.16 and 13.17, varies considerably across Wisconsin. The highest densities, which exceeded 5,000 persons per square mile in 2010, are in several neighborhoods of Milwaukee and Madison. A census block including student housing for the University of Wisconsin in Madison exceeded 10,000 persons per square mile. In contrast, vast areas of Wisconsin have densities of fewer than 2 persons per square mile. Looking at census blocks, the smallest areal unit used by the Census Bureau, one finds that many blocks consist of unpopulated

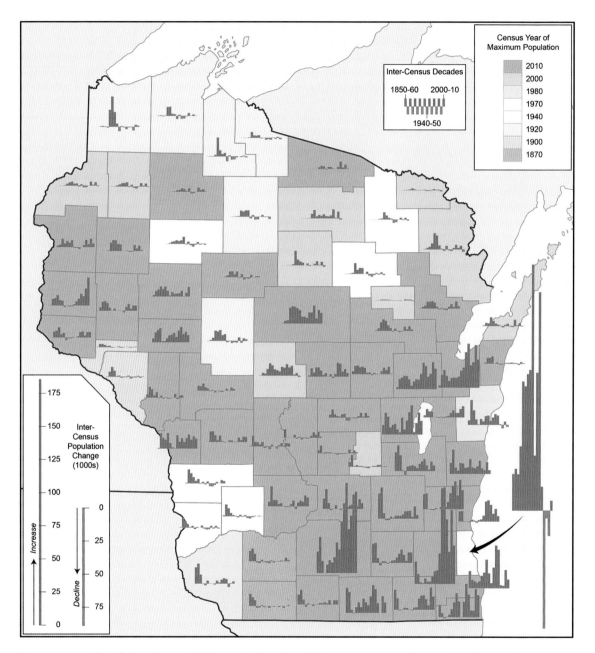

FIGURE 13.15. Population change in Wisconsin, showing which census year each county had its maximum population and graphically displaying increases and decreases in its population between each of the censuses between 1850 and 2010. Data from U.S. Censuses conducted between 1850 and 2010.

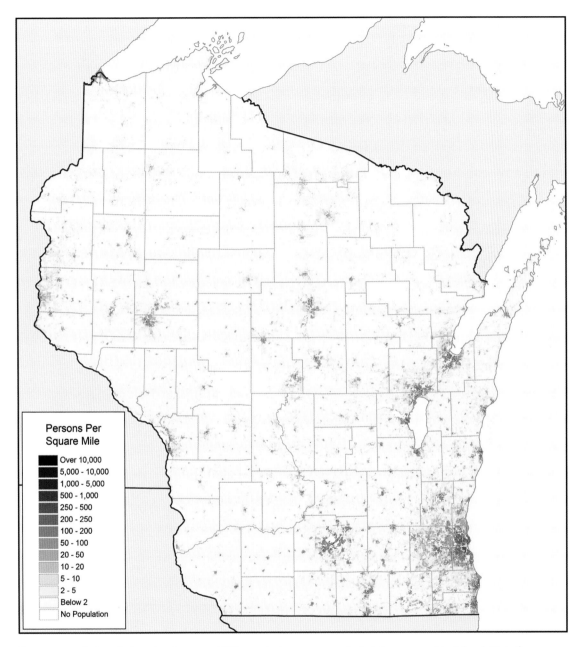

Persons Per
Square Mile

| | |
|---|---|
| ■ | Over 10,000 |
| ■ | 5,000 - 10,000 |
| ■ | 1,000 - 5,000 |
| ■ | 500 - 1,000 |
| ■ | 250 - 500 |
| ■ | 200 - 250 |
| ■ | 100 - 200 |
| ■ | 50 - 100 |
| ■ | 20 - 50 |
| ■ | 10 - 20 |
| ■ | 5 - 10 |
| ■ | 2 - 5 |
| ■ | Below 2 |
| □ | No Population |

FIGURE 13.16. Population density map of Wisconsin in 2010 compiled at the census block level. Data from USCB (2019a).

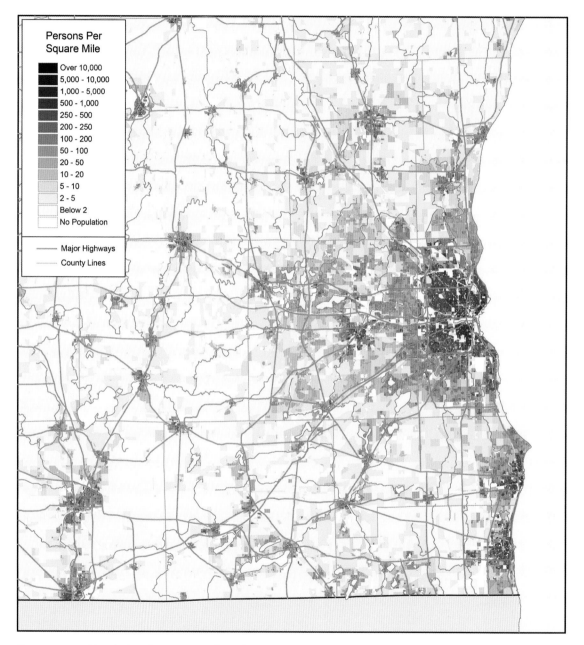

Persons Per
Square Mile

■ Over 10,000
■ 5,000 - 10,000
■ 1,000 - 5,000
■ 500 - 1,000
■ 250 - 500
■ 200 - 250
■ 100 - 200
■ 50 - 100
■ 20 - 50
■ 10 - 20
■ 5 - 10
■ 2 - 5
□ Below 2
□ No Population

—— Major Highways
—— County Lines

FIGURE 13.17. Population density map of southeast Wisconsin in 2010 compiled at the census block level. Data from USCB (2019a).

industrial areas, large farm fields that lack farmsteads, and extensive forests. Thus, census blocks with under 2 persons per square mile include 83.9 percent of Wisconsin's land area. The map distinguishes those blocks without any residents, and they cover extensive acreages in northern Wisconsin. Far more blocks are inhabited, but have fewer than 2 persons per square mile.

Considering population density at the town level, it is clear that considerable areas of northern and central Wisconsin have fewer than 2 persons per square mile, not a sufficient number of persons to meet Frederick Jackson Turner's century-old frontier definition that distinguished settled lands from wilderness. The 2010 census showed that two towns, Anderson in Iron County and Popple River in Forest County, have fewer than 1 resident per square mile. Ten towns have between 1.0 and 1.91 inhabitants per square mile. Three of these towns are large, covering nearly four Public Land Survey townships, yet their densities are under 1.5 persons per square mile of land area. They include Douglas County's Town of Dairyland, whose 184 residents extend over 140 square miles, and Sawyer County's Town of Draper, whose 204 inhabitants are spread across 136 square miles. While most of the towns with population densities under 2 persons per square mile are located in northern Wisconsin, the Town of Kingston in Juneau County has a density of 1.67 persons per square mile. An additional 17 towns have population densities between 1 and 3 persons a square mile. Two of these are in central Wisconsin, while the others are in the northern third of the state. Estimated population changes between 2000 and 2019 at the minor civil division level (Figure 13.18) indicate that a broad swath of northern Wisconsin, plus many rural areas elsewhere, continue losing a large percentage of their population.

The population of Wisconsin was more evenly distributed within the state's counties in the 1890s through the 1910s than today. Portions of Wisconsin are less settled today than they were in the nineteenth century, farms have consolidated, with many former farmhouses now gone, and those families that remain are typically much smaller than what they were a century ago. The index of territorial concentration (Figure 13.19), calculated for Wisconsin's counties, shows that since the 1920s their populations have become more clustered. It is those areas around these sparsely settled areas that often display many of the social and economic woes described earlier in the chapter. The population is often far from work, yet too dispersed to be efficiently provided with quality social, educational, transportation, and health services.

## Demographic Causes of Population Increase and Decrease

Population growth can be related to both natural increase—greater numbers of births than deaths—and to positive net migration—more arrivals than departures. As Figure 13.20 shows, both factors drove population change between 2000 and 2010, with outmigration exceeding the number of new arrivals in many of the counties that did show natural increase. If we just look at net domestic migration, excluding international movements, 44 counties had fewer arrivals than departures. International arrivals kept 6 counties from having negative net migration between 2010 and 2018. All but 2 counties (Iron and Florence) had positive international migration. Both Dane and Milwaukee Counties received over 10,000 more international arrivals than departures, while Brown County had over 2,500.

Even with these international arrivals, negative net migration and declining birthrates have put increased pressure upon many local employers. As the Wisconsin Policy Forum (WPF 2019b, 2) notes,

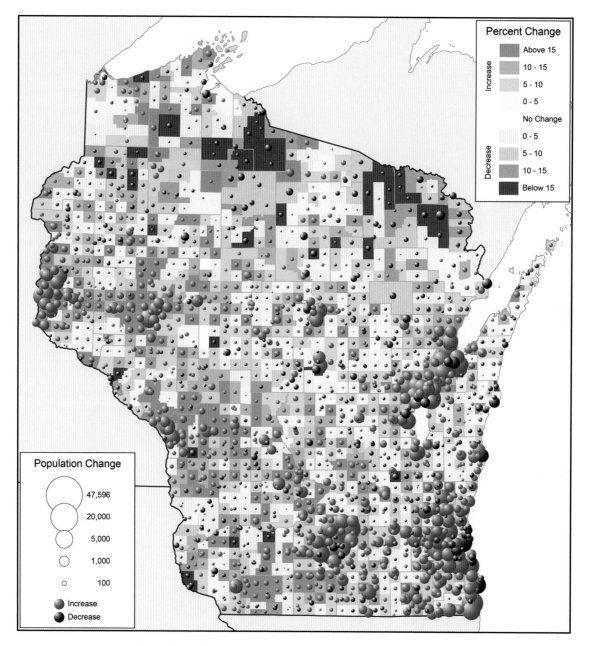

FIGURE 13.18. Estimated population change in Wisconsin between 2010 and 2019 at the minor civil division level. Data from the 2010 U.S. Census and American Community Survey estimates for 2019.

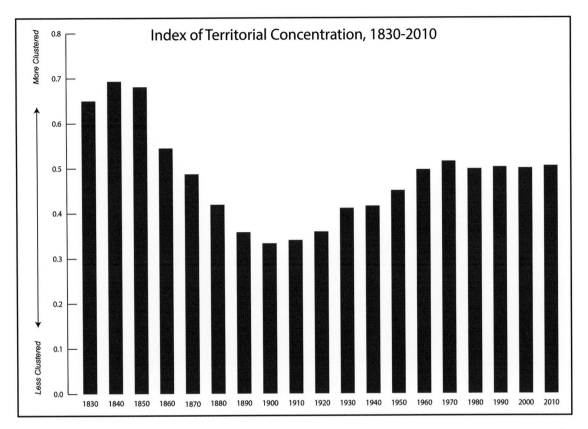

FIGURE 13.19. Chart showing index of territorial concentration in Wisconsin for each census between 1830 and 2010. Data from Schroeder (2016).

not only does the state face a decreasing working-age population and a declining youth population, who otherwise would logically comprise new workers of the future, but "more people have moved away from Wisconsin than to the state every year for more than a decade." Wood County has seen the largest proportional drop in its working-age population (5.9 percent decline), while Milwaukee, with a 1.5 percent decline, has seen the largest numerical decline (WPF 2019b). Dane, Outagamie, St. Croix, and Kenosha Counties are exceptions, with the latter having more arrivals from Illinois than departures.

## Consequences of a Growing Senior Population

At the same time Wisconsin's younger working age population has been falling, its number of residents aged 65 years or older has increased. While many older residents remain actively employed, some by financial necessity, the growing number of retirees places additional demands upon a variety of health and social services. Rural areas are often particularly at risk. For example, the 2017 Census of Agriculture indicates that 30.6 percent of Wisconsin's principal farmers are at least 65 years of age,

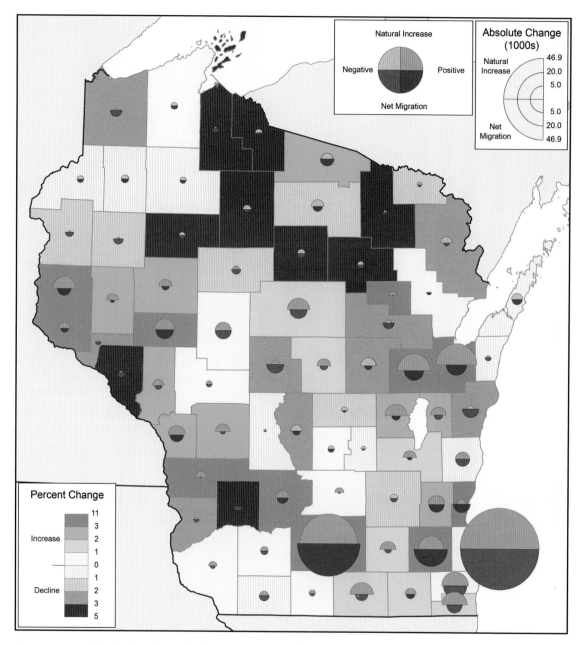

FIGURE 13.20. Population change in Wisconsin counties between 2010 and 2018 showing contributions of migration and natural increase. Data from U.S. Census Bureau. County Population Totals: 2010–2018. Population, Population Change, and Estimated Components of Population Change: April 1, 2010, to July 1, 2018 (https://www.census.gov/data/tables/time-series/demo/popest/2010s-counties-total.html).

and 9.3 percent are at least 75 years old (NASS 2019c). The lack of a suitable work force is a major hindrance to economic development in many rural counties. Yet a growing older population has kept some rural counties from posting even larger population losses.

## Impacts of Seasonal Visitors and Retirees in Rural Wisconsin

Population growth in some nonmetropolitan counties was attributed even in the 1980s to "in-migration of elderly people who have converted their lakeshore cottages into their primary homes and permanent retirement residences" (Hart 1984b, 192–193). Exploring population change at the town level, those towns that had greater than 5 percent of their area comprised by inland water had above average rates of population growth (Hart 1984a). Even though many vacation cottages are only seasonally occupied, their "increased number of visitors has generated a greater number of traditional service jobs during the longer season, and it has also increased the number of jobs, both in and out of season, in construction and repair services" (Hart 1984a, 240). In contrast, most of the rural region outside these inland lake towns and larger cities has lost population or experienced low growth.

Several areas of Wisconsin were identified in the early 2000s as naturally occurring retirement communities (NORCs), ones that formed without specific planning or marketing efforts to retirement residents. These are most conspicuous in northwestern Wisconsin, particularly within Burnett and Washburn Counties; north-central Wisconsin, especially Vilas and Oneida Counties; and central Wisconsin, extending from Juneau and Adams Counties through Portage and Waushara Counties (Hunt, Marshall, and Merrill 2002). Those naturally occurring retirement communities that are most distant from their residents' prior residence typically attracted wealthier and better-educated retirees than those who made more local moves. Thus, many older residents who obtained homes in the Northwoods with its numerous lakes came from the Chicago area, and many of those going to northwest Wisconsin moved from the Minneapolis–St. Paul metropolitan area, whereas those in central Wisconsin relocated from areas of southern Wisconsin. Yet distance is also correlated with seasonal versus permanent relocation, with those moving longer distances more likely to be seasonal versus permanent residents. Natural amenities of the area, including natural beauty, the forests, and the lakes, are far more significant in attracting long-distance retirees than other groups, who characterized the other locations. Indeed, "Amenity NORCs typically attract younger, healthier, better-educated, and more affluent migrants who move as couples to a sparsely populated rural area rich in natural amenities lakes and forests" (Hunt, Marshall, and Merrill 2002, 52).

Recent research has explored the impact of counterurbanization, the movement of affluent urbanites to "remote high-quality environments catering to the development of recreational housing as second, third, or fourth homes" (Chi and Marcouiller 2012, 48) within eight counties in northern Wisconsin, extending from Sawyer through Florence County. Using minor civil division data, Guangqing Chi and David W. Marcouiller (2012, 57) found "that the proportion of public land area was statistically significant in explaining in-migration to the frontier region," while water, wetland, lakeshore, and golf course access was not. Yet for all of the recreational use of the region, as described in the next chapter, its year-round population remains low.

## POPULATION CHANGE: REDISTRICTING AND GERRYMANDERING

Geographic differences in political affiliations have long been recognized. Rural residents have different concerns than urban dwellers, and urban residents often expect government services that cannot be practically or economically provided in sparsely settled regions. Self-employed farmers differ from factory laborers in their attitudes regarding working conditions and the provision of social services. Cities often include racial groups who are largely absent in rural areas, and cities disproportionately attract newcomers with different skills and values than the overall population. Other contrasts in attitudes can be identified and attributed to ethnicity and religious beliefs, such as the role of parochial schools, blue laws regulating work on Sunday and sales of alcohol, and provision of various family planning services. Thus, as Kazimierz J. Zaniewski and James R. Simmons (2016, 129) explain, "changes in the political map of Wisconsin are driven by . . . suburbanization, residential development in metro regions, the decline and resentment in many rural communities, and the attractiveness of cities to younger and more liberal populations."

### Wisconsin's Political Geography

In the 1910s Wisconsin elected two Democrats, eight Republicans, and one Socialist as members of the U.S. House of Representatives. The Socialist represented Milwaukee County. One Democrat came from a district that included Dodge, Washington, Ozaukee, Fond du Lac, and Sheboygan Counties, while the other Democrat represented Brown, Outagamie, Kewaunee Counties and three counties to the north. Republicans came from the remainder of the state. Since the number of Wisconsin congressional seats was reduced to nine, beginning in 1973 and continuing through 2002, Democrats occupied over half of the seats during all but two Congresses. Since then, Republicans have occupied the majority of Wisconsin's eight seats in all but two Congresses. After congressional seats were redistricted following the 2010 election, Democrats have continuously occupied three seats, those centered upon Milwaukee, south-central Wisconsin including Madison, and west-central Wisconsin including La Crosse. Republicans have controlled the other five seats. Over the past decade, "white conservative voters solidified the Republican Party's base in rural and suburban areas of the state, Democratic voting power consolidated in the cities" (Zaniewski and Simmons 2016, 131). The pronounced division between the urban Democratic strongholds and the rural and suburban counties was shown by the results of the 2018 gubernatorial election. Democratic Tony Evers won the statewide election by 29,227 votes despite having lost three-quarters of the counties, given his 150,846 vote margin in Dane County and his 138,069 plurality in Milwaukee County (WEC 2018).

### Gerrymandering in Wisconsin

Shifts in population determined by each decennial census necessitated redrawing congressional, state assembly, and state senatorial district boundaries, such that each representative or state senator is elected from a district with comparable populations. Following the release of the 2010 census data at a time the Republican Party had control of both houses of the state legislature and the governor's office, redistricting took place in a manner that favored candidates from that party. To maximize

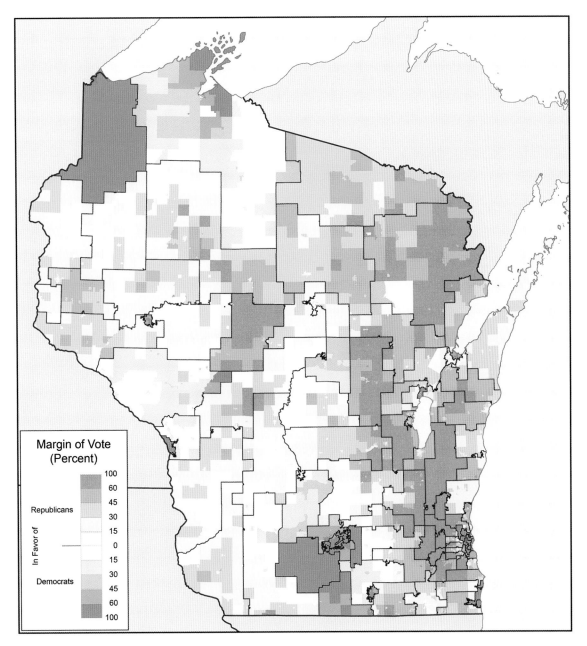

Margin of Vote
(Percent)

In Favor of

Republicans

Democrats

100
60
45
30
15
0
15
30
45
60
100

FIGURE 13.21. Proportion of votes in Wisconsin State Assembly districts won by Democratic and Republican candidates averaged over the elections of 2012, 2014, 2016, and 2018. Data from Wisconsin State Legislature's GIS Open Portal Data (https://data-ltsb.opendata.arcgis.com/search?tags=election%20data).

Republican control, the redistricting packed "Democratic voters into a few districts to dilute their vote across the state" and "pulled conservative voting blocks into Republican districts to help Republican incumbents secure future elections" (Bukowski 2017, 243–244). These actions permitted Republican candidates to win a disproportionate number of Wisconsin's congressional, state assembly, and state senatorial districts. For example, in 2012 Democrats received 51.4 percent of the votes cast statewide for the state assembly, but won 39.4 percent of the seats. In 2018 Democrats received 53 percent of the votes cast for the assembly, but won only 36 percent of the seats.

Democrats typically win their Wisconsin assembly districts by wide margins, as shown in Figure 13.21, while Republicans win many more districts by much narrower margins. This led to an unsuccessful U.S. Supreme Court challenge alleging disproportionate wasted votes between the political parties, the issue of vote efficiency. Other actions have also disenfranchised many minority voters, such as denying voting rights to convicted felons, even long after their sentences have been completed. Restrictions upon early voting and voter identification requirements have been alleged to disproportionately fall upon minority voters who would be more likely to vote for one party rather than the other. Thus, Republicans maintain political control of the state in which their party was founded in Ripon before the Civil War, even though the majority of the state's population favors the other party. Yet, Republicans occupy a greater share of the state's land area than do the Democrats. Politically, the state has long contained political opposites, such as a strong liberal and Progressive element, while simultaneously being the home of staunch conservatives. The state has also had a long political antagonism between Milwaukee and the less populated counties, such that Milwaukee, as the state's only class-one city, has unique constraints placed upon its actions and taxing authority.

## WISCONSIN'S POPULATION AND CULTURAL DIVERSIONS

Wisconsin's population is among the least racially diverse in the nation, yet it is divided on religion, on politics, and on the best ways to shape the state's future. Its urban versus rural divide is strong, shown by politics, health status, and income and wealth. Yet, as is discussed in the final chapter, Wisconsin's population is far more unified in its dislike of both the Bears and the Vikings and in its approval of the Brewers and the Bucks. Wisconsin's people share a strong affinity for their state of birth, the amenities of its outdoor environment, and being Cheeseheads and Badgers. Our final chapter focuses upon recreation and tourism in Wisconsin, activities that not only cater to Wisconsinites but underpin the state's robust tourist industry and help shape the state's image that is held by outsiders.

*Chapter Fourteen*

# Recreation and Tourism in Wisconsin

RECREATION, SUCH AS camping, boating, sports fishing, and hunting, dominates the local econ-omy in much of rural Wisconsin, yet tourism, music festivals, and sporting events are critical to the identity and economy of several metropolitan cities. Think what Green Bay would be like without the Packers and how Milwaukee mourned the departure of the Braves. Vacation cabins and retirement homes line many Wisconsin rivers and lakeshores. Many urbanites from the Chicago area have long visited Wisconsin. Resorts are prominent within certain locales, such as Lake Geneva, Wis-consin Dells, and the Door Peninsula. Different regions appeal to specific customer bases, whether judged by their affluence, their interests, or distance from home. Statewide, tourism now employs nearly 200,000 persons, and direct visitor spending totaled $13.3 billion in 2018 (WDT 2019). This chapter begins with the historical geography of Wisconsin's resort areas and then reviews the state's contemporary recreational offerings.

## HISTORICAL DEVELOPMENT OF WISCONSIN'S RESORT INDUSTRY

Railroad expansion in the nineteenth century provided wealthy Chicago and Milwaukee residents easy access to several lake-oriented resort communities that developed in southeastern Wisconsin. Other areas followed these earliest developments.

### Southeast Wisconsin

Numerous inland lakes of western Kenosha and Racine Counties, together with adjacent Lake County, Illinois, became "one of the most extensively utilized summer resort areas in the United States" (Cut-ler 1965, 84), given their location within 50 miles of Chicago. Among the best-known resort commu-nities are Lake Geneva and Oconomowoc. A century and a half ago Lake Geneva was a summer resort of wealthy Chicagoans that was known as "The Newport of Chicago society" (WPA 1954, 489). Many nineteenth-century mansions facing the lake remain today. Mail is still delivered by boat, which also transports many paying tourists, to homes along Lake Geneva. Oconomowoc likewise attracted summer visitors from near and far. "In the 1880's and 1890's wealthy families from St. Louis, New Orleans, and other Southern places came with coachmen and carriages to spend the summer at white-pillared Draper Hall . . . or at the great estates on the lakeshore. With better communications Oconomowoc developed a flourishing weekend trade that eventually turned many of the estates into summer hotels and clusters of cottages" (517). Oconomowoc remains the destination of many weekend and vacation visitors, although some of its historic buildings, such as Draper Hall, which was razed in 1969, are gone. Yet many remain (Figure 14.1), and the local historical society has prepared a walking tour for visitors. More distant, Green Lake became "widely known as a playground for well-to-do Chicago people . . . [with] many lavish summer homes, with social life centered about the Lawsonia Country Club" (364). The country club remains, several resorts face the lake, and live theatrical performances continue in its Thrasher Opera House, built in 1910.

Many smaller communities also relied upon vacationers. Within the moraine lake region of southeastern Wisconsin, which overlapped into Illinois, three distinct classes of recreational facilities were evident: upscale estate-resorts around Lake Geneva and Delavan Lake; cottage-style resorts near Twin Lakes and north of Genoa City; and summer cottages around Silver Lake eastward to Paddock Lake and Lake Shangri-La in central Kenosha County (Harper 1950). The entire moraine lake district primarily attracted visitors from the Chicago region, whereas the Oconomowoc-Waukesha Chain of Lakes Region attracted a similar class of vacationers from Milwaukee. Proximity of the privately owned recreational venues, rather than the quality of the moraine lakes that were typically small, shallow, and marshy, was its main draw. Summer visitors occupied cabins crowded along lakeshores in an otherwise agricultural zone.

### Wisconsin Dells

Kilbourn City, renamed Wisconsin Dells "in 1931 in the hope that the more descriptive name would attract tourists" (WPA 1954, 403), has long capitalized on the beauty of the rocky gorges through with the Wisconsin River flows (Figure 14.2). Never glaciated, it has many picturesque and unusual geologic features such as deep gorges, including The Narrows, Witches Gulch, and Artist's Glen, carved by glacial outwash floods, and stunning rock pillars, such as Stand Rock.

Tourists have long been attracted, particularly since the arrival of the railroad that spanned the Dells in 1857. By 1873 tourists were taken by steamboat though the Dells. Landscape images by a pioneer photographer helped promote the region. Indeed, the Milwaukee railroad "distributed more than 13,000 of Bennett's stereographs throughout the country to bolster tourism, and in 1890 the Wisconsin Central Railroad commissioned Bennett to photograph Wisconsin resorts that adjoined its lines" (Hoelscher 1998b, 552). During the last four decades of the nineteenth century, "the Dells of the Wisconsin River grew to become the most important sightseeing destination in the burgeoning Old Northwest" (Hoelscher 1997, 424).

Travelers can still arrive in Wisconsin Dells by railroad. Its Amtrak Station is just two blocks from the river excursion landing (Figure 14.3). H. H. Bennett's photography studio is now a state historic site. Yet today most tourists arrive by automobile, and river tours are just one of the resort community's many attractions, as it now touts itself as the "Waterpark Capital of the World."

FIGURE 14.1. Nineteenth-century mansion on North Lake Road in Oconomowoc, Waukesha County, with modern apartment building constructed on site of Draper Hall.

FIGURE 14.2. Eroded and tree-covered sandstone cliff and islet in Wisconsin River near Wisconsin Dells.

FIGURE 14.3. Excursion boats docked along the Wisconsin River in Wisconsin Dells for tours of the Upper Dells. The railroad bridge is in the background while Riverfront Terrace is to the left.

### Northern Wisconsin

The clear-cutting of northern Wisconsin's vast forests opened the region to tourism a century ago. Railroads provided access to former lumber camps that were converted to lodges used by wealthy sports fishermen and hunters. Larger logging communities, such as Minocqua, that boomed with the arrival of the railroad in 1887 went into a steep decline after the forests were cut. Yet "tourist trade brought a revival; saloons became dance halls and speakeasies, crowded with vacationers. By 1931, within a 40-mile radius, there were 3,000 summer homes, 218 resorts, and 32 boys' and girls' camps" (WPA 1954, 378).

With improvement of roads into the region between the two world wars, middle- and working-class vacationers came to dominate those tourists who came to Wisconsin's Northwoods, and lodge owners and recreation providers increasingly focused upon this group (Shapiro 2006). Besides the Minocqua area (Figure 14.4), Hayward also developed as an early center of Northwoods resorts, featuring lodges, cottages, and housekeeping cabins. Farther north, President Calvin Coolidge spent three summer months in 1928 on Cedar Island in the Brule River, one of five presidents beginning with Ulysses S. Grant who fished that river.

FIGURE 14.4. Marina and restaurant along Minocqua Lake in Minocqua, Oneida County.

Extensive acreage set aside as national forests, state forests and parks, and county forests and parks in Wisconsin's Northwoods replaced cutover lands and failed farms with a new sustainable and clean industry. It continues to attract large numbers of visitors to its resorts and campgrounds, while providing fishing, hiking, biking, gambling on several Indian reservations, and hunting in season. Indeed, "tourism and recreation had become major components of the state's economy by the 1950s, and in many of the northern counties they were the largest combined source of employment and income" (Thompson 1988, 288).

### Door County

Early tourists who arrived in the Door Peninsula often came by steamboat. One line ferried passengers across Green Bay from Menominee, Michigan—the nearest railroad station until 1892—while others brought vacationers directly from Chicago and Milwaukee. Resorts developed around many small ports, which "attracted affluent professional people" (Hart 2008, 71), who had longer summer vacations. Even though a geographer nearly a century ago wrote that "recreation is still in the development stage" at Ellison Bay, it had "three hotels in the village . . . and . . . fifteen cottages along the village waterfront . . . [mostly] built by summer visitors for their own use" (Platt 1928, 123). The tourism industry that developed in the Door Peninsula is more than just the lake ports, as even a century ago the county was becoming "a center of the visual and performing arts" (Hart 2008, 74). Today, it has several local groups staging live theater every summer and nearly a hundred art galleries and studios.

## Development of State Parks

Wisconsin's oldest state park, Interstate State Park, was founded in 1900 in cooperation with Minnesota to preserve the Dalles of the St. Croix River. This site, whose gorge with giant potholes was part of the glacial sluiceway carved by waters draining from Lake Superior, attracts many sightseers and campers. It is the Ice Age National Scenic Trail's western terminus and is contiguous to St. Croix National Scenic River, managed by the National Park Service. The National Park Service has only one other site in Wisconsin, the Apostle Islands National Lakeshore.

By the end of the 1920s, Wisconsin had 11 state parks, largely focused upon preserving distinctive physical landscapes. These included Devil's Lake, with its glacially formed lake situated between 500-foot quartzite bluffs in the Baraboo Range; Kohler-Andrae, which focuses upon sand dunes along its Lake Michigan beach; Pattison, which has Wisconsin's highest waterfall; Peninsula, which occupies

FIGURE 14.5. Observation tower and visitors climbing rock outcrop atop Rib Mountain, Marathon County. The rock displays an outdated plaque indicating it is the highest elevation in Wisconsin, but it is 27 feet shorter than Timm's Hill. However, Rib Mountain State Park features the greatest local relief in the state.

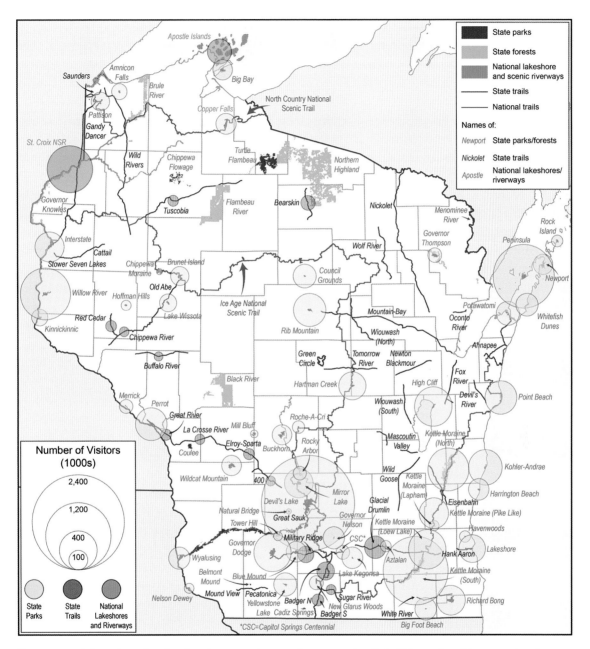

FIGURE 14.6. Location of national and state park areas, forests, and trails in Wisconsin, with proportional circles indicating the number of visitors each receives annually. Data sources: NPS (2019), WDNR (2018, 2019e), and Ziesler (2019).

shoreline and bluffs overlooking the bay of Green Bay in Door County; Perrot, extending from the Mississippi River across Brady's Bluff; Rib Mountain, which features the greatest local relief in the state (Figure 14.5); and Wyalusing, located on the bluff overlooking the confluence of the Wisconsin and Mississippi Rivers, beside Interstate State Park. Seven additional state parks were established during the 1930s. Many of Wisconsin's state parks were beneficiaries of the labor of Civilian Conservation Corps workers in the 1930s, who erected park buildings, built trails, and helped reforest cutover areas (Apps 2019).

At that time the journalist Fred Holmes (1937, 10) acclaimed, "No one can love Wisconsin deeply until a visit has been made to some of her nature shrines." The Wisconsin state parks and forests system has grown considerably since then (Figure 14.6). The state park system's strategic plan published nearly 80 years after Holmes's *Alluring Wisconsin* appeared shows a continuing focus upon accommodating visitors to Wisconsin's outdoor wonderland. "The Wisconsin State Park System is comprised of 110 parks, recreation areas, southern forests and state trails that encompass more than 150,000 acres. Each year, these properties offer the 14 million visitors diverse opportunities for recreation, education and rejuvenation. With over 5,000 campsites, numerous lakes, rivers and recreation trails it enables the public to access Wisconsin's natural and cultural wonders" (WDNR 2015b, 3). State park visitors support nearly 8,000 Wisconsin jobs, while trail users contribute another 200 (Marcouiller, Olson, and Prey 2002), indicative of the ripple effect such activity has upon the state's economy. Wisconsin has nearly 600 county parks that collectively offer many of the same types of recreational opportunities available in the state parks (Bell and Bell 2000). Although they range greatly in size and facilities, some are large, such as Petenwell County Park in Adams County, which has a campground with 500 sites adjacent to a large reservoir along the Wisconsin River.

## RECREATION AND RESORT AREAS OF WISCONSIN

Resort areas in Wisconsin were identified for 1940 and 1980 by mapping the number of summer cottages per square mile. The numbers were highest in four counties of southeast Wisconsin in 1940, while they were greatest in Walworth, Door, Vilas, and Oneida Counties in 1980 (Hart 1984b). As John Fraser Hart (1984b, 199) explains, "The data for 1940 and 1980 are not strictly comparable, but they do suggest that the number of summer cottages increased substantially in the more remote rural areas of the north while it was declining sharply on the metropolitan peripheries, where many erstwhile summer cottages apparently have been converted into year-round retirement homes or even into residences for commuters." The greatest increase in vacation housing occurred in northern Wisconsin's Burnett, Oneida, and Vilas Counties plus Adams and Door Counties (Hart 1984b). County-level data from the Census Bureau's 2014–18 American Community Survey (USCB 2019a) regarding occupied and vacant housing illustrate that this pattern has not only persisted (Figure 14.7) but has become more evident.

Vacant housing units now exceed occupied units in Adams, Burnett, Florence, Forest, Iron, Oneida, Sawyer, and Vilas Counties, and vacant units accounted for at least 30 percent of all housing within an additional 12 counties. Furthermore, the vast majority of these units, which totaled 191,564 statewide in the 2014–2018 period, are reported as being "for seasonal, recreational, or occasional use." Hart

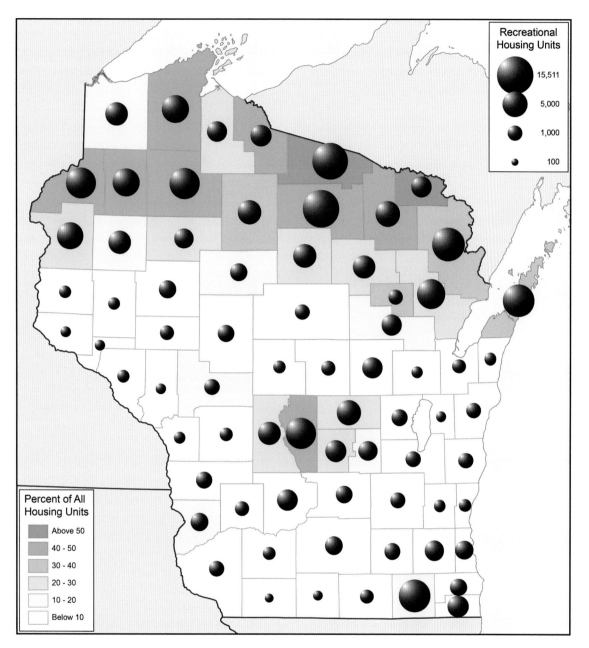

FIGURE 14.7. Vacant housing for seasonal, recreational, or occasional use by Wisconsin county, 2014–2018. Proportional circles show the number of seasonal and recreational housing units, while the shades indicate the percentage of all housing units that are normally vacant but used for these purposes. Data source: American Community Survey 5 year estimates (2014–2018) by USCB (2019a).

(1984b, 196) had assumed this category was "a reasonable surrogate for the number of summer cottages," even though some of them are more likely to be occupied at other times, such as during hunting or snowmobiling season.

Marinette, Oneida, and Vilas Counties each had over 10,000 seasonal housing units, according to the 2014–18 American Community Survey (USCB 2019a), while five counties (Burnett, Sawyer, Door, Adams, and Walworth) enumerated between 7,500 and 10,000. Their numbers have grown. While only four counties had over 7.5 summer cottages per square mile in 1980, six counties did less than four decades later. Door County led with 18.7 units per square mile during the 2014–2018 period, closely followed by Walworth County (16.6), Vilas County (15.2), Oneida County (12.4), and Adams County (12.1). Because many retirees have converted their seasonal cottages and cabins into their primary residences, the number of seasonally occupied housing units seriously underestimates the number of recreational households.

Five clusters of resort counties were identified by Hart (1984b, 202): "Sawyer and Burnett with Bayfield and Washburn in the northwest; Vilas and Oneida with Iron in the center north; Door in the northeast; Adams, Marquette, Columbia, and Sauk in the center; and Walworth in the southeast." These regions display distinctive characteristics favored by specific clientele.

> For example, some resort areas appeal to white-collar types, who prefer outdoor activities that combine uplift and mind improvement with peace and quiet: nature study, photography, hiking, sailing, canoeing, cross-country skiing. Some resort areas appeal to blue-collar types, who want speed and blood: roaring powerboats, snarling snowmobiles, the excitement of hunting and fishing. . . . Some resort areas appeal to entire families with rides, shows, and other forms of commercial entertainment. Most resort areas have some mix of all three types of activities, but one clearly is dominant and gives the area its image. (Hart 1984b, 202–203)

Door County was considered the epitome of the first category, while Wisconsin Dells and the North County characterized the second. As Hart (2008, 74) explains, "The peace and quiet of Door County did not appeal to people who hitherto had frequented frenetic commercial recreational areas, such as the Wisconsin Dells or the 'speed and blood' hunting, fishing, powerboating, and snowmobiling areas north of Rhinelander." Recreation land, even with water access, was relatively inexpensive within the Cutover Region, such that "northern or central Wisconsin attracted many people of only average means" (Thompson 1988, 290), including individuals of retirement age. In contrast, Door County and the Lake Geneva areas are more upscale.

## DESTINATION- VERSUS ACTIVITY-ORIENTED TOURISM

Today's vacation season has expanded beyond a couple of summer months to include the fall. Winter sports also attract visitors. A distinction needs to be made between those tourists who are focused upon a specific destination and will partake in those activities that it provides and those who are focused upon an activity, which may take place at a variety of locations. Obviously, there are many vacationers who focus upon a specific activity, visiting the same destination year after year. Thus, while some fishing enthusiasts may visit different streams and lakes each year, hunters who own a cabin with a hunting acreage may repeatedly return to the same location. This is also true of many who fish from their own docks or have their special site to locate their ice-fishing shanty. Conversely, there are those who focus upon visiting as many of the state park campgrounds that they can or who seek to bike all of the state rail trails, taking them to different destinations each year. We focus first upon destination tourists. Three counties each generated over $1 billion in direct visitor spending in 2018: Sauk, Milwaukee, and Dane (WDT 2019).

### Wisconsin Dells and Lake Delton

Sauk County generated $1.13 billion in direct visitor spending in 2018, the most of any nonmetropolitan county in the Wisconsin. The primary focus of its visitors is Wisconsin Dells, whose area west of the Wisconsin River is within Sauk County, and Lake Delton, located immediately to the south. These communities attract over 4 million visitors annually, generating thousands of jobs (Figure 14.8). Most hotels within this resort area are located in Sauk County, as are all but one of its major waterpark complexes. Wisconsin Dell's website claims "the indoor waterpark was pioneered here in Wisconsin Dells and we are home to the largest concentration of outdoor and indoor waterparks of any place on the planet" (WDVCB 2018).

The Dells region offers visitors a large variety of other attractions besides its over 200 waterslides, including tours of the geologic formations lining the Wisconsin River by excursion boat and amphibious "ducks," zipline tours, theatrical and music performances, and, until recently, its giant Tommy Bartlett water show among numerous other amusements and museums of oddities. Since the beginning of the twenty-first century, the Dells has experienced "a massive wave of new development," such that "themed restaurants share the landscape with condominium developments, golf courses, and escape rooms" (Gurda and Friedman 2019, 17). The Ho-Chunk Casino provides numerous ways to gamble one's money. Farther afield, yet still in Sauk County, visitors can visit Circus World Museum in Baraboo and Devil's Lake State Park just south of Baraboo. Receiving 2,674,386 visitors in fiscal year 2018, Devil's Lake is Wisconsin's most visited state park (WDNR 2018).

### Milwaukee

Milwaukee attracts many visitors who come for business and professional purposes, given the city's major role in insurance, banking, and financial software development and services. Many of these individuals devote part of their visit to enjoying a variety of amenities that the city offers. Others come just for those activities. For example, the city's Historic Third Ward has become a "destination neighborhood, a capital of chic that bears more than a passing resemblance to SoHo or the South Loop" (Gurda 2016, 19). Milwaukee County earned $2,105.3 million in direct visitor spending in 2018, the most of any county in the state and 15.8 percent of the state's total (WDT 2019).

Milwaukee has a long history of engagement with professional sports. Well before the Bucks and Brewers were on the scene, Milwaukee had a National League baseball team in 1878 and was the location where today's American League was organized in 1899 (Thompson 1988). However, between 1902 and 1952, Milwaukee's only team was the Brewers, a minor league professional baseball team.

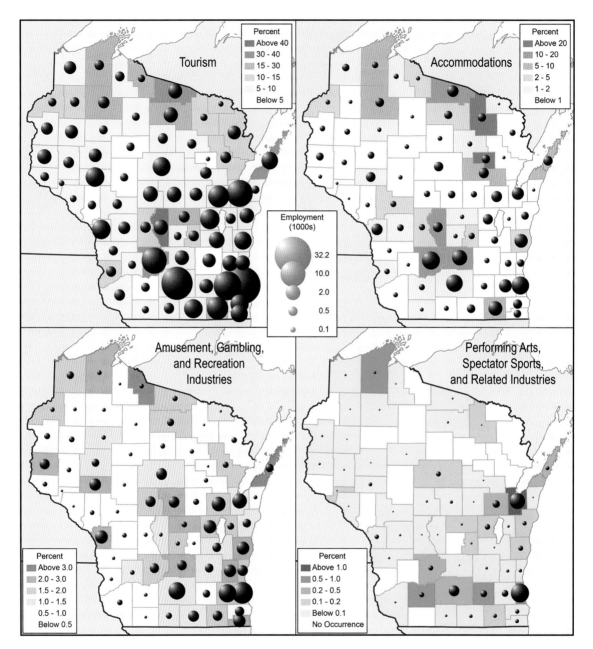

FIGURE 14.8. Employment in recreation and tourism activities in Wisconsin in 2018. Maps show total tourism-related employment, plus employment in providing accommodations; in the gambling, amusement, and recreation industries; and in the performing arts, spectator sports, and their promotion. Sources: WDT (2019) and USBLS (2019).

Yet attendance at the Borchert Field grandstand set league records, which continued with the Braves, before they departed for Atlanta. The major league Brewers, who arrived in 1970 and first played in County Stadium, now play at American Family Field, formerly Miller Park, a stadium with a retractable roof that opened in 2001.

Located in a redeveloped area of the former industrial Menomonee Valley, the Brewers' home field is one mile west of the Mitchell Park Horticultural Conservancy, which features glass domed greenhouses filled with tropical plants. It is within two miles of the Harley-Davidson Museum (Figure 14.9) and the Potawatomi Hotel and Casino, the nation's oldest off-reservation casino and an employer of nearly 3,000 workers. The Milwaukee Bucks professional basketball team, established in 1968, began playing at Fiserv Forum in 2018. This new multipurpose arena in downtown Milwaukee is home court for the Marquette University basketball team and also features various musical performances.

Summerfest, which takes place annually along Lake Michigan, claims to be the "World's largest music festival" and hosts nearly 800 bands over an 11-day run, attracting around 900,000 attendees. It takes place at the 75-acre Henry Maier Festival Park, which includes the 23,000-seat American Family Amphitheatre, the BMO Harris Pavilion that can accommodate 10,000, and 10 additional stages. Numerous ethnic festivals are also normally held annually at the site, including German Fest, Polish

FIGURE 14.9. Harley-Davidson Museum located along canal in a redeveloped area of Milwaukee's formerly industrial Menomonee Valley.

Fest, Festa Italiana, Irish Fest, and Mexican Fiesta, among others. Maier Festival Park's location near Lakeshore State Park anchors a major center of entertainment and activity for residents and visitors alike.

The futuristic Milwaukee Art Museum (Figure 14.10), with its winglike shading that adjusts its position during the day, is located four blocks north of the site of Summerfest, with Discovery World, an environmental museum and aquarium, located midway between them. Milwaukee has a long history of establishment of city parks and greenways dating from the late 1800s. These were strongly promoted by the city's Socialist leaders and Charles Whitnall, who led the city's land commission and the county's park commission a century ago, to deal with "wetland protection, flood control, stream-bank restoration, sanitation, environmental education, and public recreation" (Erickson 2004, 204). Running north from Lakeshore State Park, by the museums, is a series of greenways, through which the now 125-mile Oak Leaf Trail runs. A multipurpose recreational trail, it "is the modern version of Whitnall's figure eight around the Milwaukee metropolitan area and creates the scaffolding for the city's greenway network" (211). Extensive greenways run along the Milwaukee River, whose waters are far cleaner today than they were only a few decades ago, again being suitable for fishing and swimming (Gurda 2018). Extending west from Lakeshore State Park is the Hank Aaron State Trail, which

FIGURE 14.10. Milwaukee Art Museum's Quadracci Pavilion with its wing-like shading, reached by a pedestrian bridge over Lincoln Memorial Drive.

runs by the Harley-Davidson Museum, Potawatomi Casino, American Family Field, State Fair Park, and Milwaukee County Zoo, all major visitor attractions.

A major testament to Milwaukee's recognition as a major convention city was its selection to host the 2020 National Democratic Party Convention, although the pandemic disrupted these plans. Yet it was clear from the onset that providing accommodations for all of the delegates and press would be challenging. Given that there are approximately 5,500 hotel rooms in downtown Milwaukee, the convention attendees were to be spread throughout the metropolitan area's nearly 18,000 rooms (Hess 2019). However, Milwaukee has accommodated other large gatherings, such as the "Harley-Davidson 115th anniversary celebration [that] brought an estimated 150,000 riders to Milwaukee" (Hess 2019) in 2018.

## Madison

Badger sports fans are among the most conspicuous visitors to Madison, filling over 80,000 seats at Camp Randall Stadium on the University of Wisconsin campus seven times every football season. Built on the site of a Civil War training camp for the Union Army, the stadium is the fifth largest in the Big Ten. Until Lambeau Field in Green Bay was expanded in 2013, it was Wisconsin's largest stadium. The Badgers' basketball team plays at the Kohl Center, second only in size to the facility in Milwaukee. The University of Wisconsin brings numerous additional visitors to its campus to attend a large variety of professional and academic conferences.

Visitors, lobbyists, politicians, constituents, and protesters all descend upon Wisconsin's capitol building, located at the opposite end of State Street from the University of Wisconsin. Every Saturday during season the Dane County Farmers' Market, described as "the largest producer-only market in the country" (Destination Madison 2019), occupies the streets surrounding the capitol. The Wisconsin Historical Society's Wisconsin Historical Museum is across the street from the capitol. Numerous other activities attract visitors to the city, including the Henry Vilas Zoo, the Overture Center for the Arts, the UW Arboretum, and an extensive network of hiking and bicycling trails, running across and around the city with spokes extending south to Illinois via the Badger State Trail and west to Dodgeville via the Military Ridge State Trail. Given Madison's location amid four lakes that cover 15,000 acres, a variety of watersports attract residents and visitors. Governor Nelson State Park is located along the shore of Lake Mendota opposite Madison.

## Many Destinations Host Harvest Festivals

Many Wisconsin communities—large and small—host one or more major festivals or conventions annually, bringing large numbers of visitors. Some of these are harvest festivals, such as the three-day Warrens Cranberry Festival, touted as the world's largest celebration of this fruit for which Wisconsin is the largest producer. Door County hosts large numbers of visitors who tour the region to see the cherry blossoms in the spring and partake in its Cherry Fest in Jacksonport in early August. Gays Mills schedules a three-day Apple Festival in late September, "with parades, carnivals, arts and crafts, flea markets, music, dancing, fun run/walk, food and festivities" (Village of Gays Mills 2020). Strawberry festivals are held in Cedarburg, Sun Prairie, and Dousman. Jackson hosts a raspberry festival. Many farms schedule their own celebrations when their fields of berries or orchards of apples and cherries are ready for harvest.

Celebrations in other communities are fishy. Fond du Lac celebrates its Annual Walleye Festival in June, while Oshkosh promotes its Winter Fisheree on the first weekend of February and its Battle-On-Bago later that month, when frozen Lake Winnebago typically hosts thousands of ice fishermen and women and their ice shanties (Figure 14.11). Fond du Lac holds its annual Sturgeon Spectacular. Racine schedules its Salmon-A-Rama fishing tournament in July with prizes for winning catches and entertainment along the Lake Michigan shore. Musky Fest takes place annually in Hayward, including fishing competitions, races, a parade, and carnival. Boulder Junction claims to be the "Musky Capital of the World." Two Rivers holds a Carp Fest. These are just a few of the state's numerous harvest and fishing festivals.

Given its status as "America's Dairyland," Wisconsin hosts a variety of cheese and dairy festivals. In 2019 the three-day Great American Cheese Festival was held in Little Chute, featuring cheese—sampling, carving, and a cheese curd eating contest, along with music, a petting zoo, parade, and race. Marshfield has an annual three-day Dairyfest with cheese, music, parades, and softball games that can attract 20,000 visitors. For over a century Green County Cheese Days has been held around the

FIGURE 14.11. Ice shanties and vehicles on the frozen surface of Lake Winnebago east from Oshkosh's Menomonee Park on the opening day of sturgeon season in 2019.

courthouse square in Monroe in even numbered years. It features cheeses—particularly Swiss cheeses—beer, as well as a cow-milking contest, two parades, and arts and crafts vendors, and the organizers also promote tours of the dairy countryside to view local barn quilts. World Dairy Expo is held annually in Madison, attracting dairy farmers, dairy equipment manufacturers, and tourists. As the Expo's website claims, it is "one of the best known dairy cattle shows in the world," attracting "1,779 owners [who] exhibited 2,338 head of cattle from 37 states and 8 provinces" in 2018 (World Dairy Expo 2019). The Wisconsin Cheese Makers Association puts on an annual World Championship Cheese contest in Madison.

## Communities Host Large Conventions and Music Festivals

Several cities host activities that draw hundreds of thousands of visitors. Oshkosh, which advertises itself as Wisconsin's "event city," is the headquarters of the Experimental Aircraft Association. It holds an annual AirVenture fly-in at the city's Wittman Airport, where for one week each summer over 10,000 aircraft are parked on its grounds and over 30,000 attendees camp on the site. During that week, it is the world's busiest airport. Over a half million persons, including visitors from 80 nations, annually attend AirVenture, which features air shows, displays of vintage aircraft and the military's most modern jets, plus vendors of all things aircraft related, along with entertainment. Three major music festivals are typically held in Oshkosh every summer, during the weeks before or after AirVenture. These include Country USA, Lifefest—Wisconsin's largest festival of Christian music—and Rock USA, which together attracted 300,000 attendees before the pandemic.

Besides those in Milwaukee and Oshkosh, many other musical festivals of various genres normally take place across the state. Among these are Appleton's Mile of Music, Cadott's Country Fest, Eau Claire's Blue Ox Music Festival and Eaux Claires, Hilbert's Voices of Peace Gospel Music Fest, Kenosha's HarborPark Jazz, Rhythm & Blues Festival, Manitowoc's Metro Jam, Prairie du Chien's Prairie Dog Blues Festival, Rhinelander's Hodag Country Music Festival, Sturgeon Bay's Steel Bridge Songfest, Twin Lake's Country Thunder, Wausau's Big Bull Falls Blues Fest, and the Willard Polka Fest.

## Ethnicity and Ethnic Festivals

Ethnicity and ethnic heritage are celebrated in numerous festivals in addition to those in Milwaukee. Besides Willard, Pulaski has its annual Pulaski Polka Days celebration, which runs four days in the middle of the summer, promoting the state's official dance. La Crosse celebrates its Oktoberfest in its dedicated Oktoberfest, USA park grounds just north of the city's downtown over a four-day period, with music, food, beer, and a parade. The Travel Wisconsin website of the Wisconsin Department of Tourism listed 5 Oktoberfest celebrations besides the one in La Crosse: Madison, New Glarus, Appleton, Lake Geneva, and Germantown. Another website listed 15 such celebrations, while a blogger for Discover Wisconsin described 20 Oktoberfest celebrations in 2019.

Some communities have an annual ethnic celebration yet promote their ethnicity to tourists throughout the year. New Glarus was founded by Swiss immigrants in 1845, and while the New Glarus Hotel erected in 1853 continues to serve Swiss fare (Figure 14.12), the Swiss character of the village's other buildings is invented. Indeed, "New Glarus's history of ethnic place invention is perhaps unique in its intensity and scale" (Hoelscher 1998a, 25), as today one sees Swiss flags and banners along the downtown streets, German inscriptions on balconies and gables, window flower boxes, a few chalet-style buildings, and other hints of Swissness. Visitors seek Swiss foods at a variety of restaurants, and the community hosts its Wilhelm Tell Festival every Labor Day weekend, with a yodeling contest, music, and a street dance. Its pageant portrays Wilhelm Tell's successes in obtaining Swiss independence. At times, the celebration has drawn large number of visitors from Switzerland, to "a place more Swiss than Switzerland" (3).

Eight decades ago Fred Holmes (1944) wrote *Old World Wisconsin*, which showed how a tourist could explore the cultures of Europe by traveling across Wisconsin. It mentioned the Swissness of New Glarus, the Polish character and Italian neighborhoods of Milwaukee, a city that today revels in celebrating the heritage brought by its European immigrants as well as those groups who arrived long after Holmes's book was published. Many groups championed in his book still outwardly promote their ancestry to visitors, such as Stoughton, which lines its downtown street with Norwegian flags; Mineral Point, where Cornish pasties can be purchased; Racine, which produces kringles; and Brussels and nearby hamlets, where Belgian pies can still be purchased and booyah is served during festivals.

FIGURE 14.12. New Glarus Hotel built in 1853 in Green County's settlement founded by Swiss immigrants. While this building authentically displays its Swissness, most of the village's commercial district has been reshaped to attract tourists, by the display of Swiss flags and banners (such as the one shown on the right), decorations, and signage.

Old World Wisconsin, the Wisconsin Historical Society's nearly 600-acre outdoor living history museum near Eagle, the largest such museum in the nation, is centered around a nineteenth-century Yankee-themed Crossroads village (Figure 14.13). It is surrounded by farmsteads featuring authentic ethnic houses, barns, and other farm buildings relocated from throughout the state and gathered by ethnicity. Thus, one sees German—both Hessian and Pomeranian—Norwegian, Finnish, Danish, and Polish farmsteads, along with an African American church and cemetery.

Heritage Hill State Park, situated on a smaller site overlooking the Fox River in Allouez, provides another glimpse of Wisconsin's ethnic heritage. It features several buildings dating from the area's French fur trading era—including Wisconsin's oldest house, dating from 1776, plus a Belgian ethnic farmstead with farmhouse, barn, summer kitchen with a bakehouse, outhouse, and roadside chapel. In addition, several historic buildings were relocated from nearby Green Bay.

Persons interested in Wisconsin's Old World ethnic heritage can also travel along Wisconsin's Ethnic Settlement Trail, consisting of over 200 miles of highway indicated with Green Bay Ethnic Trail signage. Linking Kenosha with Green Bay, it passes through numerous ethnic settlements, marked by distinctive churches plus shops and restaurants featuring European-style foods and drink. Locally operated ethnic museums along the route include the Luxembourg American Cultural Center in Belgium,

FIGURE 14.13. Visitors awaiting a shuttle within the Yankee-style Crossroads Village within Old World Wisconsin, located near Eagle in Waukesha County.

Ozaukee County, and the Pinecrest Historical Village near Manitowoc. Elsewhere ethnic architectural heritage is celebrated at Norskedalem Heritage Center near Coon Valley and Knox Creek Heritage Center near Brantwood in Price County, among others.

Religion is often intimately associated with ethnicity, and religious sites often attract many visitors. For example, the National Shrine of Our Lady of Good Help in Champion, at the western edge of northeast Wisconsin's Belgian cultural region, is the "first and only Church-approved Marian Apparition Site in the United States of America" (NSOLGH 2021), attracting 150,000 visitors annually. About twice as many pilgrims annually visit the Holy Hill Basilica and National Shrine of Mary Help of Christians (Figure 14.14), erected by Bavarian Carmelites atop a glacial kame in Washington County. La Crosse draws visitors to its Shrine of Our Lady of Guadalupe, erected earlier this century as the only Marian pilgrimage shrine in western Wisconsin.

### Professional Football

Green Bay, unlike Milwaukee, is a highly unusual host city for a major professional sports team. Unlike all other National Football League teams, the Green Bay Packers are publicly owned, with any profits earmarked for a local charity. Green Bay is by far the smallest city and smallest metropolitan

FIGURE 14.14. Holy Hill Basilica and National Shrine of Mary Help of Christians, located atop a glacial kame in the Kettle Moraine west of Hubertus, Washington County.

area to have any major sport team, whether baseball, basketball, football, or hockey. Given Green Bay's population—it is 27.0 percent of that of the second-smallest metropolitan area with an NFL team—its media market is far smaller and the team is far more reliant upon rabid fans to keep the stadium full. Yet its stadium has the fourth-highest seating capacity in the NFL, and season tickets have sold out every year since 1960. There is currently a more than 30-year wait for individuals once added to the season ticket waiting list. All of the counties of Wisconsin, as well as those in Michigan's Upper Peninsula two tiers deep from the Wisconsin border, have more Packer fans than those of any other team, as do a string of counties running from northeast Iowa into northeastern Nebraska that separate Vikings territory from that dominated by Kansas City Chiefs fans (Meyer 2014). Weekends of home games see almost all hotel rooms within 50 miles of Lambeau Field fully booked and flights into both the Appleton and Green Bay airports filled.

Because of the spillover of visiting fans into adjacent counties, not all of the economic benefits go to Brown County. Nevertheless, with $696.5 million in direct visitor spending in 2018 (WDT 2019), Brown County ranks fifth of Wisconsin's counties in tourism spending, and Outagamie County is ninth. While Packer fans descend upon Green Bay in the largest numbers on game day, the team draws visitors

FIGURE 14.15. Packers Heritage Trail Plaza at Cherry and Washington Streets in downtown Green Bay. The recently renovated Hotel Northland, in the background and marked by a Packers historical marker, was at one time home to Curly Lambeau and the site of Vince Lombardi's introductory press conference.

throughout the year. The Green Bay Packers Hall of Fame and Museum at Lambeau Field draws thousands of visitors each year. The Packers Heritage Trail, focused upon downtown at Packers Heritage Trail Plaza (Figure 14.15), offers enthusiastic fans a walking tour of the city, with 25 historical markers highlighting events in the history of the franchise and key individuals associated with the team. Visitors to Green Bay also frequently patronize the Oneida Casino, located across the highway from the city's airport, and the National Railroad Museum, located in suburban Ashwaubenon.

### Minor League Professional Sports Teams

Minor league professional sports teams were found in several Wisconsin cities in 2019. Two minor league baseball teams, both Midwest League Class A teams, have Wisconsin home fields: the Beloit Snappers and Appleton's Wisconsin Timber Rattlers. Oshkosh is the home of the Wisconsin Herd, an NBA G League team, a return of professional basketball to a city that once hosted the Oshkosh All-Stars. It competed in the National Basketball League playoffs in 11 of its 12 years until 1949, when the league was merged into the National Basketball Association. A new professional women's basketball team also plays in Oshkosh's Menominee Nation Arena. Minor league ice hockey is provided by the Milwaukee Admirals, an American Hockey League team. The Green Bay Gamblers and the Madison Capitols are both U.S. Hockey League teams. Minor league soccer is provided by USL League One's Forward Madison Football Club, while Major Arena Soccer League's Milwaukee Wave plays indoor soccer. Indoor football is provided by the Green Bay Blizzard, while the Milwaukee Bombers are a Mid-American Australian Football League team. Milwaukee is also the home of a Women's Football Alliance team, the Wisconsin Dragons.

### Racing and Other Major Sporting Events

Other sporting events draw visitors to many Wisconsin communities far smaller than Green Bay and the cities hosting minor league professional sports. Elkhart Lake, long known for its historic summer resorts, hosts several major motor races during the summer. In early summer a motorcycle race at Road America determines the winner of the Dunlop Championship. In June Road America hosts amateur sports car races and the NTT IndyCar Series REV Group Grand Prix. In midsummer there are vintage race car and motorcycle exhibitions, while NASCAR XFINITY Series and production car races are held later in the summer. In total, Wisconsin has 26 racetracks and speedways. Besides Road America, the most prominent race courses include La Crosse Speedway in West Salem, the state's only NASCAR-sanctioned asphalt track that features races weekly from April through September, Great Lakes Dragway in Union Grove, and Plymouth Dirt Bike Racing.

Numerous athletic races occur across Wisconsin every year. During the winter the American Birkebeiner cross-country ski race is held between Cable and Hayward. Over 10,000 athletes from throughout the world compete on this 34-mile race and several shorter ski races, typically drawing around 20,000 spectators annually. Marathons are regularly held at many locations in Wisconsin, including Kenosha, Green Bay, and Milwaukee. During 2019, 29 marathons were scheduled in the state, ranging from La Pointe in the north to Kenosha in the south, including both large metropolitan cities and small communities such as Eagle River, Elroy, Hayward, Hurley, Luck, Minocqua, and

Norwalk. Bicycle races take place yearly, including one that circles Lake Winnebago, one that runs from Milwaukee to Madison, the Horribly Hilly Hundreds that climbs 10,000 feet running up and down hills between Blue Mounds and Black Earth, and the Peninsula Century Challenge, among many others. The Door County Century is an annual event in September that draws 3,000 participants. Another bicycle event tours Lake Geneva at a more leisurely pace, yet Wisconsin has far too many bicycle races, tours, and trails to mention them all.

## ACTIVITY-BASED OUTDOOR PURSUITS IN WISCONSIN

Rural Wisconsin hosts much of the activity-focused recreation and tourism, whose focus shifts seasonally. In the winter the focus is upon snowmobiling, cross-country skiing, downhill skiing, ice fishing, and sturgeon spearing. In the summer and early autumn we see motorists touring Wisconsin's backroads, biking its numerous rail trails, hiking in the forests and meadows, swimming in its lakes, canoeing its waterways, and sailboating or kayaking on its lakes. Fishermen and women are out during the summer and fall, and some fish through the ice in the winter. Birdwatchers are active from spring into the fall. During a more restricted season in the autumn, hunters swarm into the woods and meadows, and hikers and other sports enthusiasts flee the outdoors. Some of the activities utilize

FIGURE 14.16. Ski lift station of Granite Peak Ski Area atop Rib Mountain, Rib Mountain State Park, Marathon County.

the same locations, but during different seasons. For example, rail trails that bicyclists use in the summer are coveted by snowmobilers in the winter. Hiking trails and cross-country ski trails often follow the same route. Other locations, such as prime downhill ski slopes, are ignored once the snow is gone, except by those persons who may seek the view from their summit, such as atop Rib Mountain, overlooking the Granite Peak Ski Resort (Figure 14.16). We explore several of these activities, starting with hunting.

### Deer Hunting in Wisconsin

Deer are the most widely hunted large animal in Wisconsin. Hunting season annually draws well over 800,000 deer hunters to Wisconsin fields and forests, of which 15 percent are out-of-state residents (USFWS 2014). The gun deer season runs from the Saturday before Thanksgiving through the Sunday after Thanksgiving. Muzzle-loading gun season runs the 10 days immediately following the regular gun season. Deer hunters using bow and arrow or crossbows have a season running from mid-September into early January. In 2018, some 336,464 deer were harvested, a decline of 180,000 from 2007 (WDNR 2019c). The 2019 harvest was even lower, largely attributable to adverse weather during the main gun season. Wisconsin had the third-largest number of antlered bucks harvested among the states in 2016 (QDMA 2018). Of Wisconsin's total deer harvest, 74 percent were shot using a gun, 12 percent were taken with a bow, and 14 percent were shot with a crossbow. Hunters using a bow or crossbow are more likely to take antlered deer, while those using firearms favor antlerless deer, particularly within the central farmland zone (Dhuey and Wallenfang 2019).

Fifteen percent of Wisconsin adults hunt each year, more than in any other state except Montana and North Dakota (Winkler and Warnke 2013). Nevertheless, their numbers are declining, due to fewer younger hunters and the aging of the middle-aged "Baby Boom generation [that] has had particularly high hunting participation rates" (Winkler and Warnke 2013, 476). In addition, as noted in chapter 4, chronic wasting disease has rapidly spread in southwestern and south-central Wisconsin, reducing the herd size and discouraging hunters in that region.

Hunting occurs throughout Wisconsin, but few deer are harvested in four lakeshore counties in the Milwaukee, Racine, and Kenosha metropolitan areas. Likewise, relatively small numbers are harvested from Wisconsin's Northern Forest Zone. Counties with the most deer harvested by firearms in 2018 were Marathon with 8,759, Waupaca with 8,132, and Shawano with 8,005, all of which are in the Central Farmland Deer Management Zone (Dhuey and Wallenfang 2019). The fall deer density per square mile exceeded 50 in a swath of counties extending from Vernon County northeast through the southern half of Oconto County. Marquette County has the highest county-wide value, 74 deer per square mile (Figure 14.17), although the southernmost part of Adams County has 80. Densities within the Central Forest Deer Management Zone are markedly lower, including the northern three-quarters of Adams County, which has 41 deer per square mile. Even lower, Milwaukee County averages 4 deer per square mile and Kenosha County has 7. Iron County has the smallest inventory among those counties within the Northern Forest Deer Management Zone (WDNR 2019b). Given deer densities in areas with better browsing opportunities, including predation of farm fields, those

FIGURE 14.17. Herd of deer on snow-covered field in Marquette County west of Dalton.

counties with the greatest number of bucks harvested per unit area in 2018 (Figure 14.18) include Marquette County with 6.3 bucks per square mile and Waupaca County with 5.8 bucks per square mile. In contrast, fewer than 1 buck per square mile were harvested in Iron, Kenosha, Racine, and Milwaukee Counties.

For many deer hunters, their hunting retreat involves the entire week of Thanksgiving, providing a major cash infusion into the local economy. Most of Wisconsin's public lands are located in the forested northern third of the state, yet these lands accounted for 14.1 percent of the total deer harvest in 2018. Private lands, sometimes specifically acquired or leased for hunting purposes, account for most of the deer harvest. When hunters of other game and waterfowl are included, 23 percent of hunting takes place on public lands (USFWS 2014).

### Bear and Other Mammal Hunting in Wisconsin

Black bears are sought by far more prospective hunters in Wisconsin than the state can license. Indeed, 124,053 persons applied for one of the 12,970 black bear harvest permits in 2018, and 3,717 bear were reportedly killed (Dhuey, Walter, and Koele 2019). By far the greatest numbers of bear are shot in the northern third of the state, with Bayfield County reporting 378 and Price County 308

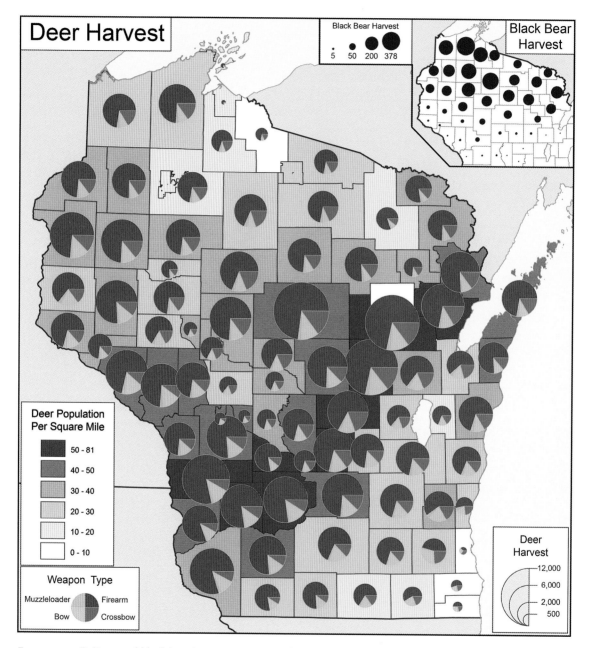

FIGURE 14.18. Deer and black bear harvests in 2018 and the density of deer population per square mile by county and deer management zone. The proportional circles indicate the share of the deer harvested by firearm, muzzleloader, bow and arrow, and crossbow. Data source: Dhuey and Wallenfang (2019).

in 2018. Four additional counties saw over 200 bear harvested: Sawyer, Ashland, Douglas, and Rusk. In contrast, within the three southernmost counties where bear were taken, Juneau County had 4 bear shot, while both Adams and Marquette Counties had 1 (Dhuey, Walter, and Koele 2019).

Bull elk can be hunted only in the Clam Lake Elk Management Zone. In contrast, small game, including cottontail rabbits and both gray and fox squirrels, are hunted statewide from the fall through midwinter, with exact dates depending upon species and location. A variety of furbearers are open to both hunting and trapping, including coyote, red and gray fox, bobcat, and raccoon. While there is a continuous open season for coyotes, the seasons for the others run from the fall through midwinter. Other furbearers, including beaver, mink, muskrat, fisher, and otter, can only be taken by trapping.

### Migratory Bird Hunting

State residents comprise 99 percent of the 105,000 migratory bird hunters in Wisconsin (USFWS 2014). Canada geese are the most widely sought migratory species. Given Canada geese's status as a migratory bird, with some nesting in northern Ontario, while others are temperate breeding populations of giant resident Canada geese, the hunting seasons' length, starting dates, and daily bag limits are controlled by the dynamics of both groups. A Wisconsin Department of Natural Resources report explains, "management of Canada and cackling geese in the Mississippi Flyway is complicated by the need to balance potentially conflicting objectives for arctic, subarctic and temperate-breeding populations" (Finger and Dhuey 2016, 9).

Hunting within the Horicon Zone, which until recently included both the Horicon National Wildlife Refuge and the Horicon Marsh State Wildlife Area, plus the northern half of Dodge County, over half of Fond du Lac County, and corners of two adjacent counties, has a disproportionate impact upon Ontario nesting geese, long necessitating tighter controls in that location. This zone has been merged into the Southern Zone (generally south of U.S. Highway 10 or southeast of Interstate and U.S. Highway 41), where the hunting season extends into early January rather than mid-December. Along the Mississippi River a slightly different season runs.

An early Canada goose harvest season targets resident geese, with an estimated 31,146 geese shot during the first half of September 2017. During the regular Canada goose harvest that followed, 78,110 geese were taken, of which 5,063 came from the former Horicon Zone (Finger, Rohrer, and Dhuey 2017). The top five counties for geese harvested in the regular season during 2016 were Dodge, Brown, Dane, Outagamie, and Waukesha (Finger and Dhuey 2016). The U.S. Fish and Wildlife Service estimates that Wisconsin had 38,400 active goose hunters in 2017 (Raftovich, Chandler, and Fleming 2018), even though twice that number of persons obtained goose hunting permits (Finger, Rohrer, and Dhuey 2017), compared with about 44,100 active duck hunters. Wisconsin's total duck harvest approximated 404,600 in 2017, with mallards accounting for 29.6 percent of the total, followed by wood ducks (21.9 percent), and blue-winged teal (9.7 percent). The typical Wisconsin duck hunter was afield 6.5 days during the hunting season, just slightly ahead of the time that hunters sought Canada geese (Raftovich, Chandler, and Fleming 2018).

### Turkey and Pheasants

Turkeys were reintroduced to Wisconsin in the 1970s, following overhunting that eliminated them in the nineteenth century. Turkeys are shot during two seasons, one in the spring and the other in the fall, with exact dates varying by region of the state. By far the largest numbers are taken in the spring, and its hunter success rate is better than in the fall. In spring 2018, a total of 212,781 turkey permits were issued, and 18 percent of the recipients obtained turkeys. That autumn 73,914 turkey hunters obtained permits for a season that ranged from 63 to 114 days depending upon location. Only 5.1 percent got their bird (Dhuey and Witecha 2019a, 2019b). The number of turkeys harvested in spring 2018, 36,811 birds, was nearly 10 times greater than the size of the fall harvest.

The reintroduction of turkeys to the wild has resulted in harvests of at least 30,000 birds every year beginning in 1999. Turkeys are taken in the greatest numbers within the two southernmost of the state's seven turkey management zones, where 20,706 birds were registered in spring 2018, compared with 1,367 turkeys within Wisconsin's northernmost two zones (Dhuey and Witecha 2019a). Less severe winters combined with abundant feeding upon crop residue in agricultural fields accounts for the greater numbers within southern and central Wisconsin. Most turkeys are hunted upon private lands, and turkeys are particularly drawn to farmland in years of low wild seed and fruit production in forests and meadows.

Pheasant hunting often involves farm-raised birds, both on private pheasant farms and public hunting grounds that are stocked with ring-necked pheasants. The Wisconsin Department of Natural Resources places about 75,000 pheasants annually on 90 properties that are either owned by the state or open to public hunting, augmenting those areas where wild pheasants are less abundant. During the 2018 harvest season, 403,766 pheasants were taken, an increase of around 100,000 from 2017. The greatest numbers were in Kenosha, Jefferson, and Waukesha Counties (Kitchell 2020).

### Economic Impact of Hunting, Fishing, and Wildlife Observation in Wisconsin

The 2011 National Survey of Fishing, Hunting, and Wildlife-Associated Recreation (USFWS 2014) reports that the average hunter in Wisconsin spent 14 days hunting annually, spending $2,833 on their effort. In total, annual hunting expenditures statewide were $2.5 billion, of which 14 percent was for food, lodging, and transportation, 9.5 percent was for guns and ammunition, with a far larger amount spent on special and auxiliary equipment, such as camping equipment, hunting clothes, boats, and vans. Hunters collectively spent nearly three-quarters of a billion dollars on licenses, permits, land leasing, and hunting land acquisition (Figure 14.19). The typical hunter of big game—mostly deer—spent an average of 11 days hunting those animals, while those hunting migratory birds spent over 12 days annually in their pursuit (USFWS 2014).

Fishing engages more Wisconsin residents and visitors than does hunting, even though fishing license sales have dropped for two decades. Twenty-three percent of Wisconsin's adult population purchase fishing licenses (Burkett and Winkler 2017). Ninety-four percent of Wisconsin's estimated 808,000 anglers fished in the state's ponds, lakes, or reservoirs other than the Great Lakes, while 354,000 Wisconsin residents fished its rivers or streams. Obviously, many anglers did both. Fifteen

FIGURE 14.19. Deer hunting stand at edge of forest along Mountain-Bay State Trail east of Eland in Shawano County.

percent of Wisconsin's fishermen and women fished in the Great Lakes. While many varieties of fish are obtained, Wisconsin anglers spend more time seeking panfish than any other type of fish, nearly three times what is spent seeking walleye (USFWS 2014). Even though sturgeon spearing attracts considerable interest during the short winter spearing season, relatively few spearers obtain their giant fish. In 2019 Lake Winnebago yielded 479 sturgeon, and the upriver lakes produced 307 (WDNR 2019d).

The typical Wisconsin angler takes 12 fishing trips each year, spending 16.7 days fishing annually. Given the popularity of waterfront resorts and cabins discussed earlier in this chapter, it is clearly evident the importance of fishing during vacation. Even though more time was spent fishing than hunting, expenditures by Wisconsin anglers averaged $1,129, less than half of what was spent by hunters (USFWS 2014).

Other outdoors enthusiasts engage in wildlife watching, and if shooting their bird or deer, only with a camera. Yet this activity involves over half of the state's residents. An estimated 1,673,000 Wisconsin residents feed birds or other wildlife near their homes, and nearly as many actively observe wildlife within a mile of their home. Over a quarter million Wisconsin residents make trips away from their home to watch wildlife, with the average wildlife watcher spending 18 days annually on such outings.

Nearly as many out-of-state residents visit Wisconsin to observe and photograph the state's wildlife, spending an average of 5 days a year on this activity (USFWS 2014). More Wisconsin residents report in-state away-from-home trips to watch songbirds and birds of prey than either large or small mammals, and these activities involve trip-related expenditures of nearly $200 million annually, with nonresidents spending over $100 million annually on their trips to observe Wisconsin wildlife. When equipment and trip expenses are summed for all wildlife observers in Wisconsin, the total exceeds $1.3 billion annually (USFWS 2014). Many of these journeys take visitors to Wisconsin's state and county parks, state and federal wildlife refuges, and its various county, state, and national forests.

## BIRD WATCHING, HIKING, BIKING, AND TOURING WISCONSIN

Just as Fred Holmes (1937, 1944) encouraged his readers to explore Wisconsin's many natural wonders and cultural landscapes over three-quarters of a century ago, Wisconsin's highways and byways remain attractive to tourists, both those arriving from other states and those taking relatively short afternoon or weekend drives from their Wisconsin homes. Many routes are signposted to guide travelers, such as the Great River Road along the Mississippi River or the historic Yellowstone Trail that runs 409 miles from Kenosha to Hudson. Wisconsin's Department of Transportation (WDOT 2018d) has designated 120 "Rustic Roads," which, according to its website, "provide hikers, bicyclists and motorists an opportunity to leisurely travel through the state's scenic countryside." These roads are lightly traveled routes in 59 counties through areas that feature "rugged terrain, native vegetation, native wildlife or include open areas with agricultural vistas." Hikers and bikers have nearly 2,000 miles of off-road trails running on abandoned railroad grades from which they can view wildlife, explore Wisconsin's forests and meadows, and appreciate both rural and urban landscapes.

## PRÉCIS

This book has provided an overview of Wisconsin's unique assemblage of physical landscapes—many of which were strongly modified by continental glaciations, its diverse vegetation along the transition zones between prairies and forests and between deciduous and coniferous forests, and its changeable seasons, creating ideal conditions for both winter and summer activities. Wisconsin's human geography is also a study of diversity. The state's population has long displayed contradictory elements of progressivism and right-wing conservatism. Its rural and urban areas battle for political control of the state, yet together they shape the state's economy and its place in the nation. In a state long known for the variety of its residents with northern, central, and eastern European ancestry, the recent arrival of Hispanics, African Americans, and the Hmong have added to the mixture, bringing diversity to the state's cities and its local farmers' markets. The economy of Wisconsin remains shaped by the geography of the state's resources, yet agriculture and manufacturing employ a far smaller share of Wisconsin residents today than in the past. What we see when traveling across the state is both a legacy of the past and a testimony of today's growing engagement in the modern information economy. Yet tourism remains a major element in that economy, just as it was a century ago, even though today we view the state from private automobiles rather than from the deck of a steamboat or window of a railroad passenger car.

# References

Acknerkecht, Erwin H. 1946. *Malaria in the Upper Mississippi Valley, 1660–1900*. Baltimore: Johns Hopkins University Press.

Adams, Barry. 2018. "Changes Come to Fox River Valley's Paper Industry." *Wisconsin State Journal*, March 10. https://www.usnews.com/news/best-states/wisconsin/articles/2018-03-10/changes-come-to-fox-river-valleys-paper-industry.

———. 2019. "A Hemp Revival Spreads across Wisconsin, Creating a New Business Model for the Stressed Farm Economy." *Wisconsin State Journal*, June 30. https://madison.com/wsj/business/a-hemp-revival-spreads-across-wisconsin-creating-a-new-business/article_75dc4eae-6b5a-5e0e-a69d-7b75b579b3ff.html.

Adams, Henry C. 1895. *Report on Transportation Business in the United States at the Eleventh Census: 1890*. Washington, DC: U.S. Census Office (Government Printing Office).

Albert, Michael D. 1993. "Wisconsin's Role in the Early Automobile Industry." *The Wisconsin Geographer* 9: 1–7.

Amato, Michael S., Sheryl Magzamen, Pamela Imm, Jeffrey A. Havlena, Henry A. Anderson, Marty S. Kanarek, and Colleen F. Moore. 2013. "Early Lead Exposure (<3 Years Old) Prospectively Predicts Fourth Grade School Suspension in Milwaukee, Wisconsin (USA)." *Environmental Research* 126: 60–65.

Angel, Jim, C. Swanston, B. M. Boustead, K. C. Conlon, K. R. Hall, J. L. Jorns, K. E. Kunkel, et al. 2018. "Midwest." In *Fourth National Climate Assessment*. Vol. 2, *Impacts, Risks, and Adaptation in the United States*, edited by D. R. Reidmiller, C. W. Avery, D. R. Easterling, K. E. Kunkel, K. L. M. Lewis, T. K. Maycock, and B. C. Stewart, 872–940. Washington, DC: U.S. Global Change Research Program.

APL (Applied Population Laboratory). 2015. *Hmong in Wisconsin—A Statistical Overview*. Madison: University of Wisconsin Applied Population Laboratory and UW Extension.

Apps, Jerry W. 2005. *Breweries of Wisconsin*. Madison: University of Wisconsin Press.

———. 2010. *Barns of Wisconsin*. Madison: Wisconsin Historical Society Press.

———. 2015. *Wisconsin Agriculture: A History*. Madison: Wisconsin Historical Society Press.

———. 2017. "From Wheat to Dairy Farming and More." *Wisconsin Magazine of History* 100 (4): 4–9.

———. 2019. *The Civilian Conservation Corps in Wisconsin: Nature's Army at Work*. Madison: Wisconsin Historical Society Press.

ASARB (Association of Statisticians of American Religious Bodies). 2019. "The ARDA: Association of Religion Data Archives." Searchable online database providing church denominational membership and adherence by county. http://www.thearda.com/mapsReports/maps/2010/WI_D469_10R.asp.

Badenhausen, Kurt. 2016. "The Best and Worst States for Business 2016." *Forbes Magazine*, November 16. https://www.forbes.com/sites/kurtbadenhausen/2016/11/16/the-best-and-worst-states-for-business-2016/#18cc149c1477.

Bardeen, Charles R. 1925. "Hospitals in Wisconsin." In *The Wisconsin Blue Book 1925*, edited by Fred L. Holmes, 235–267. Madison: Democrat Printing Company, State Printer.

Barrett, Rick, and Lee Bergquist. 2020. "Industrial Dairy Farming Is Taking Over in Wisconsin, Crowding Out Family Operations and Raising Environmental Concerns." *Milwaukee Journal Sentinel*, updated February 11. https://www.jsonline.com/in-depth/news/special-reports/dairy-crisis/2019/12/06/industrial-dairy-impacts-wisconsin-environment-family-farms/4318671002/.

Bates, John. 2018. "'The World Walked on Milwaukee Leather': Hemlock and Wisconsin's Tanning Industry." *Wisconsin Magazine of History* 101 (4): 42–49.

Bawden, Timothy. 2006. "A Geographical Perspective on Nineteenth-Century German Immigration to Wisconsin." In *Wisconsin German Land and Life*, edited by Heike Bungert, Cora Lee Kluge, and Robert C. Ostergren, 79–91. Madison: Max Kade Institute for German-American Studies, University of Wisconsin–Madison.

Beck, J. D. 1907. *The Blue Book of the State of Wisconsin*. Madison: Democratic Printing Company, State Printer.

Bell, Chet, and Jeanette Bell. 2000. *County Parks in Wisconsin*. Madison: Trails Media Group.

Benson, Mary Ellen, and Anna B. Wilson. 2015. *Frac Sand in the United States—A Geological and Industry Overview*. Open-File Report 2015-1107. Reston, VA: U.S. Geological Survey.

Berlin, Cynthia. 2002. "Cranberries: Wisconsin's Other Agribusiness." *Focus on Geography* 47 (2): 26–29.

Bernstein, Rick. 1998. Foreword to *Wisconsin's Historic Courthouses*, edited by Marv Balousek and L. Roger Turner, 7–10. Oregon, WI: Badger Books.

Billings, John S. 1895. *Report on the Social Statistics of Cities in the United States at the Eleventh Census: 1890*. Washington, DC: U.S. Census Office (Government Printing Office).

Birmingham, Robert A. 2010. *Spirits of Earth: The Effigy Mound Landscape of Madison and the Four Lakes*. Madison: University of Wisconsin Press.

Birmingham, Robert A., and Leslie E. Eisenberg. 2000. *Indian Mounds of Wisconsin*. Madison: University of Wisconsin Press.

Birmingham, Robert A., and Amy L. Rosebrough. 2017. *Indian Mounds of Wisconsin*. 2nd ed. Madison: University of Wisconsin Press.

Black, Robert F. 1974. *Geology of Ice Age National Scientific Reserve of Wisconsin*. National Park Service Scientific Monograph Series no. 2. Washington, DC: Superintendent of Documents.

Blanchard, W. O. 1924. *The Geography of Southwestern Wisconsin*. Bulletin no. 65. Madison: Wisconsin Geological and Natural History Survey.

Blumentritt, Dylan J., Herbert E. Wright Jr., and Vania Stefanova. 2009. "Formation and Early History of Lakes Pepin and St. Croix of the Upper Mississippi River." *Journal of Paleolimnology* 41 (4): 545–562.

Bockheim, James G., and Alfred E. Hartemink. 2017. *The Soils of Wisconsin*. New York: Springer.

Bonds, Anne, Judith T. Kenny, and Rebecca Nole Wolfe. 2015. "Neighborhood Revitalization without the Local: Race, Nonprofit Governance, and Community Development." *Urban Geography* 36 (7): 1064–1082.

Bosman, Julie. 2018. "School's Closed. Forever. What Happens to a Rural Town after It Loses Its Only School? Arena, Wis., Is About to Find Out." *New York Times*, June 13. https://www.nytimes.com/2018/06/13/us/arena-wisconsin-schools-empty.html.

Brewers Association. 2020. Member Directory: Breweries. https://www.brewersassociation.org/directories/breweries/.

Brinkmann, Waltraud A. R. 1985. "Severe Thunderstorm Hazard in Wisconsin." *Transactions of the Wisconsin Academy of Sciences, Arts, and Letters* 73: 1–11.

Brock, Caroline, and Bradford Barham. 2009. "Farm Structural Change of a Different Kind: Alternative Dairy Farms in Wisconsin—Graziers, Organic and Amish." *Renewable Agriculture and Food Systems* 24 (1): 25–37.

Brock, Caroline, Bradford Barham, and Jeremy D. Foltz. 2006. *Amish Dairy Farms in Southwestern Wisconsin.* PATS Research Report no. 17. Madison: University of Wisconsin–Madison Program on Agricultural Technology Studies.

Brooks, Harold E., Gregory W. Carbin, and Patrick T. Marsh. 2014. "Increased Variability of Tornado Occurrence in the United States." *Science* 346 (6207): 349–352.

Brooks, Harold E., Charles A. Doswell III, and Michael P. Kay. 2003. "Climatological Estimates of Local Daily Tornado Probability for the United States." *Weather and Forecasting* 18 (4): 626–640.

Brown, Charles E. 1916. Map Showing the Distribution of Indian Mounds in Wisconsin. Unpublished map in collection of Wisconsin Historical Society. WHS Image ID 92138.

Brown, E. A., C. H. Wu, and D. M. Mickelson. 2005. "Factors Controlling Rates of Bluff Recession at Two Sites on Lake Michigan." *Journal of Great Lakes Research* 31 (3): 306–321.

Brown, R. C. 1980. "Pine Logging in the Chippewa Valley." *Proceedings, Fifth Annual Meeting, Forest History Association of Wisconsin* 5: 16–24.

Brush, John E. 1953. "The Hierarchy of Central Places in Southwestern Wisconsin." *The Geographical Review* 43 (3): 380–402.

Buckler, William R., and Harold A. Winters. 1983. "Lake Michigan Bluff Recession." *Annals of the Association of American Geographers* 73 (1): 89–110.

Buckley, Ernest Robertson. 1898. *On the Building and Ornamental Stones of Wisconsin.* Bulletin no. 4. Madison: Wisconsin Geological and Natural History Survey.

Bukowski, Joseph W. 2017. "Redistricting Reform in Wisconsin to Curtail Gerrymandering: The Wisconsin Impartial Citizens Redistricting Commission." *Marquette Law Review* 101 (1): 233–285.

Burkett, Erin M., and Richelle L. Winkler. 2017. "Recreational Fishing in Wisconsin: Using an Age-Period-Cohort Approach to Understand Fishing Participation." Houghton: Michigan Technological University. https://www.mtu.edu/greatlakes/fishery/state-level/pdf/wisconsin-angler-demographics-2000-2014.pdf.

Caliper. 2020. Annual Average Daily Traffic (AADT) 2019. https://www.caliper.com/mapping-software-data/aadt-traffic-count-data.htm.

Cameron, Peter. 2020. "The Broadband Gap Leaves Rural Wisconsin Behind in the COVID-19 Crisis." *WisCONTEXT*, March 24. https://www.wiscontext.org/broadband-gap-leaves-rural-wisconsin-behind-covid-19-crisis.

Campo, Daniel, and Brent D. Ryan. 2008. "The Entertainment Zone: Unplanned Nightlife and the Revitalization of the American Downtown." *Journal of Urban Design* 13 (3): 291–315.

Cannon, William F., Gene L. LaBerge, John S. Klasner, and Klaus J. Schulz. 2008. "The Gogebic Iron Range—A Sample of the Northern Margin of the Penokean Fold and Thrust Belt." Professional Paper 1730. Reston, VA: U.S. Geological Survey.

Carson, Eric C., John W. Attig, and J. Elmo Rawling III. 2019. "The Glacial Record in Regions Surrounding the Driftless Area." In *The Physical Geography and Geology of the Driftless Area: The Career and Contributions of James C. Knox.* GSA Special Paper 543, edited by Eric C. Carson, J. Elmo Rawling III, J. Michael Daniels, and John W. Attig, 37–50. Boulder, CO: Geological Society of America.

Changnon, Stanley A., and David Changnon. 2006. "A Spatial and Temporal Analysis of Damaging Snowstorms in the United States." *Natural Hazards* 37 (3): 373–389.

Changnon, Stanley A., and Kenneth E. Kunkel. 2006. "Severe Storms in the Midwest." Report I/EM 2006–06. Champaign: Midwestern Regional Climate Center, Illinois State Water Survey.

Chi, Guangqing. 2012. "The Impacts of Transport Accessibility on Population Change across Rural, Suburban and Urban Areas: A Case Study of Wisconsin at Sub-county Levels." *Urban Studies* 49 (12): 2711–2731.

Chi, Guangqing, and David W. Marcouiller. 2012. "Recreational Homes and Migration to Remote Amenity-Rich Areas." *Journal of Regional Analysis & Policy* 42 (1): 47–60.

Chu, Gregory. 2001. "Ginseng Farming: A Unique Practice in Wisconsin Agriculture." *The Wisconsin Geographer* 17: 1–10.

Clayton, Jordan A., and James C. Knox. 2008. "Catastrophic Flooding from Glacial Lake Wisconsin." *Geomorphology* 93 (3–4): 384–397.

Clayton, Lee, John W. Attig, Nelson R. Ham, Mark D. Johnson, Carrie E. Jennings, and Kent M. Syverson. 2008. "Ice-Walled-Lake Plains: Implications for the Origin of Hummocky Glacial Topography in Middle North America." *Geomorphology* 97 (1–2): 237–248.

Clayton, Lee, John W. Attig, and David M. Mickelson. 2001. "Effects of Late Pleistocene Permafrost on the Landscape of Wisconsin." *Boreas* 30 (3): 174–188.

Clayton, Lee, John W. Attig, David M. Mickelson, Mark D. Johnson, and Kent M. Syverson, 2006. "Glaciation of Wisconsin." 3rd ed. Educational Series 36. Madison: Wisconsin Geological and Natural History Survey.

Cohen, Patricia. 2020. "The Struggle to Mend America's Rural Roads." *New York Times*, February 18. https://nyti.ms/39CDZ2q.

Coleman, Frank H. 1941. "Climate of Wisconsin." In *Climate and Man: Yearbook of Agriculture 1941*, U.S. Department of Agriculture, 1191–1200. Washington, DC: U.S. Government Printing Office.

Connor, Mary Roddis. 1978. "Logging in Northeastern Wisconsin." In *Some Historic Events in Wisconsin's Logging Industry: Proceedings, Third Annual Meeting, Forest History Association of Wisconsin*, edited by Ramon R. Hernandez, 31–37. Wausau: Forest History Association of Wisconsin.

Craig, William J., Randy Bixby, Mark B. Lindberg, and Tom Haight. 2015. "Mapping European Settlement in Wisconsin: Dates That Public Lands Moved to Private Ownership." Madison: Wisconsin State Cartographer's Office. https://www.sco.wisc.edu/wp-content/uploads/2015/11/Mapping%20Initial%20European%20Settlement%20in%20Wisconsin.pdf.

Crane, L. H. D. 1861. *A Manual of Customs, Precedents and Forms, in Use in the Assmenbly [sic] of Wisconsin; Together with the Rules, the Apportionments, and Other Lists and Tables for Reference, with Indices.* Madison: James Ross, State Printer.

Cravens, Stanley H. 1983. "Capitals and Capitols in Early Wisconsin." In *Wisconsin Blue Book, 1983–1984*, edited by H. Rupert Theobald and Patricia V. Robbins, 100–168. Madison: Wisconsin Legislative Reference Bureau.

Cross, John A. 1992. "1988 Drought Impacts among Wisconsin Dairy Farmers." *Transactions of the Wisconsin Academy of Sciences, Arts and Letters* 80: 21–34.

———. 1994. "Agroclimatic Hazards and Dairy Farming in Wisconsin." *The Geographical Review* 84 (3): 277–289.

———. 1995. "Dairying in an Urban Environment: The Milwaukee Metropolitan Area." *Transactions of the Wisconsin Academy of Sciences, Arts and Letters* 83: 75–86.

———. 1996. "Changing Cropland Combinations in Wisconsin, 1949–1992." *The Wisconsin Geographer* 12: 23–33.

———. 2001. "Change in America's Dairyland." *The Geographical Review* 91 (4): 702–714.

———. 2003. "Amish Surnames, Settlement Patterns, and Migration." *Names* 51 (3–4): 193–214.

———. 2004. "Expansion of Amish Dairy Farming in Wisconsin." *Journal of Cultural Geography* 21 (2): 77–101.

———. 2006. "Restructuring America's Dairy Farms." *The Geographical Review* 96 (1): 1–23.

———. 2007. "The Expanding Role of the Amish in America's Dairy Industry." *Focus on Geography* 50 (3): 7–16.

———. 2011a. "Disappearing Cheese Factories in America's Dairyland." *PAST: Pioneer America Society Transactions* 34. http://www.pioneeramerica.org/past2011/past2011artcross.html.

———. 2011b. "Twenty Years of Change in Wisconsin's Dairy Industry: 1989–2009." *The Wisconsin Geographer* 24: 51–67.

———. 2012. "Changing Patterns of Cheese Manufacturing in America's Dairyland." *The Geographical Review* 102 (4): 525–538.

———. 2014. "Continuity and Change: Amish Dairy Farming in Wisconsin over the Past Decade." *The Geographical Review* 104 (1): 52–70.

———. 2015. "Change and Sustainability Issues in America's Dairyland." *Focus on Geography* 58 (4): 173–183.

———. 2016. "Dairying Landscapes of the Amish in Wisconsin." *Material Culture* 48 (2): 16–31.

———. 2017. *Ethnic Landscapes of America.* New York: Springer Nature.

———. 2018. "Occupation Patterns of Amish Settlements in Wisconsin." *Journal of Amish and Plain Anabaptist Studies* 6 (2): 192–212.

———. 2021a. "Dairy Woes in Wisconsin: What about the Amish?" *Journal of Plain Anabaptist Communities.* 1 (2): 1–21.

———. 2021b. "Old Order Dairy Farmers in Wisconsin." *The Geographical Review*, forthcoming. https://doi.org /10.1080/00167428.2021.1889344.

Cross, John A., Bradford Barham, and Douglas Jackson-Smith. 2000. "Ethnicity and Farm Entry Behavior." *Rural Sociology* 65 (3): 461–483.

Cruz, Marcelo. 1998. "Latino Migration to Green Bay, Wisconsin in the 1990s." *The Geographical Bulletin* 40 (1): 87–101.

Curtis, John T. 1971. *The Vegetation of Wisconsin: An Ordination of Plant Communities.* Madison: University of Wisconsin Press.

Cushman, Will. 2019a. "Wisconsin Is America's Dairy Goat Land." *WisCONTEXT.* September 12. https://www .wiscontext.org/wisconsin-americas-dairy-goat-land.

———. 2019b. "Wisconsin's Place in a Bewildering Milk Pricing System." *WisCONTEXT*, April 9. https://www .wiscontext.org/wisconsins-place-bewildering-milk-pricing-system.

Cutler, Irving. 1965. *The Chicago–Milwaukee Corridor: A Geographical Study of Intermetropolitan Coalescence.* Studies in Geography no. 9. Evanston, IL: Northwestern University Department of Geography.

David, Elizabeth, and Judith Mayer. 1984. "Comparing Costs of Alternative Flood Hazard Mitigation Plans: The Case of Soldiers Grove, Wisconsin." *American Planning Association Journal* 50: 22–35.

Davis, Richard A., Jr. 2016. "Quaternary Geology of the Baraboo Area." In *Geology of the Baraboo, Wisconsin, Area*, edited by R. A. Davis Jr., R. H. Dott Jr., and I. W. D. Dalziel, 55–62. GSA Field Guide 43. Boulder, CO: Geological Society of America.

Davis, Steven M. 2013. "The Forests Nobody Wanted: The Politics of Land Management in the County Forests of the Upper Midwest." *Journal of Land Use and Environmental Law* 28 (2): 197–225.

de Julio, Mary Antoine. 1996. "The Vertefeuille House of Prairie du Chien: A Survivor from the Era of French Wisconsin." *Wisconsin Magazine of History* 80 (1): 26–56.

DeLorme. 2020. *Wisconsin Atlas and Gazetteer.* Yarmouth, ME: Garmin.

Destination Madison. 2019. "Welcome to Madison!" https://www.visitmadison.com/.

Dhuey, Brian, and Kevin Wallenfang. 2019. "2018 Wisconsin Deer Hunting Summary." In *2018 Wisconsin Big Game and Turkey Harvest Summary*, PUB-WM-284-2019, 2–38. Madison: Wisconsin Department of Natural Resources.

Dhuey, Brian, Scott Walter, and Brad Koele. 2019. "2018 Wisconsin Black Bear Harvest Report." In *2018 Wisconsin Big Game and Turkey Harvest Summary*, PUB-WM-284-2019, 47–57. Madison: Wisconsin Department of Natural Resources.

Dhuey, Brian, and Mark Witecha. 2019a. "2018 Fall Turkey Harvest Report." In *2018 Wisconsin Big Game and Turkey Harvest Summary*, PUB-WM-284-2019, 65–69. Madison: Wisconsin Department of Natural Resources.

———. 2019b. "2018 Spring Turkey Harvest Report." In *2018 Wisconsin Big Game and Turkey Harvest Summary*, PUB-WM-284-2019, 57–64. Madison: Wisconsin Department of Natural Resources.

Dolley, Thomas P. 2017a. "Silica." *2015 Minerals Yearbook*, 66.1–66.16. Reston, VA: U.S. Geological Survey.

⸻. 2017b. "Stone, Dimension." *2015 Minerals Yearbook*, 72.1–72.13. Reston, VA: U.S. Geological Survey.

Doolittle, William E. 2000. *Cultivated Landscapes of Native North America*. Oxford: Oxford University Press.

Durand, Loyal, Jr. 1943. "Dairy Barns of Southeastern Wisconsin." *Economic Geography* 19 (1): 37–44.

⸻. 1944. "The West Shawano Upland of Wisconsin: A Study of Regional Development Basic to the Problem of Part of the Great Lakes Cut-Over Region." *Annals of the Association of American Geographers* 34 (3): 135–163.

⸻. 1962. "The Retreat of Agriculture in Milwaukee County, Wisconsin." *Transactions of the Wisconsin Academy of Sciences, Arts and Letters* 51: 197–218.

Dutton, Carl E. 1971. *Geology of the Florence Area, Wisconsin and Michigan*. Professional Paper 633. Washington, DC: U.S. Geological Survey.

Ebling, Walter H., W. D. Bormuth, and F. J. Graham. 1939. *Wisconsin Dairying*. Madison: Wisconsin Crop and Livestock Reporting Service.

Ecklund, Karen, and Natasha Kassulke. 2012. "ON A MISSION to Protect Life, Property and Natural Resources." In "Facing Fire: Wildfire Prevention and Control in Wisconsin" (insert within *Wisconsin Natural Resources Magazine*). Wisconsin Department of Natural Resources. http://dnr.wi.gov/wnrmag/2012/04/wildfire%20insert.pdf.

Eichenlaub, Val. 1979. *Weather and Climate of the Great Lakes Region*. Notre Dame, IN: University of Notre Dame Press.

Eisen, Marc. 2015. "Organic Valley at the Crossroads." *The Progressive*, July 24. https://progressive.org/magazine/organic-valley-crossroads/.

Elegant, Naomi Xu. 2019. "Wisconsin Ginseng Farmers Had Been Exporting to China for a Century. Then Came the Trade War." *Fortune*, August 31. https://fortune.com/2019/08/31/trade-war-china-ginseng-farm-wisconsin/.

Ellickson, Paul B., and Paul L. E. Grieco. 2013. "Wal-Mart and the Geography of Grocery Retailing." *Journal of Urban Economics* 75: 1–14.

Emery, J. Q. 1913. *Wisconsin Cheese Factories, Creameries and Condenseries by Counties and Dairy Statistics*. Madison: Wisconsin Dairy and Food Commission.

Erickson, Donna L. 2004. "The Relationship of Historic City Form and Contemporary Greenway Implementation: A Comparison of Milwaukee, Wisconsin (USA) and Ottawa, Ontario (Canada)." *Landscape and Urban Planning* 68: 199–221.

FAA (Federal Aviation Administration). 2019. Passenger Boarding (Enplanement) and All-Cargo Data for U.S. Airports. https://www.faa.gov/airports/planning_capacity/passenger_allcargo_stats/passenger/media/cy18-all-enplanements.pdf.

Fapso, Richard J. 2001. *Norwegians in Wisconsin*. Madison: Wisconsin Historical Society Press.

Fenneman, Nevin M. 1928. "Physiographic Divisions of the United States." *Annals of the Association of American Geographers* 18 (4): 261–353.

⸻. 1938. *Physiography of Eastern United States*. New York: McGraw-Hill.

Finger, Taylor, and Brian Dhuey. 2016. *Wisconsin Canada Goose Harvest Report* 26 (3). Madison: Wisconsin Department of Natural Resources. https://dnr.wi.gov/topic/WildlifeHabitat/documents/reports/cangharv16.pdf.

Finger, Taylor, Trenton Rohrer, and Brian Dhuey. 2017. "2017 Wisconsin Canada Goose Harvest Summary." Madison: Wisconsin Department of Natural Resources. https://dnr.wi.gov/topic/WildlifeHabitat/documents/reports/cangharv17.pdf.

Finley, Robert W. 1965. *Geography of Wisconsin: A Content Outline*. Madison: College Printing & Typing.

⸻. 1976. *Original Vegetation Cover of Wisconsin: Compiled from U.S. General Land Office Notes*. St. Paul, MN: North Central Forest Experiment Station, U.S. Department of Agriculture.

Fitz, Tom. 2012. "Ironwood: The Rocks of the Penokee Range." *Wisconsin People & Ideas* 58 (2): 32–39.

Fitzpatrick, Faith A., Eric D. Dantoin, Naomi Tillison, Kara M. Watson, Robert J. Waschbusch, and James D. Blount. 2017. *Flood of July 2016 in Northern Wisconsin and the Bad River Reservation*. Scientific Investigations Report 2017-5029. Reston, VA: U.S. Geological Survey. https://doi.org/10.3133/sir20175029.

Fitzpatrick, Faith A., Marie C. Peppler, John F. Walker, William J. Rose, Robert J. Waschbusch, and James L. Kennedy. 2008. *Flood of June 2008 in Southern Wisconsin*. Scientific Investigations Report 2008-5235. Reston, VA: U.S. Geological Survey. http://pubs.usgs.gov/sir/2008/5235/pdf/sir20085235.pdf.

Fixmer, F. N. 1977. "A Short History of Industrial Forestry in Wisconsin." *Proceedings, Second Annual Meeting, Forest History Association of Wisconsin* 2: 24–29.

Flanigan, Kathy. 2020. "Beer Near: A Guide to Wisconsin Breweries." *Milwaukee Journal Sentinel*. https://projects .jsonline.com/apps/BeerNear/.

Fleming, Charles M. 1985. "Preglacial River Valleys of Marquette, Green Lake, and Waushara Counties." *Transactions of the Wisconsin Academy of Sciences, Arts, and Letters* 73: 12–25.

FMG (Fincantieri Marine Group). 2019. "First Cuts of Steel Signals Start of Construction for New Great Lakes Bulk Carrier." https://fincantierimarinegroup.com/8-14-2019/.

Foltman, Leah, and Malia Jones. 2019. "How Redlining Continues to Shape Racial Segregation in Milwaukee." *WisCONTEXT*, February 28. https://www.wiscontext.org/how-redlining-continues-shape-racial-segregation -milwaukee.

Foran, Chris. 2016. "Milwaukee Grapples with a Mid-May Blizzard—in 1990." *Milwaukee Journal Sentinel*, May 3. http://archive.jsonline.com/greensheet/milwaukee-grapples-with-a-mid-may-blizzard--in-1990-b99715810 z1-378053881.html.

Forbes. 2019. "Best States for Business." https://www.forbes.com/best-states-for-business/list/#tab:overall.

Forstall, Richard L. 1996. *Population of States and Counties of the United States: 1790 to 1990*. Washington, DC: U.S. Department of Commerce, Bureau of the Census.

Fremling, Calvin R. 2005. *Immortal River: The Upper Mississippi in Ancient and Modern Times*. Madison: University of Wisconsin Press.

Froehlich, William H. 1899. *The Blue Book of the State of Wisconsin, 1899*. Madison: Wisconsin Secretary of State.

———. 1901. *The Blue Book of the State of Wisconsin, 1901*. Madison: Wisconsin Secretary of State.

Gard, Robert. 2015. *The Romance of Wisconsin Place Names*. Madison: Wisconsin Historical Society Press.

Garrett, Ben. 2017. *Wisconsin Wildfires > 250 Acres: 1871–Present*. Wisconsin Department of Natural Resources. https://dnr.wisconsin.gov/sites/default/files/topic/ForestFire/wildfireHistoryMap.pdf.

Gartner, William Gustav. 1997. "Four Worlds without an Eden: Pre-Columbian Peoples and the Wisconsin Landscape." In Ostergren and Vale, *Wisconsin Land and Life*, 331–350.

———. 1999. "Late Woodland Landscapes of Wisconsin: Ridged Fields, Effigy Mounds and Territoriality." *Antiquity* 73 (281): 671–683.

Geisler, Ellen, Chadwick D. Rittenhouse, and Adena R. Rissman. 2016. "Logger Perceptions of Seasonal Environmental Challenges Facing Timber Operations in the Upper Midwest, USA." *Society & Natural Resources* 29 (5): 540–555.

Gess, Denise, and William Lutz. 2002. *Firestorm at Peshtigo: A Town, Its People, and the Deadliest Fire in American History*. New York: Henry Holt.

Goldstein, Andy. 2018. "Why Is Wisconsin So Terrible for Black People." *The Daily Cardinal*, November 29. https://www.dailycardinal.com/article/2018/11/why-is-wisconsin-so-terrible-for-black-people.

González, Sergio M. 2017. *Mexicans in Wisconsin*. Madison: Wisconsin Historical Society Press.

Gordon, Scott. 2017. "The Costs of Taking Broadband the Extra Mile in Wisconsin." *WisCONTEXT*, June 12. https://www.wiscontext.org/costs-taking-broadband-extra-mile-wisconsin.

Gotter, Brian. 2018. "Wind Energy in Wisconsin: How Has It Evolved?" TMJ4 Milwaukee. https://www.tmj4 .com/news/local-news/wind-energy-in-wisconsin-how-has-it-evolved-.

Guéno, Michael P. 2017. "Of a People of a Place: Reframing the Meaning(s) and Implications of Prehistoric Wisconsin Effigy Mounds." *Material Culture* 49 (2): 50–73.

Gurda, John. 1999. *The Making of Milwaukee*. Milwaukee: Milwaukee County Historical Society.

———. 2016. *Milwaukee: City of Neighborhoods*. Milwaukee: Historic Milwaukee.

———. 2018. *Milwaukee: A City Built on Water*. Madison: Wisconsin Historical Society Press.

Gurda, John, and J Tyler Friedman. 2019. "Diversions of a Different Nature: The Many-Storied Career of Wisconsin Dells." *Wisconsin People & Ideas* 65 (2): 12–18.

Gyrisco, Geoffrey M. 1997. "Victor Cordella and the Architecture of Polish and East-Slavic Identity in America." *Polish American Studies* 54 (1): 33–52.

Hadley, David W. 1976. *Shoreline Erosion in Southeastern Wisconsin*. Special Report no. 5. Madison: Wisconsin Geological and Natural History Survey.

Haines, Anna, Rebecca Roberts, and Karen Blaha. 2019. *Wisconsin Land Use Megatrends: Forests*. Stevens Point: Center for Land Use Education, University of Wisconsin–Stevens Point and University of Wisconsin–Madison Division of Extension.

Hale, Frederick. 2002. *Swedes in Wisconsin*. Revised and expanded ed. Madison: Wisconsin Historical Society Press.

———. 2007. *Swiss in Wisconsin*. Madison: Wisconsin Historical Society Press.

Hall, Alexandra. 2017. "Under Trump, Wisconsin Dairies Struggle to Keep Immigrant Workers." *The Cap Times*, March 19. https://madison.com/ct/news/local/under-trump-wisconsin-dairies-struggle-to-keep-immigrant -workers/article_198b9d16-b3b5-5557-b492-1a52e9195680.html.

Harper, Robert Alexander. 1950. *Recreational Occupance of the Moraine Lake Region of Northeastern Illinois and Southeastern Wisconsin*. Department of Geography Research Paper no. 14. Chicago: University of Chicago.

Harrison, Jill, Sarah Lloyd, and Trish O'Kane. 2009a. "Immigrant Dairy Workers in Rural Wisconsin Communities." *Briefing No. 4: Changing Hands: Hired Labor on Wisconsin Dairy Farms*. Madison: University of Wisconsin Program on Agricultural Technology Studies.

———. 2009b. "Overview of Immigrant Workers on Wisconsin Dairy Farms." *Briefing No. 1: Changing Hands: Hired Labor on Wisconsin Dairy Farms*. Madison: University of Wisconsin Program on Agricultural Technology Studies.

Hart, John Fraser. 1984a. "Population Change in the Upper Lake States." *Annals of the Association of American Geographers* 74 (2): 221–243.

———. 1984b. "Resort Areas in Wisconsin." *The Geographical Review* 74 (2): 192–217.

———. 1991. *The Land That Feeds Us*. New York: W. W. Norton.

———. 2001. "Ginseng and Entrepreneurs." *The Wisconsin Geographer* 17: 11–14.

———. 2003. *The Changing Scale of American Agriculture*. Charlottesville: University of Virginia Press.

———. 2008. *My Kind of County: Door County, Wisconsin*. Chicago: The Center for American Places at Columbia College Chicago.

Hart, John Fraser, and Mark B. Lindberg. 2014. "Kilofarms in the Agricultural Heartland." *The Geographical Review* 104 (2): 139–152.

Hartemink, Alfred E., Birl Lowery, and Carl Wacker. 2012. "Soils Maps of Wisconsin." *Geoderma* 189–190: 451–461.

Haugen, David E. 2013. *Wisconsin Timber Industry—An Assessment of Timber Product Output and Use, 2008*. Resource Bulletin NRS-78. Newtown Square, PA: U.S. Department of Agriculture, Forest Service, Northern Research Station.

———. 2017. Wisconsin Timber Industry, 2013. Resource Update FS-125. https://www.fs.fed.us/nrs/pubs/ru/ru_fs125.pdf.

———. 2020. WI Timber Products 2013 Resource Tables. Updated tables provided to the authors to replace preliminary statistics linked to Haugen 2017, shown at https://www.fs.fed.us/nrs/pubs/download/WI_2013_timber-resourcetables.pdf.

Heg, J. E., 1882. *The Blue Book of the State of Wisconsin*. Madison: Wisconsin Secretary of State.

Hernandez, Samantha. 2016. "Ephraim Dry No More." *Green Bay Press Gazette*, April 7. https://www.greenbaypressgazette.com/story/news/local/door-co/news/2016/04/05/ephraim-dry-no-more/82632760/.

Hess, Corrinne. 2019. "As DNC Convention Decision Nears, Questions over Milwaukee's Hotel Room Inventory Loom." Wisconsin Public Radio, February 26. https://www.wpr.org/dnc-convention-decision-nears-questions-over-milwaukees-hotel-room-inventory-loom.

Heyl, Allen V., Jr., Allen F. Agnew, Erwin J. Lyons, and Charles H. Behre Jr. 1959. *The Geology of the Upper Mississippi Valley Zinc-Lead District*. Professional Paper 309. Washington, DC: U.S. Geological Survey.

Heynen, Nik, Harold A. Perkins, and Parama Roy. 2006. "The Political Ecology of Uneven Urban Green Space: The Impact of Political Economy on Race and Ethnicity in Producing Environmental Inequality in Milwaukee." *Urban Affairs Review* 42 (1): 3–25.

Higgens, Dale, and Sue Reinecke. 2015. "Effects of Historical Logging on Rivers in the Lake States." *StreamNotes: The Technical Newsletter of the National Stream and Aquatic Ecology Center*. November: 1–5. https://www.fs.fed.us/biology/nsaec/assets/streamnotes2015-11.pdf.

Hildebrandt, Dirk. 2017. "Hemp: Wisconsin's Forgotten Harvest." *Wisconsin Magazine of History* 100 (3): 12–23.

Hill, James J., and Thomas J. Evans. 1981. "The Mineral Industry of Wisconsin." In *Minerals Yearbook, 1978–79*. Vol. 2, *Area Reports: Domestic*, U.S. Bureau of Mines, 571–583. Washington, DC: U.S. Government Printing Office.

Hoelscher, Steven. 1997. "A Pretty Strange Place: Nineteenth-Century Tourism in the Dells." In Ostergren and Vale, *Wisconsin Land and Life*, 424–449.

———. 1998a. *Heritage on Stage: The Invention of Ethnic Place in America's Little Switzerland*. Madison: University of Wisconsin Press.

———. 1998b. "The Photographic Construction of Tourist Space in Victorian America." *The Geographical Review* 88 (4): 548–570.

Hole, Francis D. 1976. *Soils of Wisconsin*. Madison: University of Wisconsin Press.

Holifield, Ryan, and Mick Day. 2017. "A Framework for a Critical Physical Geography of 'Sacrifice Zones': Physical Landscapes and Discursive Spaces of Frac Sand Mining in Western Wisconsin." *Geoforum* 85: 269–279.

Holmes, David G. 2004. *Irish in Wisconsin*. Madison: Wisconsin Historical Society Press.

Holmes, Fred L. 1937. *Alluring Wisconsin: The Historic Glamor and Natural Loveliness of an American Commonwealth*. Milwaukee: E. M. Hale.

———. (1944) 1990. *Old World Wisconsin: Around Europe in the Badger State*. 2nd ed. Minocqua, WI: North Word Press.

Holmes, Isiah. 2020. "The Frac Sand Industry's Migration out of Wisconsin." *Wisconsin Examiner*, January 8. https://wisconsinexaminer.com/2020/01/08/the-sand-frac-industrys-migration-out-of-wisconsin/.

Hooyer, Thomas S., and William N. Mode. 2008. *Quaternary Geology of Winnebago County, Wisconsin*. Bulletin 105. Madison: Wisconsin Geological and Natural History Survey.

Houston, Tamara G., and Stanley A. Changnon. 2007. "Freezing Rain Events: A Major Weather Hazard in the Conterminous US." *Natural Hazards* 40 (2): 485–494.

Hovde, Syd. 1995. "Christmas Trees." *Proceedings, Twentieth Annual Meeting, Forest History Association of Wisconsin* 20: 18–24.

Hubbuch, Chris. 2018. "Detour: Wisconsin Farmers Cope with New Bridge Restrictions." *La Crosse Tribune*, May 30. https://lacrossetribune.com/news/local/detour-wisconsin-farmers-cope-with-new-bridge-restrictions/article_9199f254-a493-53f0-b675-d0172ebaf5d5.html.

Hudson, John C. 1985. *Plains Country Towns*. Minneapolis: University of Minnesota Press.

———. 1997. "The Creation of Towns in Wisconsin." In Ostergren and Vale, *Wisconsin Land and Life*, 197–220.

Hunt, Michael E., Linda J. Marshall, and John L. Merrill. 2002. "Rural Areas That Affect Older Migrants." *Journal of Architectural and Planning Research* 19 (1): 44–56.

Jackson-Smith, Douglas, and Bradford Barham. 2000. *The Changing Face of Wisconsin Dairy Farms: A Summary of PATS' Research on Structural Change in the 1990s*. PATS Research Report no. 7. Madison: Program on Agricultural Technology Studies, University of Wisconsin–Madison.

Janik, Erika. 2017. "When the Mail Came to Rural Wisconsin." Wisconsin Public Radio. https://www.wpr.org/when-mail-came-rural-wisconsin.

Janus, Edward. 2011. *Creating Dairyland: How Caring for Cows Saved Our Soil, Created Our Landscape, Brought Prosperity to Our State, and Still Shapes Our Way of Life in Wisconsin*. Madison: Wisconsin Historical Society Press.

Jenney, Charles. 1895. *Report on Insurance Business in the United States at the Eleventh Census: 1890*. Part 2, *Life Insurance*. Washington, DC: U.S. Census Bureau (Government Printing Office).

Johansen, Harley E. 1971. "Diffusion of Strip Cropping in Southwestern Wisconsin." *Annals of the Association of American Geographers* 61 (4): 671–683.

Johnson, M. L. 2008. "Drought, Freeze Kill Most of Wis. Tart Cherry Crop." Fox News. http://www.foxnews.com/printer_friendly_wires/2008Jun23/0,4675,FarmSceneTartCherries,00.html.

Johnson, Neil. 2019. "Once Arrested at a Marijuana Grow Operation, He Moved Back to Rural Wisconsin and Is Growing Hemp. Now, the Biggest Risk Factor Might Be the Midwestern Winter." *The Chicago Tribune*, November 9. https://www.chicagotribune.com/midwest/ct-wisconsin-hemp-farmers-20191109-zui3b7hjmrdl5d75insvsqfd44-story.html.

Jones, Gregory V., Ryan Reid, and Aleksander Vilks. 2012. "Climate, Grapes, and Wine: Structure and Suitability in a Variable and Changing Climate." In *The Geography of Wine: Regions, Terroir and Techniques*, edited by Paul H. Dougherty, 109–133. New York: Springer Science.

Jones, Malia, and Caitlin Bourbeau. 2018. "Comparing Wisconsinites' Health in 2018, County by County." *WisContext*, November 26. https://www.wiscontext.org/comparing-wisconsinites-health-2018-county-county.

Joseph, George. 2016. "How Wisconsin Became the Home of Black Incarceration." *CityLab*, August 17. https://www.citylab.com/equity/2016/08/how-wisconsin-became-the-home-of-black-incarceration/496130/.

Kaeding, Danielle. 2019. "Changes Send Brown Co. Manure Digester Project Back for Second Look." *Wisconsin State Farmer*, November 11. https://www.wisfarmer.com/story/news/2019/11/11/psc-asks-state-agencies-take-second-look-manure-digester-project/2568931001/.

Kammel, David W. 2015. "Building Cost Estimates—Dairy Modernization." University of Wisconsin–Extension. https://fyi.uwex.edu/dairy/files/2015/11/Building-Cost-Estimates-Dairy-Modernization.pdf.

Karakis, Snejana, Barry Cameron, and William Kean. 2016. "Geology and Wine 14. Terroir of Historic Wollersheim Winery, Lake Wisconsin American Viticultural Area." *Geoscience Canada* 43: 265–282.

Kassulke, Natasha. 2009. "Who Were They and Why Did They Leave?" *Wisconsin Natural Resources*, October. https://dnr.wi.gov/wnrmag/2009/10/aztalan.htm.

Kates, Robert W. 1971. "Natural Hazard in Human Ecological Perspective: Hypotheses and Models." *Economic Geography* 47 (3): 438–451.

Kaye, Julia, Brigitte Amiri, Louise Melling, and Jennifer Dalven. 2016. *Health Care Denied*. New York: American Civil Liberties Union.

Kaysen, James P. 1978. "Railroad Logging in Wisconsin." In *Some Historic Events in Wisconsin's Logging Industry: Proceedings, Third Annual Meeting, Forest History Association of Wisconsin*, edited by Ramon R. Hernandez, 31–37. Wausau: Forest History Association of Wisconsin.

Kellogg, Louise P. 1918. "The Disputed Michigan-Wisconsin Boundary." *Wisconsin Magazine of History* 1 (3): 304–307.

———. 1925. *The French Régime in Wisconsin and the Northwest*. Madison: State Historical Society of Wisconsin.

Kelly, Kate. 2018. "Willy Wonka and the Medical Software Factory." *New York Times*, December 20. https://www.nytimes.com/2018/12/20/business/epic-systems-campus-verona-wisconsin.html.

Kenny, Judith T., and Jeffrey Zimmerman. 2004. "Constructing the 'Genuine American City': Neo-traditionalism, New Urbanism and Neo-liberalism in the Remaking of Downtown Milwaukee." *Cultural Geographies* 11 (1): 74–98.

Kitchell, Jessica. 2020. *Wisconsin Wildlife Harvest Summary, 1930–2018*. Madison: Wisconsin Department of Natural Resources.

Knipping, Mark. 2008. *Finns in Wisconsin*. Madison: Wisconsin Historical Society Press.

Knox, James C. 2019a. "Geology of the Driftless Area." In *The Physical Geography and Geology of the Driftless Area: The Career and Contributions of James C. Knox*, edited by Eric C. Carson, J. Elmo Rawling III, J. Michael Daniels, and John W. Attig, 1–35. GSA Special Paper 543. Boulder, CO: Geological Society of America.

Knox, James C. 2019b. "Lead and Zinc Mining in the Driftless Area." In *The Physical Geography and Geology of the Driftless Area: The Career and Contributions of James C. Knox*, edited by Eric C. Carson, J. Elmo Rawling III, J. Michael Daniels, and John W. Attig, 147–156. GSA Special Paper 543. Boulder, CO: Geological Society of America.

Knox, Pam Naber. 1996. *Wind Atlas of Wisconsin*. Bulletin 94. Madison: Wisconsin Geological and Natural History Survey.

Knox, Pamela Naber, and Douglas G. Norgord. 2000. *A Tornado Climatology for Wisconsin*. Bulletin 100. Madison: Wisconsin Geological and Natural History Survey.

Kochis, Nancy S., and Steven W. Leavitt. 1997. "Radon Emanation from Soil of Kenosha, Racine and Waukesha Counties, Southeastern Wisconsin." *Geoscience Wisconsin* 16: 55–62. Madison: Wisconsin Geological and Natural History Survey.

Kolinski, Dennis L. 1994. "Shrines and Crosses in Rural Central Wisconsin." *Polish American Studies* 21 (2): 33–47.

Krog, Carl E. 1996. "The Retreat of Farming and the Return of Forests in Wisconsin's Cutover." *Voyageur* 12 (2): 2–12.

Krug, W. R., and B. D. Simon. 1991. "Wisconsin Floods and Droughts." In *National Water Summary, 1988–89: Hydrologic Events and Floods and Droughts*, 567–574. Water-Supply Paper 2375. Reston, VA: U.S. Geological Survey.

Kucharik, Christopher J., Shawn P. Serbin, Steve Vavrus, Edward J. Hopkins, and Melissa M. Motew. 2010. "Patterns of Climate Change across Wisconsin from 1950 to 2006." *Physical Geography* 31 (1): 1–28.

Kurtz, Cassandra. 2018. "Forests of Wisconsin, 2017." Resource Update FS-148. Newtown Square, PA: U.S. Department of Agriculture, Forest Service, Northern Research Station. https://www.fs.fed.us/nrs/pubs/ru/ru_fs148.pdf.

Kurtz, Cassandra M., Sally E. Dahir, Andrew M. Stoltman, William H. McWilliams, Brett J. Butler, Mark D. Nelson, Randall S. Morin, et al. 2017. *Wisconsin Forests 2014*. Resource Bulletin NRS-112. Newtown Square, PA: U.S. Department of Agriculture, Forest Service, Northern Research Station.

Laatsch, William G., and Charles F. Calkins. 1992. "Belgians in Wisconsin." In *To Build in a New Land: Ethnic Landscapes in North America*, edited by Allen G. Noble, 195–210. Baltimore: Johns Hopkins University Press.

Lampard, Eric E. 1963. *The Rise of the Dairy Industry in Wisconsin: A Study in Agricultural Change, 1820–1920.* Madison: State Historical Society of Wisconsin.

Legreid, Ann Marie. 1997. "Community Building, Conflict, and Change: Geographic Perspectives on the Norwegian-American Experience in Frontier Wisconsin." In Ostergren and Vale, *Wisconsin Land and Life,* 300–319.

Lekas, Steve, Mike Gannon, and Sana Moghul. 2014. *Property Hail Claims in the United States: 2000–2013.* Jersey City, NJ: Verisk Insurance Solutions.

Lenehan, Michael. 2018. "The Grid to Nowhere: An Argument against Building Giant Transmission Lines." *Isthmus,* March 1. https://isthmus.com/news/cover-story/argument-against-building-giant-transmission-lines/.

Levine, Marc V. 2019. "Milwaukee 53206: The Anatomy of Concentrated Disadvantage in an Inner-City Neighborhood, 2000–2017." Center for Economic Development Publications 48. University of Wisconsin–Milwaukee. https://dc.uwm.edu/ced_pubs/48.

Lippelt, Irene D. 2002. *Understanding Wisconsin Township, Range, and Section Land Descriptions.* Educational Series 44. Madison: Wisconsin Geological and Natural History Survey.

Liptak, Adam. 2019. "The Supreme Court Will Soon Consider Whether the Census Will Include a Citizenship Question." *New York Times,* April 15. https://www.nytimes.com/2019/04/15/us/politics/supreme-court-citizenship-census.html?searchResultPosition=1.

Loew, Patty. 2013. *Indian Nations of Wisconsin: Histories of Endurance and Renewal.* Madison: Wisconsin Historical Society Press.

L'Roe, Andrew W. 2017. "The Effects of Industrial Forest Divestment on Land Ownership, Conservation, Recreation, and Timber Harvesting in Wisconsin's Working Forests." PhD diss., University of Wisconsin–Madison. Ann Arbor: ProQuest Dissertation Publishing no. 10254295.

Luczaj, John A. 2013. "Geology of the Niagara Escarpment in Wisconsin" *Geoscience Wisconsin* 22: part 1.

Lytwyn, Jennifer. 2010. "Remote Sensing and GIS Investigation of Glacial Features in the Region of Devil's Lake State Park, South-Central Wisconsin, USA." *Geomorphology* 123 (1–2): 46–60.

Mainville, André, and Michael R. Craymer. 2005. "Present-Day Tilting of the Great Lakes Region Based on Water Level Gauges." *Geological Society of America Bulletin* 117 (7–8): 1070–1080.

Marcouiller, Dave, Eric Olson, and Jeff Prey. 2002. *State Parks and Their Gateway Communities: Development and Recreation Planning Issues in Wisconsin* (G773). Madison: University of Wisconsin Extension.

Martin, Lawrence. 1930. "The Michigan-Wisconsin Boundary Case in the Supreme Court of the United States." *Annals of the Association of American Geographers* 20 (3): 105–163.

———. 1938. "The Second Wisconsin-Michigan Boundary Case in the Supreme Court of the United States, 1932–1936." *Annals of the Association of American Geographers* 28 (2): 77–126.

———. 1965. *The Physical Geography of Wisconsin.* 3rd ed. Madison: University of Wisconsin Press.

Mason, Joseph A. 2015. "Up in the Refrigerator: Geomorphic Response to Periglacial Environments in the Upper Mississippi River Basin, USA." *Geomorphology* 248: 363–381.

Mason, Joseph A., and James C. Knox. 1997. "Age of Colluviums Indicates Accelerated Late Wisconsinan Hillslope Erosion in the Upper Mississippi Valley." *Geology* 25 (3): 267–270.

Mason, Joseph A., Peter M. Jacobs, and David S. Leigh. 2019. "Loess, Eolian Sand, and Colluvium in the Driftless Area." In *The Physical Geography and Geology of the Driftless Area: The Career and Contributions of James C. Knox.* GSA Special Paper 543, edited by Eric C. Carson, J. Elmo Rawling III, J. Michael Daniels, and John W. Attig, 61–73. Boulder, CO: Geological Society of America.

Mather, Cotton. 1977. "Coulees and the Coulee Country of Wisconsin." *Wisconsin Academy Review* 22 (September): 22–25.

Mather, Cotton, and Matti Kaups. 1963. "The Finnish Sauna: A Cultural Index to Settlement." *Annals of the Association of American Geographers* 53 (4): 494–504.

Mauk, Alyssa. 2019. "Racine Named Second Worst Area in U.S. for Blacks, Up from Third." *Journal Times*, November 14. https://journaltimes.com/news/local/racine-named-second-worst-area-in-u-s-for-blacks/article_a469a434-ff75-5114-98c0-77060fc8f57f.html.

Mayer, Harold M. 1975a. "The Chicago–Milwaukee Corridor." In *Landscapes of Wisconsin: A Field Guide*, edited by Barbara Zakrzewska-Borowiecki, 77–97. Washington, DC: Association of American Geographers.

———. 1975b. "The City of Milwaukee." In *Landscapes of Wisconsin: A Field Guide*, edited by Barbara Zakrzewska-Borowiecki, 98–116. Washington, DC: Association of American Geographers.

McCoy, Mary Kate. 2019. "Feel Like Kwik Trips Have Exploded in Number? You're Right." Wisconsin Public Radio. https://www.wpr.org/feel-kwik-trips-have-exploded-number-youre-right.

McGee, W. J. 1891. "The Pleistocene History of Northeastern Iowa." *11th Annual Report of the Director of the U.S. Geological Survey, 1889–90*. Part 1, 189–577, 743–757.

Meine, Curt. 2015. "Of Connection and Renewal: The Historic Apple Trees of the Badger Army Ammunition Plant." *Wisconsin People & Ideas* 61 (4): 20–27.

Mervis, Jeffrey. 2019. "Researchers Object to Census Privacy Measure." *Science* 363 (6423): 114.

———. 2020. "Trump Directive on State Counts Said to Threaten Rigor of Census." *Science* 369 (6504): 611.

Meyer, Robinson. 2014. "The Geography of NFL Fandom." *The Atlantic*, September 5. https://www.theatlantic.com/technology/archive/2014/09/the-geography-of-nfl-fandom/379729/.

Mickelson, David M., and John W. Attig. 2017. *Laurentide Ice Sheet: Ice-Margin Positions in Wisconsin*. Education Series 56. Madison: Wisconsin Geological and Natural History Survey.

Mickelson, D. M., J. C. Knox, and L. Clayton. 1982. "Glaciation of the Driftless Area: An Evaluation of the Evidence." In *Quaternary History of the Driftless Area*, edited by J. C. Knox, L. Clayton, and D. M. Mickelson, 155–169. Field Trip Guidebook 5. Madison: Wisconsin Geological and Natural History Survey.

Mickelson, David M., Louis J. Maher Jr., and Susan L. Simpson. 2011. *Geology of the Ice Age National Scenic Trail*. Madison: University of Wisconsin Press.

Mickelson, David M., and Betty J. Socha. 2017. *Quaternary Geology of Calumet and Manitowoc Counties, Wisconsin*. Bulletin 108. Madison: Wisconsin Geological and Natural History Survey.

Mikoś, Susan Gibson. 2012. *Poles in Wisconsin*. Madison: Wisconsin Historical Society Press.

Millen, Timothy M., Robert E. Zartman, and Allen V. Heyl. 1995. *Lead Isotopes from the Upper Mississippi Valley District—A Regional Perspective*. Bulletin 2094-B. Washington, DC: U.S. Geological Survey.

MNI, editors. 2019. *2020 Wisconsin Manufacturers Register*. Evanston, IL: Manufacturers News.

Moertl, Frank. 1995. *Wisconsin: Its Territorial and Statehood Post Offices*. Hartland, WI: Wisconsin Postal History Society.

Moran, Joseph M. 1995. "Ice Shove! Thawing Lakes Can Be a Pile of Trouble." *Weatherwise* 48 (2): 12–15.

Moran, Joseph M., and Edward J. Hopkins. 2002. *Wisconsin's Weather and Climate*. Madison: University of Wisconsin Press.

Moranda, Scott A. 2006. "The Story of German Settlement in the Forests and on the Prairies of Wisconsin." In *Wisconsin German Land and Life*, edited by Heike Bungert, Cora Lee Kluge, and Robert C. Ostergren, 123–144. Madison: Max Kade Institute for German-American Studies, University of Wisconsin–Madison.

MRCC (Midwestern Regional Climate Center). 2019. "Midwest Climate: Climate Summaries." 1981–2010 normals, mean maximum, minimum, and mean monthly temperature data available for 89 Wisconsin stations. https://mrcc.illinois.edu/mw_climate/climateSummaries/climSummStns_temp.jsp.

Mudrey, M. G., Jr., B. A. Brown, and J. K. Greenberg. 1982. *Bedrock Geologic Map of Wisconsin*. 1:1,000,000 scale map. Madison: Wisconsin Geological and Natural History Survey.

Muldoon, Maureen A., Mark A. Borchardt, Susan K. Spencer, Randall J. Hunt, and David Owens. 2018. "Using Enteric Pathogens to Assess Sources of Fecal Contamination in the Silurian Dolomite Aquifer: Preliminary

Results." In *Karst Groundwater and Public Health*, edited by W. B. White, J. S. Herman, E. K. Herman, and M. Rutigliano, 209–213. New York: Springer International.

Mulhern, Nancy. 2009. *Population of Wisconsin, 1850–2000*. Madison: Wisconsin Historical Society Library.

NASS (National Agricultural Statistics Service). 1999. *1997 Census of Agriculture: Wisconsin State and County Data. Volume 1, Geographical Area Series, Part 49*. Washington, DC: U.S. Department of Agriculture.

———. 2004. *2002 Census of Agriculture: Wisconsin State and County Data. Volume 1, Geographical Area Series, Part 49*. Washington, DC: U.S. Department of Agriculture.

———. 2009. *2007 Census of Agriculture: Wisconsin State and County Data. Volume 1, Geographical Area Series, Part 49*. Washington, DC: U.S. Department of Agriculture.

———. 2014a. *2012 Census of Agriculture: Wisconsin State and County Data. Volume 1, Geographical Area Series, Part 49*. Washington, DC: U.S. Department of Agriculture. https://www.agcensus.usda.gov/Publications/2012/Full_Report/Volume_1,_Chapter_2_County_Level/Wisconsin/wiv1.pdf.

———. 2014b. *2012 Census of Agriculture Appendix B*. Washington, DC: USDA, National Agricultural Statistics Service.

———. 2014c. *2012 Census of Agriculture, Ag Census Web Maps*. www.agcensus.usda.gov/Publications/2012/Online_Resources/Ag_Census_Web_Maps/Overview/.

———. 2015a. *2012 Census of Agriculture: Census of Horticultural Specialties (2014), Volume 3, Special Studies, Part 3*. Washington, DC: U.S. Department of Agriculture.

———. 2015b. *2014 Organic Survey: Wisconsin*. https://www.nass.usda.gov/Statistics_by_State/Wisconsin/Publications/WI_Organic_Release.pdf.

———. 2017. *2016 Certified Organic Survey: Wisconsin*. https://www.nass.usda.gov/Surveys/Guide_to_NASS_Surveys/Organic_Production/2016_State_Publications/WI.pdf.

———. 2019a. *2017 Census of Agriculture, Ag Census Web Maps*. https://www.nass.usda.gov/Publications/AgCensus/2017/Online_Resources/Ag_Atlas_Maps/index.php.

———. 2019b. *2017 Census of Agriculture: United States Summary and State Data. Volume 1, Geographical Area Series, Part 51*. Washington, DC: U.S. Department of Agriculture. https://www.nass.usda.gov/Publications/AgCensus/2017/Full_Report/Volume_1,_Chapter_1_US/usv1.pdf.

———. 2019c. *2017 Census of Agriculture: Wisconsin State and County Data. Volume 1, Geographical Area Series, Part 49*. Washington, DC: U.S. Department of Agriculture.

NASS-CM (National Agricultural Statistics Service, Charts and Maps). 2013. *Corn for Grain 2012: Yield per Harvested Acre by County*. Previous year's harvest posted on website. https://www.nass.usda.gov/Charts_and_Maps/graphics/CR-YI-RGBChor.pdf.

———. 2015. *Corn for Grain 2014: Yield Per Harvested Acre by County*. Previous year's harvest posted on website. https://www.nass.usda.gov/Charts_and_Maps/graphics/CR-YI-RGBChor.pdf.

NASS-W (National Agricultural Statistics Service, Wisconsin Field Office). 2010a. *2010 Dairy Producer Survey*. Madison: National Agricultural Statistics Service. https://www.nass.usda.gov/Statistics_by_State/Wisconsin/Publications/Dairy/Other_Surveys/Dairy_OP_Release_10.pdf.

———. 2010b. *2010 Dairy Producer Survey—Addendum*. https://www.nass.usda.gov/Statistics_by_State/Wisconsin/Publications/Dairy/Other_Surveys/Dairy_Op_Addendum_10.pdf.

———. 2019a. Wisconsin Ag News—Chickens & Eggs. Madison: National Agricultural Statistics Service, Wisconsin Field Office. July 23. https://www.nass.usda.gov/Statistics_by_State/Wisconsin/Publications/Livestock/2019/WI-Chickens-07-19.pdf.

———. 2019b. Wisconsin Ag News—Mink. Madison: National Agricultural Statistics Service, Wisconsin Field Office. July 23. https://www.nass.usda.gov/Statistics_by_State/Wisconsin/Publications/Livestock/2019/WI-2019-Mink.pdf.

———. 2020. Milk, All, Prices Received by Farmers, Wisconsin. https://www.nass.usda.gov/Statistics_by_State/Wisconsin/Publications/Dairy/Historical_Data_Series/mkallpri.pdf.

NCDC (National Climatic Data Center). 1990. "Storm Data and Unusual Weather Phenomena—May 1990: Wisconsin." *Storm Data* 32 (5): 205.

———. 2000. "Storm Data and Unusual Weather Phenomena—May 2000: Wisconsin." *Storm Data* 42 (5): 307–317.

———. 2015. Index of /pub/data/normals/1981–2010. Asheville, NC: National Oceanic and Atmospheric Administration. https://www1.ncdc.noaa.gov/pub/data/normals/1981-2010/.

———. 2019. "National Temperature and Precipitation Maps." https://www.ncdc.noaa.gov/temp-and-precip/us-maps/.

NCEI (National Centers for Environmental Information). 2015a. "Summary of Monthly Normals, 1981–2010," for Madison Dane Co. Regional Airport. Asheville, NC: National Oceanic and Atmospheric Administration. http://www.aos.wisc.edu/~sco/clim-history/sta-data/msn/MSN-monthly/GHCND_USW00014837_2010-1-1.pdf.

———. 2015b. "Summary of Monthly Normals, 1981–2010," for Rhinelander Oneida Co. Airport. Asheville, NC: National Oceanic and Atmospheric Administration. http://www.aos.wisc.edu/~sco/clim-history/sta-data/RHI/RHI-Monthly/GHCND_USW00004803_2010-1-1.pdf.

———. 2018a. "Storm Data and Unusual Weather Phenomena—Wisconsin." *Storm Data* 60 (4): 361–367.

———. 2018b. "Storm Data and Unusual Weather Phenomena—August 2018: Wisconsin." *Storm Data* 60 (8): 501–531.

Nepveux, Michael. 2019. "How Milk Is Priced in Federal Milk Marketing Orders: A Primer." Washington, DC: Farm Bureau. https://www.fb.org/market-intel/how-milk-is-priced-in-federal-milk-marketing-orders-a-primer.

Nesbit, Robert C. 1985. *The History of Wisconsin*. Vol. 3, *Urbanization and Industrialization, 1873–1893*. Madison: State Historical Society of Wisconsin.

Newman, Judy. 2018. "Wisconsin's Biohealth Industry Is Healthy and Growing, a Report Shows." *Wisconsin State Journal*, October 21. https://madison.com/wsj/business/technology/biotech/wisconsins-biohealth-industry-is-healthy-and-growing-a-report-shows/article_d7080ae6-8c8e-51a6-8639-eb0396d6a616.html.

Noble, Allen G. 1984. *Wood, Brick, and Stone: The North American Settlement Landscape*. Vol. 2, *Barns and Farm Structures*. Amherst: University of Massachusetts Press.

Noble, Allen G., and Richard K. Cleek. 1995. *The Old Barn Book: A Field Guide to North American Barns and Other Farm Structures*. New Brunswick, NJ: Rutgers University Press.

NPS (National Park Service). 2019. "Statistical Abstract: 2018." US Department of the Interior, National Park Service. https://www.nps.gov/subjects/socialscience/visitation.htm.

NRCSA (Natural Resources Committee of State Agencies). 1956. *The Natural Resources of Wisconsin*. Madison: Wisconsin Natural Resources Committee of State Agencies.

NSOLGH (National Shrine of Our Lady of Good Help). 2021. "Adele's Story." The National Shrine of Our Lady of Good Help, Diocese of Green Bay. https://championshrine.org/the-story/.

NWS-GB (National Weather Service, Green Bay). 2017. "Wisconsin Tornado and Severe Weather Statistics." https://www.weather.gov/grb/WI_tornado_stats.

———. 2019. "Severe Storm Summary—July 19, 2019." https://www.weather.gov/grb/071919_severe_event.

———. 2020. "Final 2019 Precipitation Totals across North-Central & Northeast Wisconsin." https://www.weather.gov/grb/Final2019PrecipitationTotals.

NWS-L (National Weather Service, La Crosse). 2018. "Summary of Significant Flooding and Severe Storms August 27–28, 2018." https://www.weather.gov/arx/aug2818.

NWS-M (National Weather Service, Milwaukee/Sullivan). n.d.a. *30 Year Average Precipitation 1981–2010*. Map. https://www.weather.gov/images/mkx/climate/avg_30_year_precip.png.

———. n.d.b. *30 Year Average Snowfall 1981–2010*. Map. https://www.weather.gov/images/mkx/climate/avg_30_year_snowfall.png.

———. 2018. "August 28, 2018, Damaging Winds, Tornado and Flood Event." https://www.weather.gov/mkx/aug2818.

———. 2019a. *Wisconsin Blizzard Events: Winter 1982–83—Winter 2017–18*. Map. http://www.weather.gov/images/mkx/svr-wx-stats/Blizzard.jpg.

———. 2019b. *Wisconsin Flood Events: 1844–2018*. Map. http://www.weather.gov/images/mkx/svr-wx-stats/Flood.jpg.

———. 2019c. *Wisconsin Ice Storm Events: Winter 1982–83—Winter 2017–18*. Map. http://www.weather.gov/images/mkx/svr-wx-stats/IceStorm.jpg.

———. 2019d. *Wisconsin Lightning Events: 1982–2018*. Map. http://www.weather.gov/images/mkx/svr-wx-stats/Lightning.jpg.

———. 2019e. *Wisconsin Severe Hail Events: 1982–2016*. Map. http://www.weather.gov/images/mkx/svr-wx-stats/Hail.jpg.

———. 2019f. *Wisconsin Severe Thunderstorm Wind Events: 1844–2018*. Map. http://www.weather.gov/images/mkx/svr-wx-stats/TstormWind.jpg.

———. 2019g. *Wisconsin Tornado Events: 1844–2018*. Map. http://www.weather.gov/images/mkx/svr-wx-stats/Tornado.jpg.

NWS-TC (National Weather Service, Twin Cities). 2018. "One Year Anniversary of the 83 Mile Long Northwest Wisconsin EF3 Tornado." https://www.weather.gov/mpx/May_16th_Wisconsin_Tornado_Upgraded_to_EF3.

Ostergren, Robert C. 1997. "The Euro-American Settlement of Wisconsin, 1830–1920." In Ostergren and Vale, *Wisconsin Land and Life*, 137–162.

Ostergren, Robert C., and Thomas R. Vale, eds. 1997. *Wisconsin Land and Life*. Madison: University of Wisconsin Press.

Ottensmann, John R. 1980. "Changes in Accessibility to Employment in an Urban Area: Milwaukee, 1927–1963." *The Professional Geographer* 32 (4): 421–430.

Ottone, Gerald. 2006. "Farming in the Face of Exurbanization: The Cases of Greenville and Bellevue." *The Wisconsin Geographer* 21: 51–74.

Paddock, Susan C. 1997. "The Changing World of Wisconsin Local Government." In *Wisconsin Blue Book, 1997–1998*, edited by Lawrence S. Barish, 100–172. Madison: Wisconsin Legislative Reference Bureau.

Patterson, D. L., D. C. Endrizzi, and I. D. Lippelt. 1997. *Green Bay Area Private Claims and Williams Grant Subdivision in Brown County, Wisconsin*. Miscellaneous Map 44. Madison: Wisconsin Geological and Natural History Survey.

Pauketat, Timothy R., Robert F. Boszhardt, and Danielle M. Benden. 2015. "Trempealeau Entanglements: An Ancient Colony's Causes and Effects." *American Antiquity* 80 (2): 260–289.

Pearson, Thomas W. 2013. "Frac Sand Mining in Wisconsin: Understanding Emerging Conflicts and Community Organizing." *Culture, Agriculture, Food and Environment* 35 (1): 30–40.

———. 2017. *When the Hills Are Gone: Frac Sand Mining and the Struggle for Community*. Minneapolis: University of Minnesota Press.

Peddle, Howard. 1980. "Log Rafting on Lake Superior." *Proceedings of the Fifth Annual Meeting of the Forest History Association of Wisconsin* 5: 31–32.

Pepp, Kyle, Geoffrey Siemering, and Steve Ventura. 2019. *Digital Atlas of Historic Mining Activity in Southwestern Wisconsin*. G4177. Madison: Extension Division, University of Wisconsin–Madison.

Perrin, Richard W. E. 1981. *Historic Wisconsin Buildings: A Survey in Pioneer Architecture, 1835–1870.* Milwaukee: Milwaukee Public Museum.

Perry, Charles H. 2014. "Forests of Wisconsin, 2013." Resource Update FS-6. Newtown Square, PA: U.S. Department of Agriculture, Forest Service, Northern Research Station. https://www.fs.fed.us/nrs/pubs/ru/ru_fs6.pdf.

Perry, Charles H. (Hobie), Vern A. Everson, Ian K. Brown, Jane Cummings-Carlson, Sally E. Dahir, Edward A. Jepsen, Joe Kovach, et al. 2008. *Wisconsin's Forests, 2004.* Resource Bulletin NRS-23. Newtown Square, PA: U.S. Department of Agriculture, Forest Service, Northern Research Station.

Peterson, Eric. 2016. "Door County Cherry Blossoms in Bloom." Fox News, May 17. http://fox11online.com/news/local/lakeshore/door-county-cherry-blossoms-in-bloom.

Pew Research Center. 2019. "Religious Landscape Study: Adults in Wisconsin." Washington, DC: Pew Research Center. https://www.pewforum.org/religious-landscape-study/state/wisconsin/.

Phillip, Abby. 2014. "Wisconsin's Bar-to-Grocery Store Ratio Puts the Rest of the Country to Shame." *Washington Post,* May 29. https://www.washingtonpost.com/news/wonk/wp/2014/05/29/wisconsins-bar-to-grocery-store-ratio-puts-the-rest-of-the-country-to-shame/?utm_term=.68a9bf2508fe.

Pifer, Richard L. 1980. "The Chippewa Valley Now." *Proceedings of the Fifth Annual Meeting of the Forest History Association of Wisconsin* 5: 35–37.

Platt, Robert S. 1928. "A Detail of Regional Geography: Ellison Bay Community as an Industrial Organism." *Annals of the Association of American Geographers* 18 (1): 81–126.

Preusser, Charley. 2017. "Another Flood Arrives in Gays Mills." *Crawford County Independent.* http://www.swnews4u.com/local/another-flood-arrives-in-gays-mills/.

QDMA (Quality Deer Management Association). 2018. *Whitetail Report 2018.* https://www.qdma.com/wp-content/uploads/2018/02/Whitetail_Report_2018.pdf.

Radbruch-Hall, Dorothy H., Roger B. Colton, William E. Davies, Ivo Lucchitta, Betty A. Skipp, and David J. Varnes. 1982. *Landslide Overview Map of the Conterminous United States.* Professional Paper 1183. Reston, VA: U.S. Geological Survey.

Raftovich, R. V., S. C. Chandler, and K. K. Fleming. 2018. *Migratory Bird Hunting Activity and Harvest during the 2016–17 and 2017–18 Hunting Seasons.* Laurel, MD: U.S. Fish and Wildlife Service.

Raitz, Karl B., and Cotton Mather. 1971. "Norwegians and Tobacco in Western Wisconsin." *Annals of the Association of American Geographers* 61 (4): 684–696.

Raney, William F. 1936. "The Building of Wisconsin Railroads." *The Wisconsin Magazine of History* 19 (4): 387–403.

Rast, Joel. 2007. "Annexation Policy in Milwaukee: An Historical Institutionalist Approach." *Polity* 39 (1): 55–78.

Ratcliffe, Michael, Charlynn Burd, Kelly Holder, and Alison Fields. 2016. "Defining Rural at the U.S. Census Bureau." American Community Survey and Geography Brief ACSGEO-1. Washington, DC: U.S. Census Bureau.

Rawling, J. E., III, P. R. Hanson, A. R. Young, and J. W. Attig. 2008. "Late Pleistocene Dune Construction in the Central Sand Plain of Wisconsin, USA." *Geomorphology* 100 (3–4): 494–505.

Reading, William H., IV, and James W. Whipple. 2007. *Wisconsin Timber Industry: An Assessment of Timber Product Output and Use in 2003.* Resource Bulletin NRS-19. Newtown Square, PA: U.S. Department of Agriculture, Forest Service, Northern Research Station.

RFA (Renewable Fuels Association). 2014. *Fueling a Nation—Feeding the World: The Role of the U.S. Ethanol Industry in Food and Feed Production.* http://www.ethanolrfa.org/wp-content/uploads/2015/09/9864ff506e6519057b_t5m6brouu.pdf.

Rhemtulla, Jeanine M., David J. Mladenoff, and Murray A. Clayton. 2009. "Legacies of Historical Land Use on Regional Forest Composition and Structure in Wisconsin, USA (mid-1800s–1930s–2000s)." *Ecological Applications* 19 (4): 1061–1078.

Riemer, Jeffrey W. 2003. "Grass-Roots Power through Internet Technology—The Case of the Crandon Mine." *Society & Natural Resources* 16 (10): 853–868.

Rogerson, Peter A. 2015. "A New Method for Finding Geographic Centers, with Application to U.S. States." *Professional Geographer* 67 (4): 686–694.

Rohe, Randall E. 1997. "Lumbering: Wisconsin's Northern Urban Frontier." In Ostergren and Vale, *Wisconsin Land and Life*, 221–240.

———. 1999. "Lumbering's Lingering Impact on Oshkosh." *Oshkosh—Sawdust City, Then and Now: Proceedings, Twenty-Fourth Annual Meeting, Forest History Association of Wisconsin* 24: 10–16.

———. 2002. *Ghosts of the Forest: Vanished Lumber Towns of Wisconsin*. Marinette, WI: Forest History Association of Wisconsin.

Roth, Filibert. 1898. *On the Forestry Conditions of Northern Wisconsin*. Bulletin no. 1. Madison: Wisconsin Geological and Natural History Survey.

Ruefle, William, William H. Ross, and Diane Mandell. 1992. "Attitudes toward Southeast Asian Immigrants in a Wisconsin Community." *The International Migration Review* 26 (3): 877–898.

Salamon, Sonya. 1992. *Prairie Patrimony: Family, Farming, and Community in the Midwest*. Chapel Hill: University of North Carolina Press.

Schafer, Joseph. 1927. *Four Wisconsin Counties: Prairie and Forest*. Wisconsin Domesday Book, General Studies, vol. 2. Madison: State Historical Society of Wisconsin.

———. 1932. *The Wisconsin Lead Region*. Wisconsin Domesday Book, General Studies, vol. 3. Madison: State Historical Society of Wisconsin.

Schlemper, M. Beth. 2004. "The Regional Construction of Identity and Scale in Wisconsin's Holy Land." *Journal of Cultural Geography* 22 (1): 51–81.

———. 2006. "The Borders of the Holyland of East-Central Wisconsin." In *Wisconsin German Land and Life*, edited by Heike Bungert, Cora Lee Kluge, and Robert C. Ostergren, 189–205. Madison: Max Kade Institute for German-American Studies, University of Wisconsin–Madison.

———. 2007. "From the Rhenish Prussian Eifel to the Wisconsin Holyland: Immigration, Identity and Acculturation at the Regional Scale." *Journal of Historical Geography* 33 (2): 377–402.

Schmid, John. 2012. "Paper Cuts: Wisconsin's Paper Industry Battles the Threat of Digital, China as a Paper Power." *Milwaukee Journal Sentinel*. http://archive.jsonline.com/business/paper-industry-digital-china-wis consin-181832171.html.

———. 2013. "Wisconsin Economy Stuck in Old-Growth Industries." *Milwaukee Journal Sentinel*, June 17. http://archive.jsonline.com/business/wisconsin-economy-stuck-in-old-growth-industries-b99341002I-211923141 .html/.

———. 2014. "Wisconsin's Paper Industry Braces for Uncertainty." *Milwaukee Journal Sentinel*, June 7. http:// archive.jsonline.com/business/wisconsins-paper-industry-braces-for-uncertainty-b992841872I-262261441 .html/.

Schmidt, Silke. 2016. "Despite State Efforts, Arsenic Continues to Poison Many Private Wells in Wisconsin." *Wisconsin Watch*, January 24. http://wisconsinwatch.org/2016/01/despite-state-efforts-arsenic-continues-to -poison-many-private-wells-in-wisconsin/#.

Schockel, Bernard H. 1916. "Settlement and Development of Jo Daviess County." Bulletin no. 26, 173–228. Urbana: Illinois Geological Survey.

Schreuder, Yda. 1997. "Americans by Choice and Circumstance: Dutch Protestant and Dutch Catholic Immigrants in Wisconsin, 1850–1905." In Ostergren and Vale, *Wisconsin Land and Life*, 320–330.

Schroeder, Jonathan P. 2016. "Historical Population Estimates for 2010 U.S. States, Counties and Metro/Micro Areas, 1790–2010." Data Repository for the University of Minnesota. http://doi.org/10.13020/D6XW2H.

Schubring, Selma Langenhan. 1926. "A Statistical Study of Lead and Zinc Mining in Wisconsin." *Transactions of the Wisconsin Academy of Sciences, Arts, and Letters* 22: 9–98.

Schultz, Gwen. 2004. *Wisconsin's Foundations: A Review of the State's Geology and Its Influence on Geography and Human Activity*. Madison: University of Wisconsin Press.

Scull, Peter, and Randall J. Schaetzl. 2011. "Using PCA to Characterize and Differentiate Loess Deposits in Wisconsin and Upper Michigan, USA." *Geomorphology* 127 (3–4): 143–155.

Shapiro, Aaron. 2006. "Up North on Vacation: Tourism and Resorts in Wisconsin's North Woods 1900–1945." *Wisconsin Magazine of History* 89 (4): 2–13.

Sharma, Sapna, Kevin Blagrave, and John J. Magnuson, et al. 2019. "Widespread Loss of Lake Ice around the Northern Hemisphere in a Warming World." *Nature Climate Change* 9 (3): 227–231.

Shastri, Devi. 2017. "Mosquito Capable of Spreading Zika Found in Wisconsin for First Time." *Milwaukee Journal Sentinel*, July 19. http://www.jsonline.com/story/news/health/2017/07/19/mosquito-capable-spreading-zika-found-wisconsin-first-time/490086001/.

Shears, Andrew. 2014. "Local to National and Back Again: Beer, Wisconsin & Scale." In *The Geography of Beer: Regions, Environment, and Societies*, edited by Mark Patterson and Nancy Hoalst-Pullen, 45–56. New York: Springer.

Silva, Erin, Fengxia Dong, Paul Mitchell, and John Hendrickson. 2014. "Impact of Marketing Channels on Perceptions of Quality of Life and Profitability for Wisconsin's Organic Vegetable Farmers." *Renewable Agriculture and Food Systems* 30 (5): 428–438.

Simon, Roger D. 1978. "The City-Building Process: Housing and Services in New Milwaukee Neighborhoods, 1880–1910." *Transactions of the American Philosophical Society* 68 (5).

Smith, Alice E. 1973. "Courts and Judges in Wisconsin Territory." *Wisconsin Magazine of History* 56 (3): 179–188.

Smith, Guy-Harold. 1929. "The Settlement and the Distribution of the Population in Wisconsin." *Transactions, Wisconsin Academy of Sciences, Arts and Letters* 24: 56–107.

Sorden, L. G. 1979. "The Northern Wisconsin Settler Relocation Project, 1934–1940." *Proceedings of the Fourth Forest History Association of Wisconsin Conference* 4: 14–17.

Steinberger, Florence. 2018. "For Hmong Families, Growing Food—and Selling It at Markets—Is a Proud Tradition." *Milwaukee Journal Sentinel*, September 18. https://www.jsonline.com/story/life/food/2018/09/18/hmong-families-milwaukee-make-most-land-farmers-markets/1277906002/.

Stoltman, James B., and Robert F. Boszhardt. 2019. "An Overview of Driftless Area Prehistory." In *The Physical Geography and Geology of the Driftless Area: The Career and Contributions of James C. Knox*, edited by Eric C. Carson, J. Elmo Rawling III, J. Michael Daniels, and John W. Attig, 93–118. GSA Special Paper 543. Boulder, CO: Geological Society of America.

Stover, C. W., B. G. Reagor, and S. T. Algermissen. 1980. *Seismicity Map of the State of Wisconsin*. Miscellaneous Field Studies Map MF-1229. Reston, VA: U.S. Geological Survey.

Suchan, Trudy A., Marc J. Perry, James D. Fitzsimmons, Anika E. Juhn, Alexander M. Tait, and Cynthia A. Brewer. 2007. *Census Atlas of the Unites States*. Series CENSR-29. Washington, DC: U.S. Census Bureau.

Syverson, Kent M., and Patrick M. Colgan. 2004. "The Quaternary of Wisconsin: A Review of Stratigraphy and Glaciations History." In *Quaternary Glaciations—Extent and Chronology*. Part 2, *North America*, edited by J. Ehlers and P. L. Gibbard, 295–311. Amsterdam: Elsevier.

Taschler, Joe. 2018. "Three Years and $300 Million Later, Kroger Continues to Transform Wisconsin Stores." *Milwaukee Journal Sentinel*, November 23. https://www.jsonline.com/story/money/business/2018/11/23/kroger-continues-transform-its-wisconsin-stores/1982535002/.

Tetzlaff, Trent. 2018. "Kwik Trip Takeover: Rise of the Midwest Powerhouse Resets Convenience Store Marketplace." *Appleton Post-Crescent*, December 10. https://www.postcrescent.com/story/news/2018/12/10/small-convenience-stores-work-keep-up-kwik-trip/2068606002/.

Thaller, Michael L., and Curtis W. Richards. 1975. "The Port of Milwaukee." In *Landscapes of Wisconsin: A Field Guide*, edited by Barbara Zakrzewska-Borowiecki, 117–122. Washington, DC: Association of American Geographers.

Thompson, William F. 1988. *The History of Wisconsin*. Vol. 6, *Continuity and Change, 1940–1965*. Madison: State Historical Society of Wisconsin.

Thornbury, William D. 1965. *Regional Geomorphology of the United States*. New York: John Wiley & Sons.

Thwaites, Reuben Gold. 1888. "The Boundaries of Wisconsin." *Wisconsin Historical Collections* 11: 451–501.

Timme, Ernst G., ed. 1887. *The Blue Book of the State of Wisconsin*. Madison: Wisconsin Secretary of State.

Tishler, William H. 1992. "Norwegians in Wisconsin." In *To Build in a New Land: Ethnic Landscapes in North America*, edited by Allen G. Noble, 226–241. Baltimore: Johns Hopkins University Press.

Trewartha, Glenn T. 1943. "The Unincorporated Hamlet: One Element of the American Settlement Fabric." *Annals of the Association of American Geographers* 33 (1): 32–81.

Trotta, L. C., and R. D. Cotter. 1973. *Depth to Bedrock in Wisconsin*. 1:1,000,000 scale map. Madison: Wisconsin Geological and Natural History Survey.

Turner, Frederick Jackson. 1921. *The Frontier in American History*. New York: Henry Holt.

Urban Land Institute. 2015. "Former Chrysler Engine Plant Site Redevelopment: Kenosha, Wisconsin." ULI Advisory Services Panel Report. Washington, DC: Urban Land Institute.

USACE (U.S. Army Corps of Engineers). 2019. *Waterborne Commerce of the United States*. Part 2, *Gulf Coast, Mississippi River System, Puerto Rico, and Virgin Islands*. Part 3, *Great Lakes*. USACE Navigation Data Center. https://publibrary.planusace.us/#/series/Waterborne%20Commerce%20of%20the%20United%20States.

———. 2020. "FINAL 2019 and Long-Term (1918–2019) Mean, Max & Min Monthly Mean Water Levels." https://www.lre.usace.army.mil/Portals/69/docs/GreatLakesInfo/docs/WaterLevels/LTA_GLWL-English _2019.pdf?ver=2020-02-04-152044-737.

USBC (U.S. Bureau of the Census). 1904. *Occupations at the Twelfth Census*. Special Reports. Washington, DC: U.S. Department of Commerce and Labor.

———. 1912. *Thirteenth Census of the United States Taken in the Year 1910*. Vol. 9, *Manufactures, 1909: Reports by States, with Statistics for Principal Cities*. Washington, DC: U.S. Department of Commerce and Labor.

———. 1914a. *Thirteenth Census of the United States Taken in the Year 1910*. Vol. 3, *Population 1910: Nebraska-Wyoming*. Washington, DC: U.S. Department of Commerce.

———. 1914b. *Thirteenth Census of the United States Taken in the Year 1910*. Vol. 4, *Population 1910: Occupation Statistics*. Washington, DC: U.S. Department of Commerce.

———. 1914c. *Thirteenth Census of the United States Taken in the Year 1910*. Vol. 5, *Agriculture 1909 and 1910: General Report and Analysis*. Washington, DC: U.S. Department of Commerce.

———. 1914d. *Thirteenth Census of the United States Taken in the Year 1910*. Vol. 7, *Agriculture 1909 and 1910: Reports by States, with Statistics for Counties—Nebraska to Wyoming*. Washington, DC: U.S. Department of Commerce.

———. 1923. *Fourteenth Census of the United States Taken in the Year 1920*. Vol. 9, *Manufacturers 1919: Reports by States, with Statistics for Principal Cities*. Washington, DC: U.S. Department of Commerce.

———. 1952a. *United States Census of Agriculture: 1950, Counties and State Economic Areas: Wisconsin*. Vol. 1, pt. 7. Washington, DC: U.S. Department of Commerce.

———. 1952b. *United States Census of Agriculture: 1950, Ranking Agricultural Counties*. Vol. 5, *Special Reports*, pt. 7. Washington, DC: U.S. Department of Commerce.

———. 1952c. *United States Census of Population: 1950*. Vol. 1, *Number of Inhabitants*. Washington, DC: U.S. Department of Commerce.

USBLS (U.S. Bureau of Labor Statistics). 2019. "Occupational Employment Statistics: May 2018." Excel worksheets: MSA_M2018_dl and state_M2018_dl. https://www.bls.gov/oes/tables.htm.

———. 2020. "Table 3. Employees on Nonfarm Payrolls by State and Selected Industry Sector, Seasonally Adjusted." December 2019. Revised March 27. https://www.bls.gov/news.release/laus.t03.htm.

USC (U.S. Census). 1853. *The Seventh Census of the United States: 1850*. Washington, DC: Robert Armstrong, Public Printer.

———. 1864a. *Agriculture of the United States in 1860; Compiled from the Original Returns of the Eighth Census*. Washington, DC: Government Printing Office.

———. 1864b. *Population of the United States in 1860; Compiled from the Original Returns of the Eighth Census*. Washington, DC: Government Printing Office.

———. 1883a. *Report on the Productions of Agriculture as Returned by the Tenth Census (June 1, 1880)*. Washington, DC: Government Printing Office.

———. 1883b. *Statistics of the Population of the United States at the Tenth Census (June 1, 1880)*. Washington, DC: Government Printing Office.

USCB (U.S. Census Bureau). 2005. *Wisconsin: 2002 Economic Census, Manufacturing*. EC02-31A-WI (RV). Washington, DC: U.S. Department of Commerce.

———. 2018. "2016 County Business Patterns & Nonemployer Statistics Combined Report." Dataset: CBP2016. Excel file: combine16wi. https://www.census.gov/data/tables/2016/econ/cbp/2016-combined-report.html.

———. 2019a. American FactFinder. (This online site, removed March 31, 2020, was utilized to access data at the county and minor civil division level [town, city, and village] from the 1990, 2000, and 2010 U.S. Censuses; Economic, Manufacturing, and Commerce Censuses conducted between 2002 and 2017; and American Community Surveys conducted annually, with results expressed both annually and for five-year periods, at https://factfinder.census.gov/faces/nav/jsf/pages/index.xhtml).

———. 2019b. "2017 County Business Patterns & Nonemployer Statistics Combined Report." Dataset: CBP2017. Excel file: combine17wi. https://www.census.gov/data/tables/2017/econ/cbp/2017-combined-report.html.

USCO (U.S. Census Office). 1883. *Population of the United States at the Tenth Census (June 1, 1880) Embracing Extended Tables of the Population of States, Counties, and Minor Civil Divisions with Distinction of Race, Sex, Age, Nativity, and Occupations*. Washington, DC: Government Printing Office.

———. 1895. *Report on Manufacturing Industries in the United States at the Eleventh Census: 1890*. Washington, DC: Government Printing Office.

USDA (U.S. Department of Agriculture). 2019. *Local Food Directories: National Farmers Market Directory*. https://www.ams.usda.gov/local-food-directories/farmersmarkets.

USDA-NASS (U.S. Department of Agriculture, National Agricultural Statistics Service). 2019a. *Wisconsin Crop Progress & Condition*. 19 (10): week ending June 2.

———. 2019b. *Wisconsin Crop Progress & Condition*. 19 (12): week ending June 16.

———. 2019c. *Wisconsin Crop Progress & Condition*. 19 (37): week ending December 8.

USDA-NCSS (U.S. Department of Agriculture, National Cooperative Soil Survey). 2009. Official Series Description—Antigo Series. https://soilseries.sc.egov.usda.gov/OSD_Docs/A/ANTIGO.html.

USDA-RMA (U.S. Department of Agriculture, Risk Management Agency). 2013. "RMA Indemnities (as of 08/05/2013)." https://legacy.rma.usda.gov/data/indemnity/2012/80513table.pdf.

———. 2017. "RMA Indemnities (as of 07/03/2017)." https://legacy.rma.usda.gov/data/indemnity/2017/070317table.pdf.

———. 2020. "RMA Indemnities (as of 04/27/2020)." https://www.rma.usda.gov/-/media/RMA/Maps/Total-Crop-Indemnity-Maps/Crop-Year-2019/042720table.ashx?la=en.

USDA-WFO (U.S. Department of Agriculture, National Agricultural Statistics Service, Wisconsin Field Office). 2018a. "Wisconsin Ag News—2017 Corn County Estimates." https://www.nass.usda.gov/Statistics_by_State/Wisconsin/Publications/County_Estimates/WI-CtyEst-Corn-16-17.pdf. Posting removed at end of 2020.

———. 2018b. "Wisconsin Ag News—2017 Oats County Estimates." https://www.nass.usda.gov/Statistics_by_State/Wisconsin/Publications/County_Estimates/WI_CE_Oats_2017.pdf. Posting removed at end of 2020.

———. 2018c. "Wisconsin Ag News—2017 Soybean County Estimates." https://www.nass.usda.gov/Statistics_by_State/Wisconsin/Publications/County_Estimates/WI-CtyEst-Soybean-16-17.pdf. Posting removed at end of 2020.

———. 2018d. "Wisconsin Ag News—2017 Winter Wheat County Estimates." https://www.nass.usda.gov/Statistics_by_State/Wisconsin/Publications/County_Estimates/WI_CE_Winter_Wheat_2017.pdf. Posting removed at end of 2020.

USEIA (U.S. Energy Information Administration). 2018. "Wisconsin State Energy Profile." https://www.eia.gov/state/print.php?sid=WI.

USFS (U.S. Forest Service). 2019. EVALIDator Version 1.8.0.01. https://apps.fs.usda.gov/Evalidator/evalidator.jsp.

USFWS (U.S. Fish and Wildlife Service and U.S. Census Bureau). 2014. *2011 National Survey of Fishing, Hunting, and Wildlife-Associated Recreation: Wisconsin.* FHW/11-WI (RV). https://www.census.gov/prod/2012pubs/fhw11-nat.pdf.

USGS (U.S. Geological Survey). 1998. "The Mineral Industry of Wisconsin." *1997 Minerals Yearbook.* https://minerals.usgs.gov/minerals/pubs/state/985598.pdf.

———. 2017. "The Mineral Industry of Wisconsin." *2012–13 Minerals Yearbook.* https://s3-us-west-2.amazonaws.com/prd-wret/assets/palladium/production/mineral-pubs/state/2012_13/myb2-2012_13-wi.pdf.

UWPHI (University of Wisconsin Population Health Institute). 2019a. County Health Rankings and Roadmaps: Wisconsin Dentists. https://www.countyhealthrankings.org/app/wisconsin/2019/measure/factors/88/data.

———. 2019b. County Health Rankings and Roadmaps: Wisconsin Primary Care Physicians. https://www.countyhealthrankings.org/app/wisconsin/2019/measure/factors/4/data.

———. 2019c. County Health Rankings and Roadmaps: Wisconsin Uninsured. https://www.countyhealthrankings.org/app/wisconsin/2019/measure/factors/85/data.

———. 2019d. *Wisconsin County Health Rankings State Report.* Madison: University of Wisconsin.

Veldman, Jan Willem. 2018. "Interview with Award Winning US Dairy Holsum Dairies." *DairyGlobal,* April 19. https://www.dairyglobal.net/Farm-trends/Articles/2018/4/Interview-with-awarding-winning-US-dairy-Holsum-Dairies-274758E/.

Vias, Alexander C. 2004. "Bigger Stores, More Stores, or No Stores: Paths of Retail Restructuring in Rural America." *Journal of Rural Studies* 20 (3): 303–318.

Village of Gays Mills. 2020. "Gays Mills Apple Festival." http://www.gaysmills.org/applefest.html.

Vogel, Virgil J. (1965) 1980. "Wisconsin's Name: A Linguistic Puzzle." *Wisconsin Magazine of History* 48 (3): 181–186. Reprinted in *Wisconsin Stories* series of the State Historical Society of Wisconsin.

Vogeler, Ingolf. 1986. *Wisconsin: A Geography.* Boulder, CO: Westview Press.

Vogl, Richard J. 1970. "Fire and the Northern Wisconsin Pine Barrens." In *Proceedings: 10th Tall Timbers Fire Ecology Conference,* 175–209. Tallahassee, FL: Tall Timbers Research Station.

Vue, Mai Zong. 2020. *Hmong in Wisconsin.* Madison: Wisconsin Historical Society Press.

Wangemann, William F. 2005. "The Wood Industry in and around Sheboygan, Wisconsin." *The Historic Sheboygan Area: Proceedings, Thirtieth Annual Meeting, Forest History Association of Wisconsin* 30: 12–15.

Ward, Kevin. 2007. "'Creating a Personality for Downtown': Business Improvement Districts in Milwaukee." *Urban Geography* 28 (8): 781–808.

Waring, George E., Jr. 1887. *Report on the Social Statistics of Cities.* Washington, DC: U.S. Census Office, Government Printing Office.

Warner, Hans B. 1880. *The Blue Book of the State of Wisconsin, 1880.* Madison: Wisconsin Secretary of State.

WASAL (Wisconsin Academy of Sciences, Arts and Letters). 2007. *The Future of Farming and Rural Life in Wisconsin: Findings, Recommendations, Steps to a Healthy Future.* Madison: Wisconsin Academy of Sciences, Arts and Letters.

WASS (Wisconsin Agricultural Statistics Service). 1999. *1999 Wisconsin Agricultural Statistics.* Madison: Wisconsin Department of Agriculture, Trade and Consumer Protection.

———. 2002. *2002 Wisconsin Agricultural Statistics.* Madison: Wisconsin Department of Agriculture, Trade and Consumer Protection.

———. 2012. *2012 Wisconsin Agricultural Statistics.* Madison: Wisconsin Field Office, National Agricultural Statistics Service.

———. 2017. *2017 Wisconsin Agricultural Statistics.* Madison: Wisconsin Field Office, National Agricultural Statistics Service.

———. 2018. *2018 Wisconsin Agricultural Statistics.* Madison: Wisconsin Field Office, National Agricultural Statistics Service.

———. 2019. *2019 Wisconsin Agricultural Statistics.* Madison: Wisconsin Field Office, National Agricultural Statistics Service.

Watson, G. E., 1952. *Thirty Fifth Report of the Superintendent of Public Instruction of the State of Wisconsin.* Madison: Wisconsin Superintendent of Public Instruction.

WBO (Wisconsin Broadband Office). 2018. County Coverage Atlas. Maps of all 72 counties showing broadband connectivity speed using December 31, 2016, data. Madison: Public Service Commission of Wisconsin. https://psc.wi.gov/Documents/broadband/CoverageAtlas/County/CountyCoverageAtlas.pdf.

WCFA (Wisconsin County Forests Association). 2013. "Forests Acres." http://www.wisconsincountyforests.com/forest-resources/forest-acres/.

WCG (Wisconsin Cartographers' Guild). 1998. *Wisconsin's Past and Present: A Historical Atlas.* Madison: University of Wisconsin Press.

WCGA (Wisconsin Corn Growers Association). 2018. "Wisconsin Corn Facts." http://wicorn.org/resources/corn-facts/.

WDA (Wisconsin Department of Administration). 1975. *Major Areas of Potential Flood Hazard and Steep Slope.* 1:500,000 map. Madison: State Planning Office, Wisconsin Department of Administration.

WDATCP (Wisconsin Department of Agriculture, Trade and Consumer Protection). 1989. Unpublished computer tape listing all Wisconsin dairy farmers whose herds had undergone Brucellosis Ring Testing, required for milk sales.

———. 2009. *April, 2009 Dairy Producers.* Compact Disk of Active Dairy Producers—Comma delimited file. Madison: WDATCP.

———. 2019a. *March, 2019 Dairy Producers.* Compact Disk of Active Dairy Producers—Comma delimited file. Madison: WDATCP.

———. 2019b. *November, 2019 Dairy Producers.* Compact Disk of Active Dairy Producers—Comma delimited file. Madison: WDATCP.

———. 2019c. *Wisconsin Dairy Plant Directory 2019–2020.* Madison: WDATCP.

———. 2019d. *Wisconsin Dairy Task Force 2.0 Final Report: Toward a Vibrant Wisconsin Dairy Industry: Issues, Insights and Recommendations.* Madison: WDATCP. https://datcp.wi.gov/Documents/DairyTaskForce20Final Report.pdf.

WDHS (Wisconsin Department of Health Services). 2017a. Refugee Arrivals in Wisconsin by Destination County—FFY 2017. Madison: WDHS. https://www.wistatedocuments.org/digital/collection/p267601coll4/search/searchterm/987856042/field/dmoclcno.

———. 2017b. "West Nile Virus." https://www.dhs.wisconsin.gov/arboviral/westnilevirus.htm.

———. 2019a. *Annual Wisconsin Death Report, 2017* (P-01170-19). Madison: WDHS, Division of Public Health, Office of Health Informatics.

———. 2019b. "Lyme—Incidence—Crude Rates per 100,000." https://dhsgis.wi.gov/DHS/EPHTracker/#/map/Lyme%20Disease/lymeIndex/NOTRACT/Incidence/lymeCtyIndex2.

———. 2019c. "Trauma Care System." https://www.dhs.wisconsin.gov/trauma/index.htm.

WDMA (Wisconsin Department of Military Affairs). 2002. *Hazard Analysis for the State of Wisconsin*. Madison: Wisconsin Emergency Management, WDMA.

———. 2008. *State of Wisconsin Hazard Mitigation Plan*. Madison: Wisconsin Emergency Management, WDMA.

———. 2017a. *State of Wisconsin Hazard Mitigation Plan*. Madison: Wisconsin Emergency Management, WDMA.

———. 2017b. *State of Wisconsin Threat Hazard Identification and Risk Assessment (THIRA & SPR)*. Madison: Homeland Security Council and Wisconsin Emergency Management, WDMA.

WDNR (Wisconsin Department of Natural Resources). 1993. *The Floods of 1993: The Wisconsin Experience*. Madison: Bureau of Water Regulation and Zoning.

———. 2014. "Ferrous Mining in Wisconsin." PUB-WA-1721. Madison: WDNR.

———. 2015a. *The Ecological Landscapes of Wisconsin: An Assessment of Ecological Resources and a Guide to Planning Sustainable Management*. PUB-SS-1131. Madison: WDNR.

———. 2015b. *Wisconsin State Park System, 2015–2020: Strategic Directions*. Madison: WDNR. https://dnr.wi.gov/topic/parks/documents/StrategicDirections2015.pdf.

———. 2016. "Forestry and the Wisconsin Economy." Individual fact sheets for each county are accessible at https://dnr.wi.gov/topic/forestbusinesses/factsheets.html. Factsheets on website annually replaced with updated statistics.

———. 2017a. "Arsenic in Drinking Water." PUB-DG-062. https://dnr.wi.gov/files/PDF/pubs/DG/DG0062.pdf.

———. 2017b. *Industrial Sand Mining in Wisconsin: Strategic Analysis*. https://dnr.wi.gov/topic/EIA/documents/ISMSA/ISMSA2017.06.12.pdf.

———. 2017c. "Wisconsin's Managed Forest Law: A Program Summary." PUB-FR-295. https://woodlandinfo.org/files/2017/09/FR-295.pdf.

———. 2018. "Estimated Attendance by Year (Fiscal Years 2002–2018)." https://dnr.wi.gov/topic/parks/documents/WSPSattendance.pdf.

———. 2019a. "CWD Prevalence in Wisconsin." https://dnr.wi.gov/topic/wildlifehabitat/prevalence.html.

———. 2019b. "Deer Abundance and Densities in Wisconsin Deer Management Units." https://dnr.wi.gov/topic/hunt/maps.html/falldeerperdr.pdf.

———. 2019c. "Preliminary 9-Day Gun Deer Season Comparison, 2017 vs. 2018." Madison: WDNR. https://dnr.wi.gov/topic/wildlifehabitat/documents/deer_Prelim9day_2017v2018.pdf.

———. 2019d. "Sturgeon Spearing: Daily Sturgeon Harvest Sheet." https://dnr.wi.gov/topic/fishing/documents/sturgeon/HarvestFeb242019.pdf.

———. 2019e. *Visitor Guide: Wisconsin State Park System*. Madison: Bureau of Parks and Recreation. PUB-PR-002. https://dnr.wi.gov/topic/parks/documents/WSPSVisitorGuide.pdf.

———. 2020. "Lake Winnebago Opening Day." https://dnr.wi.gov/topic/fishing/documents/sturgeon/2020OpeningDayHarvestTotalsShantyCounts.pdf.

WDNR-RM (Wisconsin Department of Natural Resources, Runoff Management). 2019–2020. DNR Runoff Management CAFO Permittees. https://dnr.wi.gov/topic/AgBusiness/data/CAFO/index.asp. Active file continuously updated, with permits searchable through the index.

WDOT (Wisconsin Department of Transportation). 2009. *Connections 2030: Statewide Long-Range Transportation Plan*. Madison: WDOT. https://wisconsindot.gov/Pages/projects/multimodal/conn2030.aspx.

———. 2014. *Wisconsin Rail Plan 2030 Final.* Madison: WDOT. https://wisconsindot.gov/Pages/projects/multi modal/railplan/2030.aspx.

———. 2015. *Wisconsin State Airport System Plan 2030.* Madison: WDOT. https://wisconsindot.gov/Pages/proj ects/multimodal/sasp/air2030-chap.aspx.

———. 2018a. *Wisconsin State Freight Plan.* Madison: WDOT. https://wisconsindot.gov/Documents/projects/ sfp/plan.pdf.

———. 2018b. *Wisconsin Northwoods Freight Rail Study.* Madison: WDOT. https://wisconsindot.gov/Documents/ projects/multimodal/rail/northwoods2018.pdf.

———. 2018c. "Wisconsin Get-Around Guide: Intercity Public Transportation Information." Madison: WDOT. http://wisconsindot.gov/Documents/travel/pub-transit/get-around.pdf.

———. 2018d. *Wisconsin Rustic Roads.* Madison: WDOT. https://wisconsindot.gov/Pages/travel/road/rustic -roads/default.aspx.

———. 2019. *Wisconsin Aviation Activity—2018.* Madison: WDOT. https://wisconsindot.gov/Documents/doing -bus/aeronautics/resources/2018activity.pdf.

———. 2020. "Wisconsin Railroads and Harbors 2020." Map. https://wisconsindot.gov/Documents/travel/rail/ railmap.pdf.

WDPI (Wisconsin Department of Public Instruction). 2019. "WISEdash Data Files by Topic." https://dpi.wi.gov/ wisedash/download-files/type?field_wisedash_upload_type_value=Enrollment-Private-School&field_wise dash_data_view_value=All.

WDT (Wisconsin Department of Tourism). 2019. "Total Tourism Impacts 2018: Wisconsin and Counties." https://www.wpr.org/sites/default/files/48b75ff3-8218-4c83-ad4b-b69a9fb80211-county-economic-impact -table-2018-final_2.pdf.

Weaver, John C. 1954. "Crop-Combination Regions in the Middle West." *The Geographical Review* 44 (2): 175–200.

WEC (Wisconsin Election Commission). 2018. "WEC Canvass Reporting System, County by County Report, 2018 General Election." State Constitutional Offices. https://elections.wi.gov/sites/electionsuat.wi.gov/files/ County%20by%20County%20Report-2018%20Gen%20Election-State%20Constitution%20Offices.pdf.

Wehrwein, George S. 1935. "Town Government in Wisconsin." In *Wisconsin Blue Book 1935,* edited by Leone G. Bryhan, 95–107. Madison: Democrat Printing Company, State Printer.

Wehrwein, George S., and J. A. Baker. 1937. "The Cost of Isolated Settlement in Northern Wisconsin." *Rural Sociology* 2 (3): 253–265.

Weichelt, Katie L. 2016. "A Historical Geography of the Paper Industry in the Wisconsin River Valley." PhD diss., University of Kansas.

Weidman, Samuel. 1904. *The Baraboo Iron-Bearing District of Wisconsin.* Bulletin 13. Madison: Wisconsin Geo- logical and Natural History Survey.

Weiland, Andrew. 2016. "Wisconsin Has 10 Firms on 2016 Fortune 500 List: Up from 8 in 2018." *BizTimes: Milwaukee Business,* June 6. https://biztimes.com/wisconsin-has-10-firms-on-2016-fortune-500-list/.

———. 2019. "More Wisconsin Companies Make Fortune 1000." *BizTimes: Milwaukee Business News,* May 22. https://biztimes.com/more-wisconsin-companies-make-fortune-1000/.

WGNHS (Wisconsin Geological and Natural History Survey). 2005. *Bedrock Geology of Wisconsin.* Small-scale map. Madison: WGNHS.

———. 2014. "Frac Sand in Wisconsin." Factsheet 5. https://wgnhs.wisc.edu/pubs/fs05/.

Whitbeck, Ray Hughes. 1913. *The Geography and Industries of Wisconsin.* Bulletin no. 26. Madison: Wisconsin Geological and Natural History Survey.

———. 1915. *The Geography of the Fox-Winnebago Valley.* Bulletin no. 42. Madison: Wisconsin Geological and Natural History Survey.

———. 1921. *The Geography and Economic Development of Southeastern Wisconsin*. Bulletin no. 58. Madison: Wisconsin Geological and Natural History Survey.

WHS (Wisconsin Historical Society). 2017. "Wisconsin's Name: Where It Came from and What It Means." http://www.wisconsinhistory.org/Content.aspx?dsNav=N:4294963828-4294963805&dsRecordDetails=R:CS3663.

WICCI (Wisconsin Initiative on Climate Change Impacts). 2011. *Wisconsin's Changing Climate: Impacts and Adaptation*. Madison: University of Wisconsin Nelson Institute for Environmental Studies and Wisconsin Department of Natural Resources.

Wikipedia. 2019. "List of U.S. States and Territories by Elevation." https://en.wikipedia.org/wiki/List_of_U.S._states_by_elevation.

Wilkerson, Miranda E., and Joseph Salmons. 2008. "'Good Old Immigrants of Yesteryear' Who Didn't Learn English: Germans in Wisconsin." *American Speech* 83 (3): 259–283.

Wines, Michael, and Emily Bazelon. 2020. "Flaws in Census Count Imperil Trump Plan to Exclude Undocumented Immigrants." *New York Times*, December 4. https://www.nytimes.com/2020/12/04/us/census-trump.html.

Winkler, Richelle, and Keith Warnke. 2013. "The Future of Hunting: An Age-Period-Cohort Analysis of Deer Hunter Decline." *Population and Environment* 34 (4): 460–480.

Wisconsin Assembly. 1853. *Manual for the Use of the Assembly of the State of Wisconsin for the Year 1853 Prepared Pursuant to a Resolution of the Assembly*. Madison: Brown & Carpenter, Printers.

WDVCB (Wisconsin Dells Visitor & Convention Bureau). 2018. "The Waterpark Capital of the World!" https://www.wisdells.com/media/facts/water-parks.htm.

WIST (Wisconsin Institute for Sustainable Technology). 2019. *An Assessment of the Economic Contribution of Pulp, Paper and Converting to the State of Wisconsin*. Madison: Wisconsin Economic Development Corporation. https://wedc.org/wp-content/uploads/2019/08/2019-WI-Paper-Industry-Report.pdf.

WLRB (Wisconsin Legislative Reference Bureau). 2015. *State of Wisconsin 2015–2016 Blue Book*. Madison: Wisconsin Department of Administration.

———. 2017. *State of Wisconsin 2017–2018 Blue Book*. Madison: Wisconsin Department of Administration.

———. 2019. *State of Wisconsin 2019–2020 Blue Book*. Madison: Wisconsin Department of Administration.

WLRL (Wisconsin Legislative Reference Library). 1950. *The Wisconsin Blue Book 1950*. Madison: State of Wisconsin.

Wokatsch, Randy. 2018. "Dairy Farm Crisis Grows Critical for Many." *Wisconsin Agriculturalist*, April 30. https://www.farmprogress.com/dairy/dairy-farm-crisis-grows-critical-many.

World Dairy Expo. 2019. https://worlddairyexpo.com/.

WPA (Work Projects Administration). 1954. *Wisconsin: A Guide to the Badger State*. New York: Hastings House.

———. 1979. *The Story of Mineral Point, 1827–1941*. Mineral Point, WI: Mineral Point Historical Society.

WPF (Wisconsin Policy Forum). 2019a. "Wisconsin's Brain Drain Problem." *Focus*, no. 10. https://wispolicyforum.org/wp-content/uploads/2019/05/Focus_19_10.pdf.

———. 2019b. "Wisconsin's Workforce Challenges Intensify. *Focus*, no. 17. https://wispolicyforum.org/wp-content/uploads/2019/09/Focus_19_17.pdf.

WSCaO (Wisconsin State Cartographer's Office). 2017. "Wisconsin High Points." https://www.sco.wisc.edu/wisconsin/high-points/.

WSCGA (Wisconsin State Cranberry Growers Association). 2013. "Interesting Cranberry Facts." http://www.wiscran.org/media/1346/2013interestingcranberryfacts.pdf.

WSCO (Wisconsin State Climatology Office). n.d. *Length of "Frost Free" Season*. Map. http://www.aos.wisc.edu/~sco/clim-history/sta-data/WI-growing_season/WI-32degfreezedates-1981-2010-50pseasonlength-c.gif.

———. 2017. "Wisconsin Wind Data: Monthly Winds at Milwaukee (WI)." http://www.aos.wisc.edu/~sco/clim-history/stations/mke/milwind.html.

———. 2020. "History of Freezing and Thawing of Lake Mendota, 1852–53 to 2019/20." http://www.aos.wisc.edu/~sco/lakes/Mendota-ice.html.

WTC (Wisconsin Technology Council). 2015. "'Rust Belt' No More: Traveling the 'I-Q Corridor.'" http://wisconsintechnologycouncil.com/wp-content/uploads/2015/10/I-Q_Corridor_Final.pdf.

Wylie, Robert C. 1992. "The Evolution of Logging Dams in the Wisconsin River Flow Management System." *Proceedings, Seventeenth Annual Meeting, Forest History Association of Wisconsin* 17: 27–33.

Wyman, Mark. 1998. *The Wisconsin Frontier*. Bloomington: Indiana University Press.

Young Center. 2019. "Amish Population, 2019." Young Center for Anabaptist and Pietist Studies, Elizabethtown College. http://groups.etown.edu/amishstudies/statistics/population-2019/.

Zakrzewska-Borowiecki, Barbara. 1975. "Land Form of Southeastern Wisconsin." In *Landscapes of Wisconsin: A Field Guide*, edited by Barbara Zakrzewska-Borowiecki, 30–51. Washington, DC: Association of American Geographers.

Zaniewski, Kazimierz J., and Carol J. Rosen. 1998. *The Atlas of Ethnic Diversity in Wisconsin*. Madison: University of Wisconsin Press.

Zaniewski, Kazimierz J., and James R. Simmons. 2016. "Divided Wisconsin: Partisan Spatial Electoral Realignment." *The Geography Teacher* 13 (3): 128–133.

Zeitlin, Richard H. 2000. *Germans in Wisconsin*. Madison: State Historical Society of Wisconsin.

Ziesler, Pamela S. 2019. *Statistical Abstract: 2018*. Natural Resource Data Series. NPS/NRSS/EQD/NRDS—2019/1219. Fort Collins, CO: National Park Service. https://irma.nps.gov/Datastore/Reference/Profile/2259799.

# Index

Page numbers for illustrations and captions are in *italics*.